Supramolecular Systems for Gene and Drug Delivery

Supramolecular Systems for Gene and Drug Delivery

Editors

Francisco José Ostos
José Antonio Lebrón
Pilar López-Cornejo

MDPI • Basel • Beijing • Wuhan • Barcelona • Belgrade • Manchester • Tokyo • Cluj • Tianjin

Editors
Francisco José Ostos
University of Seville
Spain

José Antonio Lebrón
University of Seville
Spain

Pilar López-Cornejo
University of Seville
Spain

Editorial Office
MDPI
St. Alban-Anlage 66
4052 Basel, Switzerland

This is a reprint of articles from the Special Issue published online in the open access journal *Pharmaceutics* (ISSN 1999-4923) (available at: https://www.mdpi.com/journal/pharmaceutics/special_issues/supramolecular_gene_drug_delivery).

For citation purposes, cite each article independently as indicated on the article page online and as indicated below:

LastName, A.A.; LastName, B.B.; LastName, C.C. Article Title. *Journal Name* **Year**, *Volume Number*, Page Range.

ISBN 978-3-0365-3377-3 (Hbk)
ISBN 978-3-0365-3378-0 (PDF)

Contents

About the Editors

Francisco José Ostos, Ph.D., is a postdoctoral fellowship at the Clinical Unit of Infectious Diseases, Microbiology, and Preventive Medicine from the Institute of Biomedicine of Seville (IBiS) and School of Medicine from the University of Seville (since 2021). His research focuses on developing an innovative multidisciplinary approach at the interface of nanotechnology, supramolecular chemistry, cell biology, and immunology in order to eradicate viral infections. His research period resulted in the publication of impressive works in journals with a high impact factor, including 18 research articles, 2 book chapters, and 2 books. Furthermore, he presented inventions in 30 international/national meetings, and participated in 7 research projects. He received some awards (One secondary award from the Spanish Ministry of Education, Culture and Sport (XIV Certamen Universitario Arquímedes), and three awards as best publication from the University of Seville.

José Antonio Lebrón has a research contract funded by the "Fundación ONCE" in its "Oportunidad al Talento" program at the "Professor Rodríguez Velasco Kinetics Group (FQM 206)" of the University of Seville. His research has led him to publish papers in high-impact-factor journals, including 19 research articles. He has also participated in six research projects and two book chapters, and has published two books of his own. He has contributed 33 communications to national/international congresses. His research focuses on the synthesis and development of new drug nanocarriers (drugs or genetic material) for the treatment of diseases. The candidate has also participated in the organization of two congresses, one national 2017 "J2IFAM2017" and another international 2013 "(8th ICCK 2013)".

Pilar López-Cornejo, Professor of Physical Chemistry at the University of Seville. Her research is related to restricted geometry systems, macromolecules, colloids, and ligand–receptor interactions. Her research focuses on the development of biocompatible nanocarriers based on the use of calixarenes, metallomicelles, liposomes, CNTs, MOFs, and polymers for encapsulating nucleic acids, antineoplastic drugs, peptides, proteins, antiretrovirals, and toll-like receptors. Publications include 2 books, 8 book chapters, 95 research articles, and >50 meetings. She has participated in more than 20 projects, leading some of them. She is also the supervisor of three doctoral theses (+1 under progress) and is the reviewer of several scientific journals of high impact. She is also responsible for bilateral agreements, including Sevilla-Pisa and Sevilla-Hamburgo (Erasmus Program), and Seville-Tunisia (Erasmus +). As a member of the Colloids and Interface Group and the RSEQ, she has collaborated with different national and international research groups.

 *pharmaceutics*

Editorial

Supramolecular Systems for Gene and Drug Delivery

José A. Lebrón [1], Pilar López-Cornejo [1,*] and Francisco J. Ostos [2,3,*]

[1] Department of Physical Chemistry, Faculty of Chemistry, University of Seville, C/Profesor García González 1, 41012 Seville, Spain; jlebron@us.es

[2] Clinical Unit of Infectious Diseases, Microbiology and Preventive Medicine, Institute of Biomedicine of Seville (IBiS), Virgen del Rocío University Hospital, CSIC, University of Seville, 41013 Seville, Spain

[3] Department of Medical Biochemistry, Molecular Biology, and Immunology, School of Medicine, University of Seville, 41009 Seville, Spain

* Correspondence: pcornejo@us.es (P.L.-C.); fostos@us.es (F.J.O.)

Citation: Lebrón, J.A.; López-Cornejo, P.; Ostos, F.J. Supramolecular Systems for Gene and Drug Delivery. *Pharmaceutics* 2022, 14, 471. https://doi.org/10.3390/pharmaceutics14030471

Received: 18 February 2022
Accepted: 21 February 2022
Published: 22 February 2022

Publisher's Note: MDPI stays neutral with regard to jurisdictional claims in published maps and institutional affiliations.

Several biomaterial-based supramolecular systems (cyclodextrins [1], calixarenes [2,3], polymers [4], carbon nanotubes [5], nanoparticles [6,7], liposomes [3,8], nanogels [9], and nanocomplexes [10], among others) have been widely used for biomedical applications, such as gene and drug delivery. Numerous researchers have developed novel supramolecular systems for enhancing their biocompatibility and pharmacological activity, thus increasing their therapeutic properties. These nanosystems are considered to be promising platforms in gene therapy and drug delivery due to their higher transfection (or encapsulation) efficiency and low cytotoxicity.

This Special Issue, "Supramolecular Systems for Gene and Drug Delivery", brings together the latest research articles, published in *Pharmaceutics*. Noticeably, 10 original research articles were published by authors from 12 different countries on what is a hot topic in this research field.

I. Asela et al. [1] prepared nanosponges based on β-cyclodextrin (βCDNS), which were loaded with the drugs phenylethylamine (PhEA) and 2-amino-4-(4-chlorophenyl)-thiazole (AT). Subsequently, the supramolecular βCDNS drug complexes were functionalized with gold nanoparticles (AuNPs), forming the βCDNS-PhEA-AuNP and βCDNS-AT-AuNP systems. The drug-loading capacity was higher for the βCDNS and βCDNS-drug-AuNP systems than with native βCD.

B. Gómez-González et al. [2] studied the formation of inclusion complexes between alkyl sulfonate guests and a cationic pillar [5] arene receptor in water using NMR and ITC measurements. The results demonstrated the formation of host–guest complexes stabilized by electrostatic interactions and hydrophobic effects.

J. A. Lebrón et al. [3] studied the formation of calixarene-based liposomes. Four amphiphilic calixarenes were used. The lipid bilayer was formed with one calixarene and with the phospholipid 1,2-dioleoyl-sn-glycero-3-phosphoethanolamine (DOPE). The liposomes containing the least cytotoxic calixarene (TEAC$_{12}$)$_4$ were used as nanocarriers of both nucleic acids and the antineoplastic drug doxorubicin (DOX). The results showed that (TEAC$_{12}$)$_4$/DOPE/p-EGFP-C1 lipoplexes, of a given composition, can transfect the genetic material, although the transfection efficiency substantially increases in the presence of an additional amount of DOPE as coadjuvant. On the other hand, the (TEAC$_{12}$)$_4$/DOPE liposomes showed a high doxorubicin encapsulation efficiency and a slow controlled release, which could diminish the side effects of the drug.

V. Karava et al. [4] prepared microparticles (MPs) based on newly synthesized poly(l-lactic acid)-co-poly(butylene adipate) (PLA/PBAd) block copolymers for the preparation of aripiprazole (ARI)-loaded long-acting injectable (LAI) formulations. In terms of in vitro dissolution profile, results suggested that the newly synthesized PLA/PBAd block copolymers can successfully control the release rate and extent of the API's release from the prepared MPs, indicating that, probably, under in vivo conditions, their use may lead to new formulations that will be able to maintain a continuous therapeutic level for an extended period of time, with reduced lag time compared to the currently marketed ARI LAI product.

L. Tang et al. [5] successfully prepared a multi-walled carbon nanotube (MWNT)-based drug delivery system with the synergistic effect of PTT photothermal therapy and chemotherapy for efficient tumor removal. The integration of photothermal agents ICG-NH2 to MWNT was achieved by linking hyaluronic acid (HA). To realize the synergistic therapeutic effect of chemotherapy and phototherapy, DOX was attached on the wall of MWNT via a π–π interaction to obtain the final MWNT-HA-ICG/DOX nanocomplexes. Both in vitro and in vivo experiments verified the great therapeutic efficacy of MWNT-HA-ICG/DOX nanocomplexes.

L. S. Mbatha et al. [6] formulated folic acid (FA)-modified, poly-amidoamine-generation-5 (PAMAM G5D)-grafted gold nanoparticles (AuNPs) and evaluated their cytotoxicity and transfection efficiency using the luciferase reporter gene (FLuc-mRNA) in vitro. These nanosystems showed low cytotoxicity and good transfection efficiency.

S. Yin et al. [7] prepared NPs based on the insertion of two types of functional peptides, half-life extension peptide PAS and tumor-targeting peptide RGDK (Arg-Gly-Asp-Lys), into human heavy-chain ferritin (HFn) at the C-terminal through flexible linkers with two distinct enzyme-cleavable sites. RGDK peptide enhanced the internalization efficiency of HFn and showed a significant increase in growth inhibition. Pharmacokinetic study in vivo demonstrated that PAS peptides extended ferritin half-life. RGDK peptides greatly enhanced drug accumulation in the tumor site, rather than in other organs, in a biodistribution analysis. Drug-loaded, PAS-RGDK-functionalized HFns curbed tumor growth with significantly greater efficacies in comparison with drug-loaded HFn.

C. E. Torres et al. [8] prepared magnetoliposomes (MLP), which are liposomes that contain magnetite nanoparticles (MNP) inside. This study presents a low-cost microfluidic approach for the synthesis and purification of MLPs to improve their biocompatibility, with functional testing via hemolysis, platelet aggregation, cytocompatibility, internalization, and endosomal escape assays to determine their potential application in gastrointestinal delivery. In addition, the authors achieved encapsulation efficiencies between 20% and 90% by varying the total flow rates (TFRs), flow rate ratios (FRRs), and MNP concentrations.

F. Bintang Ilhami et al. [9] developed a new concept in cooperative adenine–uracil (A-U) hydrogen bonding interactions between anticancer drugs and nanocarrier complexes, which was successfully demonstrated by invoking the co-assembly of water-soluble, uracil end-capped polyethylene glycol polymer (BU-PEG) upon association with the hydrophobic drug adenine-modified rhodamine (A-R6G). This concept holds promise as a smart and versatile drug delivery system, which leads to the formation of self-assembled A-R6G/BU-PEG nanogels in aqueous solution, for the achievement of targeted, more efficient cancer chemotherapy.

A. Jagusiak et al. [10] described the Congo red–doxorubicin (CR-DOX) complexes, analyzed their interaction with some proteins, and explained the mechanism of this interaction. This kind of interaction between CR-DOX and the described proteins may in future become an important therapeutic system, with the possibility of targeted drug transport and delivery. Supramolecular ribbon-like CR complexed with doxorubicin is a promising system in the treatment of cancers and may open new avenues for novel treatment strategies.

We would like to thank all the authors and reviewers of this Special Issue. We also acknowledge the Assistant Editor, Ms. Daisy Tu, for her tremendous efforts in ensuring its implementation. In addition, authors are encouraged to submit original research articles and reviews in the next Special Issue, "Supramolecular Systems for Gene and Drug Delivery (Volume II)", led by us.

Funding: F. J. Ostos thanks the Junta de Andalucía for the postdoctoral grant (PAIDI-DOCTOR, DOC_00963). J. A. Lebrón also thanks the Fundación ONCE funded by the Fondo Social Europeo.

References

1. Asela, I.; Donoso-González, O.; Yutronic, N.; Sierpe, R. β-Cyclodextrin-Based Nanosponges Functionalized with Drugs and Gold Nanoparticles. *Pharmaceutics* **2021**, *13*, 513. [CrossRef] [PubMed]
2. Gómez-González, B.; García-Río, L.; Basílio, N.; Mejuto, J.C.; Simal-Gandara, J. Molecular Recognition by Pillar[5]arenes: Evidence for Simultaneous Electrostatic and Hydrophobic Interactions. *Pharmaceutics* **2021**, *14*, 60. [CrossRef] [PubMed]
3. Lebrón, J.A.; López-López, M.; García-Calderón, C.B.; Rosado, I.V.; Balestra, F.R.; Huertas, P.; Rodik, R.V.; Kalchenko, V.I.; Bernal, E.; Moyá, M.L.; et al. Multivalent Calixarene-Based Liposomes as Platforms for Gene and Drug Delivery. *Pharmaceutics* **2021**, *13*, 1250. [CrossRef] [PubMed]
4. Karava, V.; Siamidi, A.; Vlachou, M.; Christodoulou, E.; Bikiaris, N.D.; Zamboulis, A.; Kostoglou, M.; Gounari, E.; Barmpalexis, P. Poly(l-Lactic Acid)-co-poly(Butylene Adipate) New Block Copolymers for the Preparation of Drug-Loaded Long Acting Injectable Microparticles. *Pharmaceutics* **2021**, *13*, 930. [CrossRef] [PubMed]
5. Tang, L.; Zhang, A.; Mei, Y.; Xiao, Q.; Xu, X.; Wang, W. NIR Light-Triggered Chemo-Phototherapy by ICG Functionalized MWNTs for Synergistic Tumor-Targeted Delivery. *Pharmaceutics* **2021**, *13*, 2145. [CrossRef] [PubMed]
6. Mbatha, L.S.; Maiyo, F.; Daniels, A.; Singh, M. Dendrimer-Coated Gold Nanoparticles for Efficient Folate-Targeted mRNA Delivery In Vitro. *Pharmaceutics* **2021**, *13*, 900. [CrossRef] [PubMed]
7. Yin, S.; Wang, Y.; Zhang, B.; Qu, Y.; Liu, Y.; Dai, S.; Zhang, Y.; Wang, Y.; Bi, J. Engineered Human Heavy-Chain Ferritin with Half-Life Extension and Tumor Targeting by PAS and RGDK Peptide Functionalization. *Pharmaceutics* **2021**, *13*, 521. [CrossRef]
8. Torres, C.E.; Cifuentes, J.; Gómez, S.C.; Quezada, V.; Giraldo, K.A.; Puentes, P.R.; Rueda-Gensini, L.; Serna, J.A.; Muñoz-Camargo, C.; Reyes, L.H.; et al. Microfluidic Synthesis and Purification of Magnetoliposomes for Potential Applications in the Gastrointestinal Delivery of Difficult-to-Transport Drugs. *Pharmaceutics* **2022**, *14*, 315. [CrossRef]
9. Ilhami, F.B.; Bayle, E.A.; Cheng, C.-C. Complementary Nucleobase Interactions Drive Co-Assembly of Drugs and Nanocarriers for Selective Cancer Chemotherapy. *Pharmaceutics* **2021**, *13*, 1929. [CrossRef]
10. Jagusiak, A.; Chłopaś, K.; Zemanek, G.; Kościk, I.; Roterman, I. Interaction of Supramolecular Congo Red and Congo Red-Doxorubicin Complexes with Proteins for Drug Carrier Design. *Pharmaceutics* **2021**, *13*, 2027. [CrossRef]

 *pharmaceutics*

Article

β-Cyclodextrin-Based Nanosponges Functionalized with Drugs and Gold Nanoparticles

Isabel Asela [1,†], Orlando Donoso-González [1,2,†], Nicolás Yutronic [1,*] and Rodrigo Sierpe [1,2,3,*]

[1] Laboratorio de Nanoquímica y Química Supramolecular, Departamento de Química, Facultad de Ciencias, Universidad de Chile, Las Palmeras 3425, Ñuñoa, 7800003 Santiago, Chile; i.asela.m@gmail.com (I.A.); orlando.donoso@ug.uchile.cl (O.D.-G.)

[2] Laboratorio de Nanobiotecnología y Nanotoxicología, Departamento de Química Farmacológica y Toxicológica, Facultad de Ciencias Químicas y Farmacéuticas, Universidad de Chile, Santos Dumont 964, Independencia, 8380000 Santiago, Chile

[3] Laboratorio de Biosensores, Departamento de Química Farmacológica y Toxicológica, Facultad de Ciencias Químicas y Farmacéuticas, Universidad de Chile, Santos Dumont 964, Independencia, 8380000 Santiago, Chile

* Correspondence: nyutroni@uchile.cl (N.Y.); rsierpe@ciq.uchile.cl (R.S.); Tel.: +562-29787255 (N.Y.); +562-29782885 (R.S.)

† Main Authors.

Citation: Asela, I.; Donoso-González, O.; Yutronic, N.; Sierpe, R. β-Cyclodextrin-Based Nanosponges Functionalized with Drugs and Gold Nanoparticles. *Pharmaceutics* **2021**, *13*, 513. https://doi.org/10.3390/pharmaceutics13040513

Academic Editors: Francisco José Ostos, José Antonio Lebrón and Pilar López-Cornejo

Received: 17 February 2021
Accepted: 4 April 2021
Published: 8 April 2021

Abstract: Drugs are widely used as therapeutic agents; however, they may present some limitations. To overcome some of the therapeutic disadvantages of drugs, the use of β-cyclodextrin-based nanosponges (βCDNS) constitutes a promising strategy. βCDNS are matrices that contain multiple hydrophobic cavities, increasing the loading capacity, association, and stability of the included drugs. On the other hand, gold nanoparticles (AuNPs) are also used as therapeutic and diagnostic agents due to their unique properties and high chemical reactivity. In this work, we developed a new nanomaterial based on βCDNS and two therapeutic agents, drugs and AuNPs. First, the drugs phenylethylamine (PhEA) and 2-amino-4-(4-chlorophenyl)-thiazole (AT) were loaded on βCDNS. Later, the βCDNS–drug supramolecular complexes were functionalized with AuNPs, forming the βCDNS–PhEA–AuNP and βCDNS–AT–AuNP systems. The success of the formation of βCDNS and the loading of PhEA, AT, and AuNPs was demonstrated using different characterization techniques. The loading capacities of PhEA and AT in βCDNS were 90% and 150%, respectively, which is eight times higher than that with native βCD. The functional groups SH and NH_2 of the drugs remained exposed and allowed the stabilization of the AuNPs, 85% of which were immobilized. These unique systems can be versatile materials with an efficient loading capacity for potential applications in the transport of therapeutic agents.

Keywords: β-cyclodextrin-based nanosponge; phenylethylamine; 2-amino-4-(4-chlorophenyl)thiazole (AT); gold nanoparticles; carrier of therapeutic agents

1. Introduction

β-cyclodextrin (βCD) is a cyclic oligosaccharide approved by the FDA (Food and Drug Administration) that has been widely used as a pharmaceutical excipient in food products, textiles, cosmetics, and medical products [1]. In modern drug delivery investigations, βCD has been used as a host molecule for the preparation of drug carrier systems in diverse forms, such as vesicles, hydrogels, micelles, metal–organic systems, and nanoparticles [2–6]. Structural modifications of native βCD have been shown to increase its inclusion capacity and solubility and have allowed bioapplications of a large number of guest biomolecules [7–9]. An innovative modification to βCD recently studied was the synthesis of a polymeric cross-linked network, forming a highly porous and branched matrix of nanometric dimensions called the β-cyclodextrin-based nanosponge (βCDNS) [10,11]. This nanostructure contains multiple lipophilic cavities and carbonate bridges, leading to a network of hydrophilic channels [12], which allows βCDNS to serve as a polymeric conjugate, increasing the loading capacity, association, and stability of the included drugs [7,13–19]. Notably, a high loading capacity is a characteristic feature of βCDNS since they can interact with different molecules of suitable dimensions, using either the cavities of βCD or the multiple pores generated in the crosslinking [7,11,12]. Due to the above, studies on βCDNS applied to drug administration have been reported.

Drugs are some of the most widely used therapeutic agents; however, they may present some limitations, such as early instability, poor aqueous solubility, and low bioavailability. Therefore, strategies for the inclusion of drugs in matrices of native or modified βCD have been an excellent alternative for solving these disadvantages. In this work, the loading of the drugs phenylethylamine (PhEA) and 2-amino-4-(4-chlorophenyl)-thiazole (AT) on βCDNS was studied, which led to formation of two new systems: βCDNS–PhEA and βCDNS–AT. PhEA is a psychoactive stimulant that is used as an antidepressant without inducing tolerance; however, it is rapidly metabolized in organisms by the MAO-B enzyme, making it difficult to reach the site of action [20,21]. AT is a thiazole derivative that is currently used as an antimicrobial and anti-inflammatory agent but is rapidly degraded and has a poor aqueous solubility [22–24]. Our group previously studied complex formation between native βCD and these drugs. An increase was reported in the aqueous solubility and stability of PhEA over time due to its inclusion; in addition, the drug was released from the βCD–PhEA complex using laser irradiation and gold nanoparticles (AuNPs) present in the medium [25]. The inclusion of AT in βCD increased its aqueous solubility, allowing the application of higher doses in in vitro studies of permeability and antibacterial activity. Finally, it was demonstrated that the βCD–AT complex maintained its antibacterial activity against six strains of clinical relevance [26]. In this sense, the incorporation of βCDNS could show novel results, increasing the loading capacity or solubility, among other advantages [7,10,11,16]. Notably, AuNPs could also be added as a remarkable second therapeutic agent.

AuNPs have been widely employed in nanobiotechnology due to their unique properties, which allow them to be incorporated into new nanomaterials [27]. The main characteristics of AuNPs include their optoelectronic properties, as surface plasmon resonance (SPR), which are related to their shape, size, and large surface-to-volume ratio; besides its excellent biocompatibility and low toxicity [28]. The chemical reactivity of the surface atoms of AuNPs allows their functionalization and assembly with various chemical species, enabling their application in chemical and biological sensing, imaging, therapeutics, detection and diagnostics, biolabeling, and drug delivery [29–35]. Notably, AuNPs have been used as therapeutic and diagnostic agents, even in hard-to-reach places, such as the brain, since they can cross the blood–brain barrier [36–40]. Due to their photothermal properties, AuNPs can release compounds that are attached or close to its surface, such as drugs, at specific sites of action in a controlled manner due to the generation of thermal energy when excited by a laser specifically tuned to the SPR frequency [25,41,42]. Furthermore, AuNPs can accumulate passively in sites with an immature vasculature and with extensive fenestrations, such as tumor tissues, or in injured sites where an immune response develops. This is called the enhanced permeability and retention effect (EPR effect) [43–45]. It has been shown that the EPR effect combined with a longer blood circulation time of some types of nanoparticles can increase drug concentrations in tumors by 10 to 100 times compared to the use of free drugs [46]. In recent years, a series of complexes based on βCD have allowed the stabilization of AuNPs, building systems with promising applications in biological and chemical areas [25,26,47–53]. Therefore, if properly designed, βCDNS loaded with drugs and AuNPs could be used as new systems with biomedical applications, acting synergistically in nanotherapy.

In this work, we propose the development of a new system based on βCDNS for the transport of two therapeutic agents, drugs and AuNPs. For this, inclusions of PhEA and AT were studied separately. Later, the complexes were functionalized with AuNPs, forming the βCDNS–PhEA–AuNP and βCDNS–AT–AuNP systems. We believe that these unique systems, based on βCDNS, drugs, and AuNPs, can be versatile materials with potential applications in the therapy and diagnosis of diseases.

2. Materials and Methods

2.1. Material

Anhydrous βCD ($C_{42}H_{70}O_{35}$) $\geq$ 97%, 1134.98 g/mol; diphenylcarbonate (DPC, $C_6H_5O)_2CO$) 99%, 214.22 g/mol; PhEA ($C_8H_{11}N$) $\geq$ 99.5%, 121.18 g/mol, density (δ): 0.962 g/mL; AT ($C_9H_7ClN_2S$) $\geq$ 98%, 210.68 g/mol; sodium hydroxide (NaOH) $\geq$ 97%, 40.00 g/mol; tetrachloroauric acid ($HAuCl_4$) $\geq$ 99.9%, 393.83 g/mol; and sodium citrate ($Na_3C_6H_5O_7$) $\geq$ 99%, 294.10 g/mol were provided by Merck (Merck, Darmstadt, Germany). Ultrapure water (18 $MWcm^{-1}$) was obtained from a Milli-Q water system (Synergy UV equipment, Merck, Darmstadt, Germany).

2.2. Synthesis of β-Cyclodextrin Nanosponges

For βCDNS synthesis, anhydrous βCD and DPC were used as precursors. Synthesis was carried out by adapting Patel's protocol [54]. βCD (0.189 g) and 0.143 g of DPC were mixed in solid state at a 1:4 molar ratio βCD:DPC. A round-bottom flask with the mixture was heated inside an oil bath on a heating plate, with constant stirring for 5 h at 100 °C, observing its melting. The solid mixture obtained was ground in a mortar, washed with distilled water, and filtered under vacuum. The product was washed in a Soxhlet apparatus with acetone for 24 h, to remove phenol by-product. Later, it was moistened with water and dried for 2 h in a vacuum system using a Buchner funnel connected to a Kitasato flask to remove trace βCD. Finally, the product was dried for 72 h at 65 °C and stored.

2.3. β-Cyclodextrin Nanosponges Loading with Drugs

To load βCDNS with PhEA and AT, the saturated solutions method [55] was used with minor modifications. βCDNS were dispersed in a NaOH 0.1 M solution at room temperature, while the drugs were dissolved in ethanol. The solutions were mixed under constant agitation for 15 min and then left without agitation for 24 h. The resulting solution was centrifuged, and the supernatant was lyophilized and reserved [56,57]. The loading capacity of the βCDNS–PhEA and βCDNS–AT systems was calculated from the weights of βCD and drugs obtained using Equation (1) [58].

$$\text{Loading capacity} = \frac{\text{Weight of drug in βCDNS}}{\text{Weight of βCD in βCDNS}} \times 100 \tag{1}$$

2.4. Association Constant, K_a

For both drugs, studies were performed following the Higuchi and Connors method [59]. First, known concentrations (C) of each drug were measured by UV-Vis. From the A_{max} vs. C graph, the slope corresponded to the ε of each drug. Then, the βCDNS concentration versus the loaded drug concentration (calculated by Beer–Lambert law) was plotted. The value of the slope of the graphs related the amount of βCDNS added to the amount of solubilized drug, indicating the degree of solubilization. Degree of solubilization was used to calculate the association constant (K_a) and complexation efficiency of each system using Equations (2) and (3), respectively.

$$K_{a\,(1:1)} = \frac{\text{Degree of solubilization}}{[C_o](1 - \text{Degree of solubilization})} \tag{2}$$

$$\text{Complexation efficiency} = K_{a\,(1:1)}[C_o] = \frac{\text{Degree of solubilization}}{(1 - \text{Degree of solubilization})} \tag{3}$$

$[C_o]$ corresponds to the concentration of the free drug in the absence of βCDNS.

2.5. Synthesis of Gold Nanoparticles and Their Immobilization on β-Cyclodextrin Nanosponges–Drug Systems

Synthesis of AuNPs was performed using the Turkevich method [60]. A reflux system on a round-bottom flask (with three necks) was mounted by placing a thermometer, a condenser, and a rubber stopper on each neck. Here, 0.474 mL of $HAuCl_4$ was added with 18 mL of water. Sodium citrate (22.8 mg) was dissolved in 2.0 mL of water and heated at 60 °C for approximately 5 min. When aqueous solution of $HAuCl_4$ was refluxed and the gas–liquid equilibrium stabilized at a temperature of 186 °C, the citrate solution (at 60 °C) was added through the neck with the stopper. The reflux was continued under constant agitation ($6\times g$) for 30 min until a deep red solution was obtained. Later, the solution was cooled slowly to room temperature. The obtained AuNPs were filtered, diluted, set to pH 8.8 using an NaOH solution, and stored at 4 °C.

Immobilization was carried out via solubilization of the βCDNS–drug supramolecular complexes in an alkaline environment of AuNPs, setting the pH to 8.8 using NaOH. These mixtures formed homogeneous colloidal solutions that were centrifuged to decant only the βCDNS–drug systems interacting with AuNPs. Once the systems βCDNS–drug–AuNP were separated from the supernatant, they were resuspended in a new aqueous solution, forming the systems βCDNS–AT–AuNP and βCDNS–PhEA–AuNP. The concentration of AuNPs was calculated using UV-Vis spectroscopy. The molar extinction coefficient was obtained from the literature [61,62], and it was applied together with the Beer–Lambert equation.

2.6. Analysis by Nuclear Magnetic Resonance of Protons, ^{1}H-NMR

All the samples were dissolved in deuterated dimethylsulfoxide (DMSO)-d_6.

2.7. Preparation of Samples for Studies by Scanning and Transmission Electron Microscopy, SEM and TEM

For SEM studies, the βCD and βCDNS samples were prepared directly depositing the solid material onto carbon tape, then a gold coating was applied using magneton sputtering (pressure 0.5 mbar, Ar atmosphere, current 25 mA over 15 s). βCDNS–drug–AuNP samples were prepared by dropping aliquots on carbon tape, allowing them to dry overnight. The AuNPs immobilized on βCDNS–drug systems allowed the conductivity of these samples.

For TEM studies, the βCD and βCDNS samples were dissolved in ethanol (20% v/v), then mixed, sonicated, and dripped onto a copper grid with a continuous Formvar film. The βCDNS–drug samples were dissolved in ethanol (20% v/v), then mixed, sonicated, and dripped onto a holey carbon grid. Finally, all these samples were stained with phosphotungstic acid. The AuNPs samples were deposited directly on the grid with a continuous Formvar film.

2.8. Preparation of the Samples for Studies by Dynamic Light Scattering (DLS) and ζ Potential

βCDNS and βCDNS–drugs were redispersed to measurements. To determine the size distribution of the samples, the results were retrieved from the intensity distribution values using the cumulant method. The measurement conditions were set for organic βCD-based samples (refraction index: 1.49 and k: 0).

AuNPs and AuNPs with βCDNS–drug were diluted $10\times$ for measurements. Sonication and filtration were performed through a 0.45 μm filter. To determine the size distribution of AuNPs on the samples, the results were retrieved from the intensity distribution values using the cumulant method. On the other hand, the Smoluchowski approximation was used to calculate the ζ potentials from the measured electrophoretic mobility. The measurement conditions were set for colloidal gold samples (refraction index: 1.33 and k: 0.20).

2.9. Equipment Used for Characterization of the Samples

2.9.1. Nuclear Magnetic Resonance of Protons, ^{1}H-NMR

^{1}H-NMR characterizations of the βCDNS, PhEA, AT, and βCDNS–drug samples were performed in a Bruker Advance 400 MHz instrument (Bruker, Billerica, MA, USA) at 30 °C using TMS as an internal reference. The MestreNova program was used for data processing.

2.9.2. Infrared Spectroscopy, IR

The analyses were performed on a Jasco FT/IR-4600 instrument (Jasco, Easton, PA, USA). Spectral resolution: 1 cm^{-1}, number of scans: 4. CO_2 and H_2O correction through the software of the equipment was made. Baseline correction of KBr was performed.

2.9.3. Thermogravimetric Analysis, TGA

Analyses were performed on Perkin-Elmer model 4000 equipment (Perkin-Elmer, Waltham, MA, USA) over a temperature range from 0 °C to 800 °C with a rate of 10 °C/min under an air atmosphere with a flow of 20 mL/min.

2.9.4. Scanning and Transmission Electron Microscopy, SEM and TEM

For both characterizations, Inspect F50 HR-SEM instrument (Fei Company, Hillsboro, OR, USA) was used. For the scanning electron microscopy (SEM) images, an Everhart-Thornley detector was used, while for the transmission electron microscopy (TEM) images, the detector was scanning transmission electron microscope (STEM). An acceleration voltage of 10.0 kV, pressure of 9.71×10^{-8} Pa, and observation magnitudes of 16,000× and 100,000× were used.

2.9.5. UV-Visible Spectrophotometry

A Shimadzu UV-2450 instrument (Shimadzu, Kyoto, Japan) was employed to obtain the absorbance spectra. Measurements were made in 1.0 cm diameter quartz cuvettes between 200 and 800 nm using water at pH 8.8 as the reference. The UVProve program, version 1.10, was used for data processing.

2.9.6. Dynamic Light Scattering (DLS) and ζ Potential

The samples were measured on a Malvern Zetasizer Nano ZS instrument (Malvern, Malvern, UK).

2.9.7. Lyophilization of the Samples

BenchTop Pro, Omnitronic team equipment, SP Scientific (Omnitronic team, Gardiner, NY, USA) was used.

For data processing and graphic design, OriginPro 8.0 software (OriginLab, Northampton, MA, USA) was used.

2.9.8. Metallization of the Samples

PELCO SC-6 magnetron sputtering (PELCO, Fresno, CA, USA) was used. A gold foil was placed inside the vacuum chamber at 0.5 mbar, under inert atmosphere of argon. To begin the process, a current of 25 mA was used over 15 s to ionize the gas, hitting the metal foil and releasing Au atoms. These Au atoms were deposited over the βCD and βCDNS systems.

3. Results and Discussion

3.1. Synthesis and Characterization of β-Cyclodextrin-Based Nanosponges

Different synthesis routes have been reported for βCDNS formation, and they use ultrasonic baths; heating plates; solvents, such as ethanol or acetone for the washing stages; and even different molar ratios of βCD and DPC [7,10]. For this reason, different methodologies were evaluated to optimize the synthesis of βCDNS, eliminate byproducts, and increase yield. For the ultrasonic bath (A) and heating plate (B) methods, the use of acetone and a 1:4 molar ratio showed yields greater than 60%, as shown in Figure A1 (Appendix A). Considering the reproducibility of the synthesis and the lower amount of generated byproducts exhibited by method B relative to method A, method B with a heating plate was selected.

βCDNS formation was confirmed using ^{1}H-NMR. The technique allowed us to compare the chemical shifts of the signals for βCD protons in βCDNS and in native βCD. Figure 1 shows the spectra of (A) βCD, (B) DPC, and (C) βCDNS with a scheme showing the proton assignments for βCD and DPC. Table 1 shows the proton assignment for βCD and their respective chemical shifts and integrations in the ^{1}H-NMR spectra. The shifts of the signals are due to the change in the chemical environment of the βCD matrices when they are linked to form βCDNS. Notably, the greatest changes were observed in the integration delta ($\Delta\int$) of the hydroxyl groups, because they react with DPC to form linkers between βCD matrices, strongly suggesting βCDNS formation.

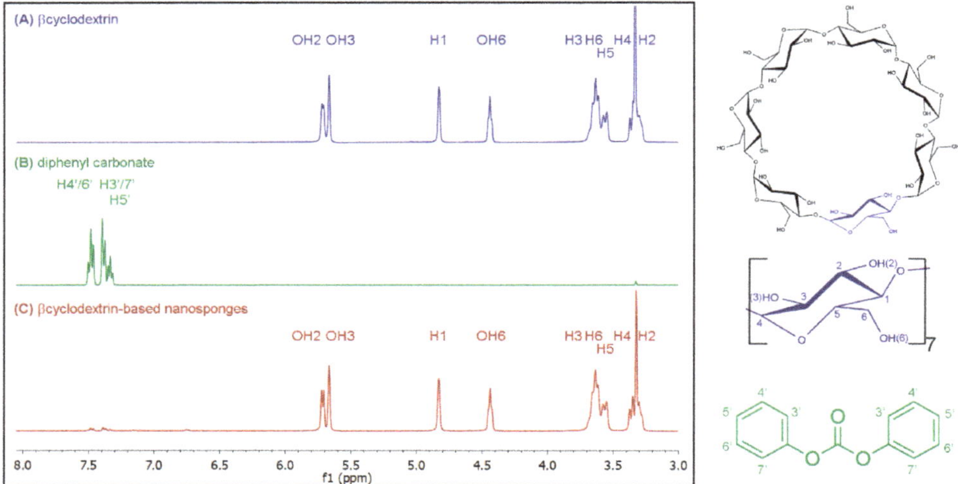

Figure 1. ^{1}H-NMR spectra of (**A**) native β-cyclodextrin (βCD), (**B**) diphenylcarbonate (DPC), and (**C**) β-cyclodextrin-based nanosponge (βCDNS) synthesized (**left**) together with the molecular structures and the assignments of the protons of βCD and DPC (**right**).

Table 1. Proton assignments, ^{1}H-NMR chemical shifts, and integrations of the β-cyclodextrin (βCD) and β-cyclodextrin-based nanosponge (βCDNS) signals.

Protons	δβCD (ppm)	δβCDNS (ppm)	$\lvert\Delta\delta\rvert$ (ppm)	βCD Integration ($\int$)	βCDNS Integration ($\int$)	$\lvert\Delta\int\rvert$
H1	4.825	4.827	0.002	7.00	7.00	0
H2	3.311	3.301	0.010	6.99	6.98	0.01
H3	3.669	3.655	0.014	6.99	6.98	0.01
H4	3.370	3.361	0.009	7.14	7.15	0.01
H5	3.566	3.562	0.004	7.04	7.04	0
H6	3.621	3.624	0.003	13.83	13.84	0.01
OH2	5.706	5.714	0.008	7.01	6.68	0.33
OH3	5.661	5.665	0.004	7.02	6.57	0.45
OH6	4.435	4.437	0.002	7.15	6.43	0.72

βCDNS formation was also characterized using IR vibrational spectroscopy. Commonly, this study focuses on comparing the signals of native βCD and βCD forming nanosponges and recognizing the vibration signal of the carbonyl group, which is an indicator of βCD crosslinking. Figure 2A shows the IR spectra of (A) βCD, (B) DPC, and (C) βCDNS. Characteristic peaks of βCD are observed at 3363 cm^{-1} (O-H alcohol stretching), 2924 cm^{-1} (C-H stretching), 1417 cm^{-1}, 1368 cm^{-1}, 1157 cm^{-1} (O-H bending), 1080 cm^{-1}, and 1029 cm^{-1} (C-O stretching). These data are consistent with literature data [63–65]. For βCDNS, the characteristic peaks are located mostly in the same regions observed for βCD, but with shifts or variations in intensity due to changes in the chemical environment. These were observed at 3366 cm^{-1} (O-H alcohol stretching), 2928 cm^{-1} (C-H stretching), 1645 cm^{-1} (C=O stretching), 1367, 1234, and 1155 cm^{-1} (O-H bending), and 1079 cm^{-1} and 1030 cm^{-1} (C-O stretching). Notably, the appearance of peaks at 1783, 1715, and 1235 cm^{-1} derived from signals present in DPC confirm the crosslinking of βCD forming nanosponges. The peak at 1760 cm^{-1} (C=O stretching) of DPC is masked by a peak in the βCDNS spectrum.

Thermogravimetry was performed to analyze and confirm the formation of βCDNS, differentiating it from its precursors through changes in their thermal decomposition, as is typically observed in the synthesis of polymeric materials [66]. Figure 2B shows thermograms of (A) βCD, (B) DPC, and (C) βCDNS. The loss of hydration water was observed in the first decomposition at temperatures up to 100 °C, with the percentage of mass loss being 11.5% for βCD and 2.7% for βCDNS of the total mass samples. Decomposition of 100% of the mass of DPC was observed in the range 130 to 250 °C. A second range of decomposition in βCD was observed between 300 and 350 °C, corresponding to a loss of 71% of the sample mass. For βCDNS, this second range was between 210 and 350 °C, consuming 70% of the total mass. The decrease in the temperature at the beginning of thermal degradation suggests that DPC, a crosslinker molecule, binds to the primary OH groups of βCDs, forming the nanopolymer through carbonyl groups. Changes in the peaks of the TGA curves (see Figure A2 in Appendix A) from 337 °C (βCD) to 327 (βCDNS) are typically observed in the formation of polymeric materials due to changes in chemical structure [67–69]. Finally, the oxidation interval for βCD ranged from 350 to 700 °C, encompassing 17.5% of the mass. However, βCDNS oxidation ranges from 350 to 580 °C, encompassing 27.3% of the mass. This also suggests modifications in the reactive structure of the polymer relative to native βCD.

To explain the change in the beginning of the range of thermal degradation for βCDNS, the average between the beginning temperatures for βCD and DPC, which were 300 and 130 °C, respectively, was evaluated. The calculated average temperature was 215 °C, which coincided with the value of the beginning of thermal degradation observed in the βCDNS thermogram, fulfilling the "eutectic mixture" criterion [70]. In addition, the high value of the degradation interval for βCDNS supports its thermal stability.

To obtain information on the morphology and size of βCDNS, the material was characterized using electron microscopy techniques and DLS. Figure 3 shows micrographs obtained by FE-SEM of native βCD (Figure 3A) and βCDNS (Figure 3B), directly revealing the morphological differences between both. βCD has irregular crystalline structures, while βCDNS has a characteristic porous appearance. TEM images were obtained to determine the average diameter of βCDNS, which were previously dispersed by sonication. Figure 3C,D shows the βCDNS and the resulting histogram, respectively. The average diameter, obtained from the count of more than 450 nanoparticles seen in various TEM images, was 146 ± 54 nm (see more images in Figure A3 in Appendix A). The staining of the βCDNS sample revealed some βCD crystals, which was verified by obtaining TEM images of native βCD with the same dispersion and staining protocol described for βCDNS (see Figure A3 in Appendix A). In addition, a hydrodynamic diameter of 133.9 ± 66.9 nm was found for βCDNS using DLS. These size data are concordant and strongly suggest the nanometric dimensions of the system studied (see more details in Appendix C).

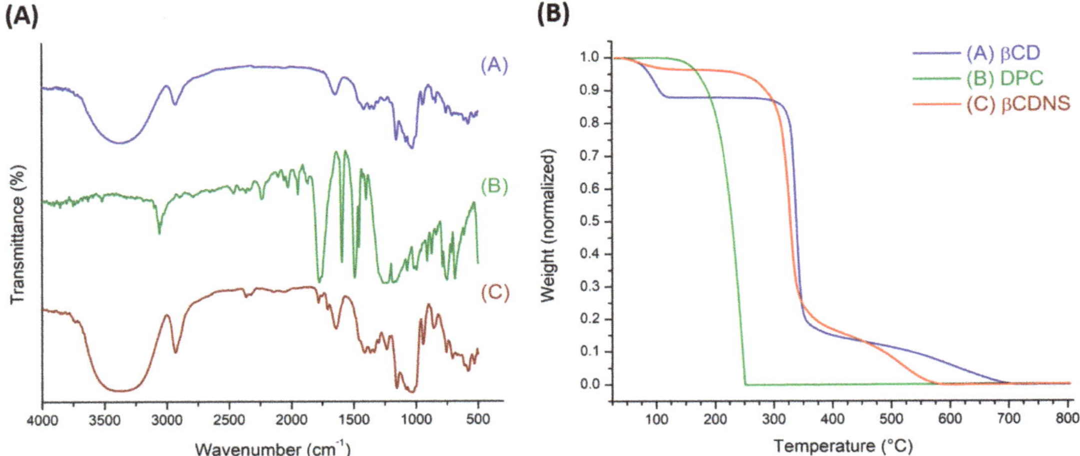

Figure 2. (**A**) FT-IR spectra of (A) βCD, (B) DPC, and (C) βCDNS; (**B**) normalized thermograms of (A) βCD, (B) DPC, and (C) βCDNS.

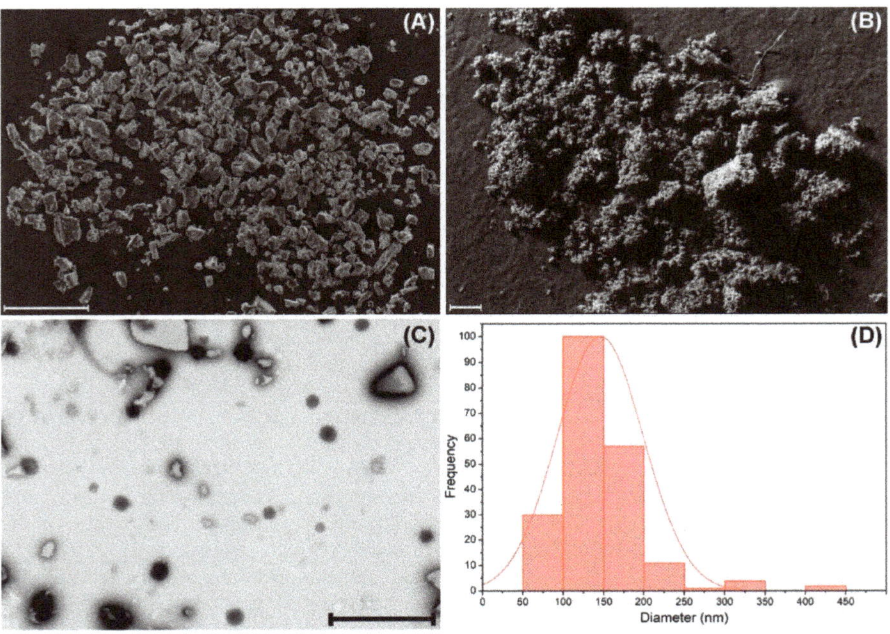

Figure 3. SEM micrographs of (**A**) βCD and (**B**) βCDNS. (**C**) transmission electron microscopy (TEM) micrograph of βCDNS and (**D**) the size distribution observed in TEM micrographs of βCDNS. Scale bar for figure (**A**) and (**B**) is 200 μm; scale bar for figure (**C**) is 1000 nm.

3.2. Loading of β-Cyclodextrin-Based Nanosponges with Drugs

The βCDNS obtained was loaded with two drugs separately, forming the βCDNS–PhEA and βCDNS–AT systems. Once each supramolecular complex was formed in the solubilized phase of the aqueous solution, the effective inclusion of the drugs and the stoichiometric relationship of both systems were analyzed using ^{1}H-NMR.

Figure 4 shows the spectra of βCDNS–PhEA (A) and βCDNS–AT (B) with their molecular structures and proton assignments for the respective drug. The loading of PhEA to form the βCDNS–PhEA system (A) and the loading of AT to form the βCDNS–AT system (B) were confirmed with the respective assignments of protons in the molecular structures of PhEA and AT (see full spectra, Figures A4 and A5, in Appendix B). Tables 2 and 3 show the chemical shifts and integrals recorded for the protons of βCDNS and of the PhEA and AT drugs resulting from the inclusion process.

For the βCDNS–PhEA system, Table 2, the largest chemical shifts for βCDNS were observed for the internal protons H3 and H5 and the hydroxyl groups OH2 and OH3, probably due to preferential inclusion in the widest zone of the βCD cavity. In addition, chemical shifts for all the βCDNS protons were observed, mainly towards lower fields, which demonstrates the effective loading of PhEA within βCD cavities and in the multiple interstitial spaces of the interstitial βCDNS produced by crosslinking. Analyzing the chemical shifts of the PhEA protons, a change in the chemical environment due to inclusion was also evidenced, consistent with that reported in the literature [25,71].

For the βCDNS–AT system, Table 3, chemical shifts were observed in all the βCDNS protons oriented towards the interior and exterior of the cavity due to the change in the chemical environment of βCDNS resulting from AT loading. This finding shows that the inclusion of the drug occurs in βCD cavities and between the formed interstitial spaces. Chemical shifts towards higher fields were observed in the protons NH$_2$b, Hb′/f′, and Hc′/e′ of AT, which demonstrates the electronic shielding effect of the drug due to its inclusion in the nanosponges, in accordance with that reported in the literature [26].

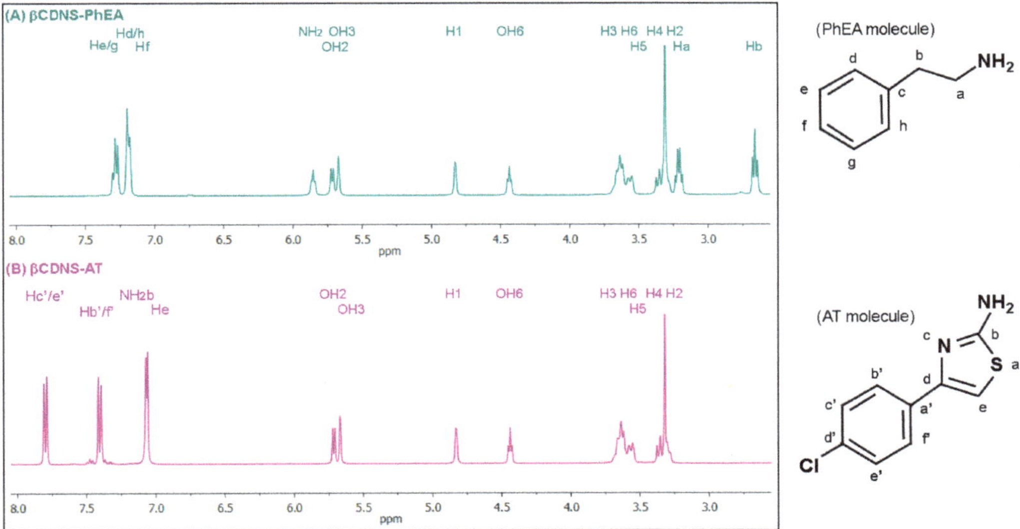

Figure 4. ^{1}H-NMR spectra of (**A**) phenylethylamine (PhEA) loaded in βCDNS (βCDNS–PhEA) and (**B**) 2-amino-4-(4-chlorophenyl)-thiazole (AT) loaded in βCDNS (βCDNS–AT) (**left**) together with the molecular structures and the assigning protons with respect to PhEA and AT (**right**).

Table 2. Proton assignments, ^{1}H-NMR chemical shifts, and integrations of the βCDNS, phenylethylamine (PhEA), and PhEA loaded in βCDNS (βCDNS–PhEA) signals.

Proton	δβCDNS (ppm)	δβCDNS–PhEA (ppm)	|Δδ| (ppm)	βCDNS Integration (∫)	βCDNS–PhEA Integration (∫)
H1	4.827	4.829	0.002	7.00	7.00
H2	3.301	-	-	6.98	-
H3	3.655	3.660	0.005	6.98	6.96
H4	3.361	3.363	0.002	7.15	7.03
H5	3.562	3.565	0.003	7.04	7.05
H6	3.624	3.626	0.002	13.84	13.93
OH2	5.713	5.719	0.006	6.68	6.67
OH3	5.665	5.671	0.006	6.57	6.64
OH6	4.437	4.436	0.001	6.43	6.51
Proton	**δPhEA (ppm)**	**δβCDNS–PhEA (ppm)**	**|Δδ| (ppm)**	**PhEA Integration (∫)**	**βCDNS–PhEA Integration (∫)**
NH$_2$	1.385	5.856	4.471	2.05	7.07
Ha	2.798	3.211	0.413	2.05	15.11
Hb	2.658	2.660	0.002	2.07	16.05
Hd/h	7.205	7.198	0.007	1.97	16.00
He/g	7.285	7.284	0.001	2.00	16.02
Hf	7.172	7.179	0.007	0.96	7.70

Table 3. Proton assignments, ^{1}H-NMR chemical shifts, and integrations of the βCDNS, 2-amino-4-(4-chlorophenyl)-thiazole (AT) and AT loaded in βCDNS (βCDNS–AT) signals.

Proton	δβCDNS (ppm)	δβCDNS–AT (ppm)	|Δδ| (ppm)	βCDNS Integration (∫)	βCDNS–AT Integration (∫)
H1	4.827	4.832	0.005	7.00	7.00
H2	3.301	-	-	6.98	-
H3	3.655	3.664	0.009	6.98	7.01
H4	3.361	3.367	0.006	7.15	7.09
H5	3.562	3.568	0.006	7.04	7.04
H6	3.624	3.629	0.005	13.84	13.86
OH2	5.713	5.717	0.004	6.68	6.70
OH3	5.665	5.669	0.004	6.57	6.53
OH6	4.437	4.440	0.003	6.43	6.53
Proton	**δAT (ppm)**	**δβCDNS–AT (ppm)**	**|Δδ| (ppm)**	**AT Integration (∫)**	**βCDNS–AT Integration (∫)**
He	7.061	7.061	0.000	0.99	8.49
NH2 b	7.079	7.073	0.006	2.00	16.10
Hb'/f'	7.408	7.407	0.001	1.95	16.09
Hc'/e'	7.803	7.801	0.002	2.00	16.52

Notably, the integration of the βCDNS protons and the protons of each drug in their respective ^{1}H-NMR spectra, Tables 2 and 3, showed a stoichiometric βCD:drug ratio of 1:8 in both systems, which is an amount of drug eight times greater than those reported for βCD–PhEA [25] and βCD–AT [26], each of which exhibits a 1:1 stoichiometry. This amount is equivalent to 0.9 mg of PhEA loading per 1 mg of βCD unit in βCDNS, and on the other hand, to 1.5 mg of AT loading per 1 mg of βCD unit in βCDNS. Applying Equation (1) [58] (Section 2, Material and Methods), the loading capacity in βCDNS is 90% for PhEA and 150% for AT, which is higher than the loading capacity of 11% for PhEA and 19% for AT in βCD native, according to reported data [25,26]. These results show that the drug loading of the βCDNS formed is higher than that of native βCD and that βCDNS could be used as a more efficient drug carrier than native βCD (see the details in the Appendix B).

The loading of drugs into βCDNS was also analyzed by FT-IR spectroscopy by comparing peaks for vibrations before and after the inclusion process. Figure 5 shows the spectra of (A) PhEA, (B) βCDNS-PhEA, (C) AT, and (D) βCDNS–AT.

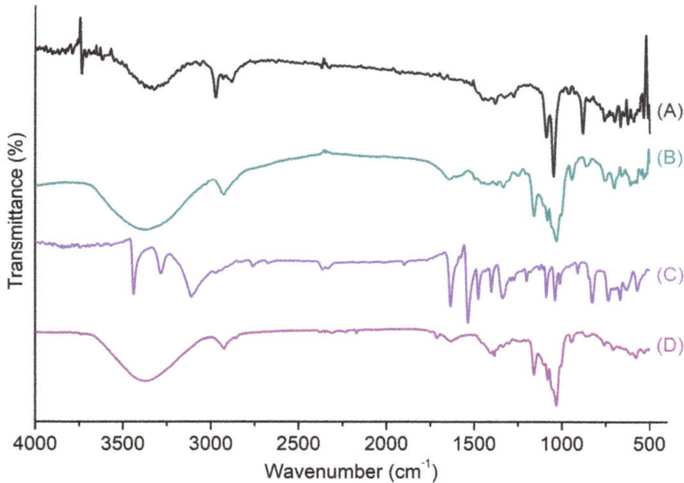

Figure 5. IR spectra of (**A**) phenylethylamine (PhEA), (**B**) PhEA loaded in βCDNS (βCDNS–PhEA), (**C**) 2-amino-4-(4-chlorophenyl)-thiazole (AT), and (**D**) AT loaded in βCDNS–AT.

In the vibrational analysis of the βCDNS–PhEA system, the βCDNS peaks at 3570 cm^{-1} and 3170 cm^{-1} corresponding to O-H alcohol stretching and N-H primary amine asymmetric and symmetric stretching, respectively, were identified. The peaks at 2926 cm^{-1} corresponding to C-H stretching, at 1642 cm^{-1} corresponding to C=O stretching, at 1333 cm^{-1} and 1157 cm^{-1} corresponding to O-H bending, and at 1081 cm^{-1} and 1029 cm^{-1} corresponding to C-O stretching were also identified. These vibrations remain unchanged in comparison to those of the βCDNS spectrum without loaded drugs. The peak from PhEA found for the βCDNS–PhEA system corresponding to N-H symmetric stretching was observed at 2950 cm^{-1}, while the peak at 745 cm^{-1} corresponding to C-H aromatics was masked due to the inclusion process.

In the case of the βCDNS–AT system, decreases in the intensity of some peaks with respect to those of βCDNS were observed. However, the characteristic peaks were located in the same regions of the spectra. O-H alcohol stretching, and N-H primary amine asymmetric and symmetric stretching vibrations were observed at 3170 cm^{-1} and 3570 cm^{-1}, respectively. C-H stretching appeared at 2924 cm^{-1}, C=O stretching at 1637 cm^{-1}, O-H group bending at 1384 cm^{-1} and 1157 cm^{-1}, and finally, C-O stretching appeared at 1079 cm^{-1} and 1029 cm^{-1}. The characteristic peaks of AT at 1476 cm^{-1}, corresponding to C=C aromatics, and at 3438 cm^{-1}, corresponding to N-H aromatic stretching, were masked in βCDNS–AT due to the inclusion in βCDNS.

The changes in the intensity and definition of the βCDNS peaks observed in the IR spectra suggested a change in their conformations due to drug loading, which was also corroborated by DLS and TEM. The hydrodynamic diameters of βCDNS–PhEA and βCDNS–AT were 270.5 ± 48.0 nm and 335.5 ± 150.5 nm, respectively, observing an increase in the size of both systems with respect to βCDNS (see more details in Appendix C). Figure 6 shows TEM images of βCDNS loaded with PhEA (A–E) and AT (F–I). Changes in the shapes of the systems with respect to that of βCDNS were also observed; in addition, the average diameter calculated using TEM images of these systems increased to 252 ± 39 nm with respect to βCDNS. The loading of the drugs PhEA and AT could promote a process of association and intermolecular interactions between different βCDNS. This would explain the increase in size observed using TEM and DLS.

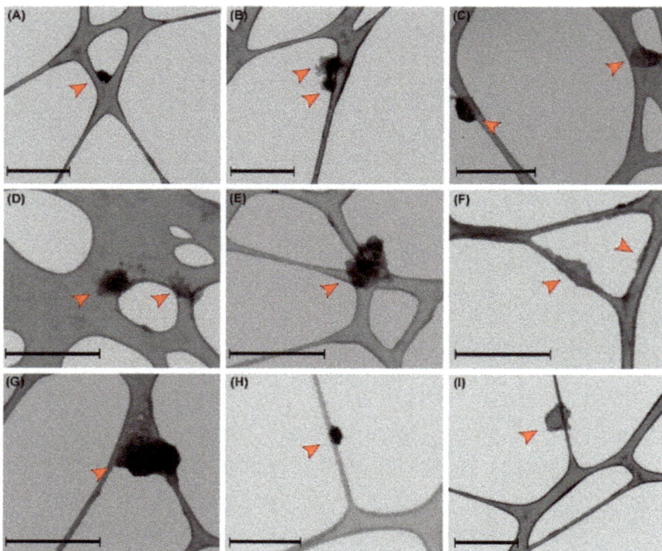

Figure 6. TEM micrographs of (**A–E**) βCDNS–PhEA and (**F–I**) βCDNS–AT. Scale bar for all images is 1000 nm (Red arrows highlight the nanosystems in the micrographs).

The degree of solubilization indicates the tendency to increase the aqueous solubility of the drugs due to the action of βCDNS, while the complexation efficiency corresponds to the concentration of drug included versus the concentration of drug initially used in the process. This is directly related to the effectiveness of βCDNS and intermolecular interactions to keep drugs entrapped in the complex. The degree of solubilization of the drugs, the K_a, and the complexation efficiency for the βCDNS–PhEA and βCDNS–AT systems were calculated using phase solubility studies (Equations (2) and (3), Section 2, Material and Methods) [59] and are shown in Table 4. Additionally, they were compared with the results obtained for the complexation of PhEA and AT using native βCD [25,26].

An increase in the aqueous solubility of PhEA and AT using βCDNS was observed, when they were compared to the solubility of free drugs (see Figures A7 and A8, Appendix B). Notably, the degree of solubilization achieved by the presence of βCDNS was more than 1.3 times higher for PhEA and 5 times higher for AT than with native βCD. This is especially relevant in therapy since drugs to be pharmacologically active must be soluble in water. The K_a values are 1318 M^{-1} and 484 M^{-1} for the βCDNS–PhEA and βCDNS–AT systems, respectively. These results indicate that the interactions that allow inclusion are strong, forming two highly stable systems over time due to the incorporation of βCDNS. The complexation efficiency values obtained for both systems show that the complexation using βCDNS is optimal, being the same for PhEA in native βCD and seven times greater for AT in native βCD. The above findings are in accordance with the previous discussion given by stoichiometry studies and loading capacity calculated using NMR (more details in the Appendix B).

Table 4. Comparative data on the degree of solubilization, association constants, and complexation efficiency of the drugs PhEA and AT included in βCDNS versus native βCD.

System	Degree of Solubilization	K_a (M^{-1})	Complexation Efficiency
βCDNS–PhEA	0.035	1318	0.037
βCDNS–AT	0.297	484	0.422
βCD–PhEA *	0.028	760	0.038
βCD–AT *	0.051	970	0.054

* Reference values obtained from the literature [25,26].

In general, the K_a values of the βCD complexes vary between 50 and 2000 M^{-1}. Lower values at 50 M^{-1} indicate a limitation in the pharmaceutical formulation since they have low stability and do not release the drug at its site of action [25,72–74]. On the other hand, K_a values greater than 2000 M^{-1} also present limitations, such as poor pharmacokinetics, since the drug release rates can be affected [72,73]. This is why the use of a strategy for the controlled release of the drugs included in βCDNS becomes relevant. AuNPs can release absorbed energy in the form of heat and can release molecules near their surface as a result of the photothermal effect [28,75,76]. This was demonstrated for a drug in AuNP- and βCD-based systems using laser irradiation [25,47]. In this sense, the incorporation of AuNPs into the two systems could, in addition to acting as a therapeutic agent, promote the controlled release of the drugs.

3.3. Synthesis and Immobilization of Gold Nanoparticles on Drug-Loaded β-Cyclodextrin-Based Nanosponges

Once the βCDNS–drug systems were obtained, the interactions with colloidal AuNPs were studied to load another therapeutic agent and form the βCDNS–PhEA–AuNP and βCDNS–AT–AuNP systems. AuNPs were synthesized following the Turkevich method at pH 5.5. These AuNPs were then stabilized at pH 8.8 to facilitate their immobilization on drug-loaded βCDNS. Figure 7A shows the absorbance spectra of AuNPs at pH 5.5 and 8.8, and Figure 7B shows a representative TEM micrograph of spherical AuNPs with an average diameter of 18 ± 4 nm (see histogram in Figure A9, Appendix C) AuNPs with diameters between 4 and 100 nm do not present cytotoxic effects [77], which would allow possible drug delivery applications.

Figure 7C,D shows the UV-Vis spectra of the βCDNS–PhEA–AuNP and βCD–AT–AuNP systems, respectively, in addition to those of the initial AuNP solution and the supernatant resulting from the functionalization of each mixture. The recorded plasmon bands demonstrate a preferential interaction of AuNPs with βCDNS–drug, with an immobilization of 85%, maintaining the main characteristics of the plasmon band and indicating that AuNPs remain stable in both systems.

Table 5 shows the intensities and the maximum wavelengths from the absorbance spectra. In addition, the hydrodynamic diameter and surface charge of βCDNS–PhEA–AuNP and βCDNS–AT–AuNP in aqueous solution are shown. These analyses represent the behavior of AuNPs in the different systems, because Au is highly efficient to absorb and scatter light, being superior to the organic material present.

A shift in the wavelength of the maximum absorbances with respect to those for the as-synthesized AuNPs occurred for both systems due to the interparticle coupling caused by the increased proximity between these nanostructures when immobilized; in turn, the permanence of the plasmon bands was evidence of the stability achieved and that the aggregation of AuNPs did not occur. In turn, increases in hydrodynamic diameters from 33.9 ± 13.2 nm for AuNPs with citrate to 51.2 ± 24.7 nm for AuNPs in the βCDNS–PhEA–AuNP system and up to 114.0 ± 42.2 nm for AuNPs in the βCDNS–AT system were observed due to the proximity between the immobilized AuNPs and the presence of βCDNS–drug complexes. Furthermore, this behavior was consistent with the increase in size of the βCDNS when they were loaded with the drugs. The reported partial and dynamic inclusion of AT in βCD could explain the greater hydrodynamic diameter of the

AuNPs on βCDNS–AT with respect to βCDNS–PhEA. The two functional groups, NH$_2$ and SH, of AT are exposed [26], facilitating its interaction with AuNPs, while PhEA only has one NH$_2$ group that is completely included within βCD [25,71].

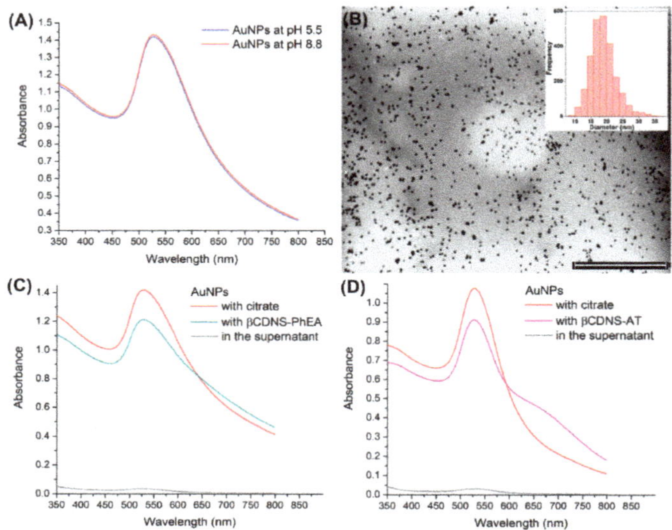

Figure 7. (**A**) UV-Vis spectra of AuNPs at pH 5.5 and pH 8.8; (**B**) TEM micrograph of AuNPs with their size histogram inserted (scale bar of 1000 nm); (**C**) UV-Vis spectra of AuNPs with citrate and with βCDNS–PhEA, including supernatant of the functionalization; and (**D**) UV-Vis spectra of AuNPs with citrate and with βCDNS–AT, including supernatant of the functionalization.

The registered surface charge of the AuNPs was −51.4 ± 7.9 mV due to the stabilizing citrate ions, which changed to −33.0 ± 5.3 mV and −38.4 ± 6.9 mV for AuNPs in the βCDNS–PhEA–AuNP and in the βCDNS–AT–AuNP systems, respectively, due to the replacement of a fraction of citrate molecules by neutral supramolecular complexes. As a control, a drug-free βCDNS solution was subjected to the same mixing protocol with colloidal AuNPs, confirming through different characterization techniques that the interaction between βCDNS and AuNPs does not occur (see the details in the Appendix C).

Table 5. Data obtained from the UV-VIS spectra, dynamic light scattering (DLS), and ζ potentials of the as-synthesized gold nanoparticles (AuNPs) and AuNPs immobilized on the βCDNS–PhEA and βCDNS–AT supramolecular systems.

System	Intensity (a.u.) of A$_{max}$	Wavelength (nm) of A$_{max}$	Immobilized AuNPs (%)	Hydrodynamic Diameter (nm)	PDI	Superficial Charge (mV)
AuNPs–citrate	1.419	528	0	33.9 ± 13.2	0.537	−51.4 ± 7.9
βCDNS–PhEA–AuNP	1.214	529	85.5	51.2 ± 24.7	0.571	−33.0 ± 5.3
βCDNS–AT–AuNP	0.912	531	84.7	114.0 ± 42.2	0.663	−38.4 ± 6.9

Figure 8A,B shows SEM micrographs of the βCDNS–PhEA–AuNP (A) and βCDNS–AT–AuNP (B) systems, respectively. The images clearly show the AuNPs immobilized on the βCDNS–drug supramolecular complexes. In addition, an irregular morphology was observed, probably due to the process of functionalization of βCDNS, as suggested by the TEM images (Figure 6).

Various characterization techniques and direct observation using electron microscopy confirmed the simultaneous loading of βCDNS with two therapeutic agents, drugs and AuNPs, forming the βCDNS–PhEA–AuNP and βCDNS–AT–AuNP systems. If properly designed, that is, by establishing parameters for the colloidal stability, concentration, surface charge, and size, among others, βCDNS and AuNPs could be considered nontoxic and used in therapy without generating adverse effects in the organism. In this sense, in the design and formation of these two new systems, the established parameters were realized.

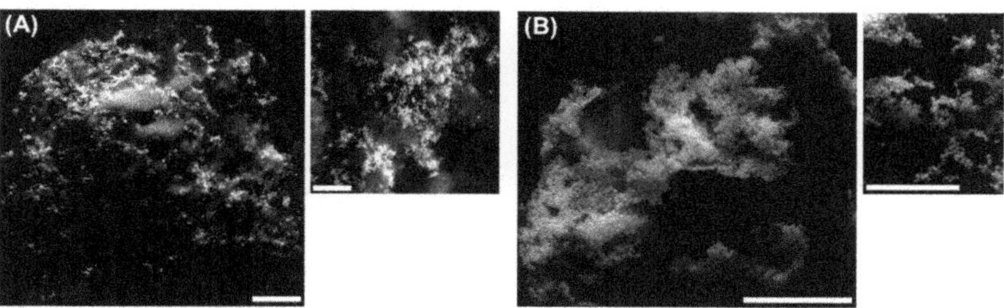

Figure 8. SEM micrographs of (**A**) βCDNS–PhEA with gold nanoparticles (AuNPs) immobilized on the surface and a zoomed view, with bar scales of 2000 nm and 500 nm; (**B**) βCDNS–AT with AuNPs immobilized on its surface and a zoomed view, with bar scales of 1000 nm and 500 nm.

4. Conclusions

The formation of βCDNS was confirmed by different techniques that indicated its polymeric characteristics and nanometric dimensions. Therapeutic agents PhEA and AT were successfully included in the multiple cavities of the nanostructures, forming the βCDNS–PhEA and βCDNS–AT systems. The loading capacity of βCDNS was 90% for PhEA and 150% for AT, being eight times higher than with native βCD. An increase in the aqueous solubility of PhEA and AT when complexed with βCDNS was demonstrated. In addition, a higher degree of solubilization and complexation efficiency of both drugs was obtained with βCDNS than with native βCD. The synthesized AuNPs were also loaded into each system, reaching an immobilization percentage of 85%. The hydrodynamic diameter and surface charge of AuNPs were 51 nm and −33 mV in the βCDNS–PhEA–AuNP system and 114 nm and −38 mV in the βCDNS–AT–AuNP system, respectively, which are relevant parameters for biological studies. βCDNS loaded with the two therapeutic agents (drug and AuNP) were observed directly by SEM images, showing the porous morphologies of the nanosponges and the nanoparticles immobilized on their surfaces due to the SH and NH_2 functional groups of the drugs. We believe that these unique systems, based on βCDNS, drugs, and AuNPs, can be versatile materials with an efficient loading capacity for potential applications in the transport of therapeutic agents. Finally, to continue researching in the field of drug delivery, studies that demonstrate the controlled release of PhEA and AT from βCDNS–drug–AuNP using laser irradiation are required and this, together with studies of cell permeability, toxicity, and pharmacological activity, has been considered in a future perspective.

Author Contributions: Conceptualization, I.A., N.Y. and R.S.; methodology, I.A., N.Y. and R.S.; validation, I.A. and R.S.; formal analysis, I.A., O.D.-G. and R.S.; investigation, I.A. and R.S.; resources, I.A., O.D.-G., N.Y. and R.S.; data curation, I.A. and R.S.; writing—original draft preparation, I.A., O.D.-G. and R.S.; writing—review and editing, I.A., O.D.-G., N.Y. and R.S.; visualization, I.A. and R.S.; supervision, I.A. and R.S.; project administration, I.A., N.Y. and R.S.; funding acquisition, I.A., O.D.-G., N.Y. and R.S. All authors have read and agreed to the published version of the manuscript.

Funding: Orlando Donoso-González gives thanks for financing of ANID doctoral scholarship No. 21180548. Rodrigo Sierpe gives thanks for financing to ANID-FONDECYT for postdoctoral research grant No. 3180706. Orlando Donoso and Rodrigo Sierpe acknowledge the financing of ANID-FONDAP No. 15130011.

Institutional Review Board Statement: Not applicable.

Informed Consent Statement: Not applicable.

Data Availability Statement: Not applicable.

Conflicts of Interest: The authors declare no conflict of interest.

Appendix A

Appendix A.1. β-cyclodextrin Nanosponge Synthesis

The methods for the synthesis of βCDNS were:

Appendix A.1.1. Method A. (Patel, 2014)

In a round-bottomed flask, βCD was mixed with DPC, and two different molar ratios were studied: 1:4 and 1:8. The flask was then placed in an ultrasonic bath at 90 °C for 5 h. The solid product obtained was then repeatedly washed with distilled water and vacuum filtered for 2 h using a Kitasato flask with a Büchner funnel with filter paper. The product was washed for 24 h in a Soxhlet system, and two different solvents, ethanol and acetone, were studied for this process; the product was finally dried and stored in an amber flask with a Teflon seal.

Appendix A.1.2. Method B. (Modified from Patel)

In a round-bottomed flask, βCD was mixed with DPC, and two different molar ratios were studied: 1:4 and 1:8. The flask was heated in an oil bath on a heating plate with constant agitation for 5 h at 100 °C. The mouth of the flask was covered with a septum, and a syringe was introduced into it to let the phenol gas released from the reaction pass through. The solid mixture obtained was extracted from the flask, a small amount of water was added and an ultrasonic bath was used to release residues from the walls. Once all the solid was extracted, it was ground in a mortar, washed with distilled water, and then vacuum filtered for 2 h. The product was washed for 24 h, and two different solvents, ethanol and acetone, were studied for this process: ethanol and acetone. Finally, the product was dried for between 48 and 72 h at 65 °C and stored.

Figure A1 shows a bar graph with the mass yields of different types of βCDNS syntheses with changes in the heating method, the solvents used in the washing steps and the βCD:DPC molar ratio.

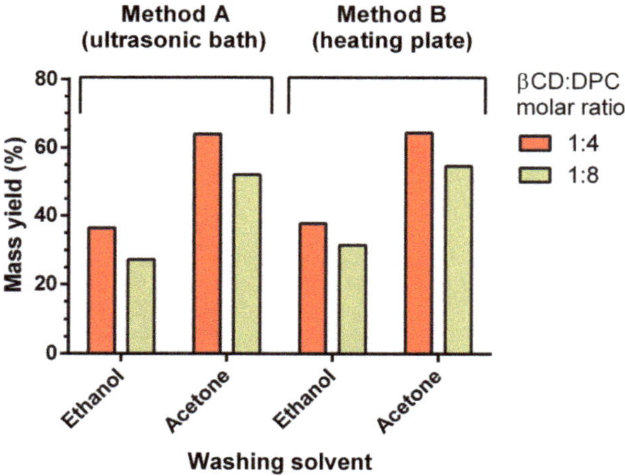

Figure A1. Mass yield (in percentage) of βCDNS versus the washing solvent, heating method and βCD:DPC molar ratio.

Appendix A.2. Derivative Curves from TGA Characterization of β-Cyclodextrin Nanosponge

Figure A2 shows the derivative curves from TGA (Figure 2B), analysing the differences in the peaks of the thermogravimetric decomposition curves.

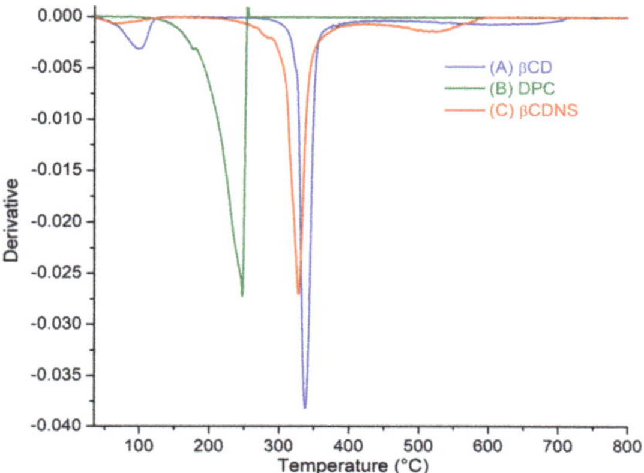

Figure A2. TGA derivative curves of (A) βCD, (B) DPC and (C) βCD.

Appendix A.3. TEM Characterization of Native β-Cyclodextrin

Figure A3 shows a TEM image of Figure A3A–C βCDNS to obtain its mean size; and Figure A3D native βCD with the same dispersion and staining protocol described for βCDNS.

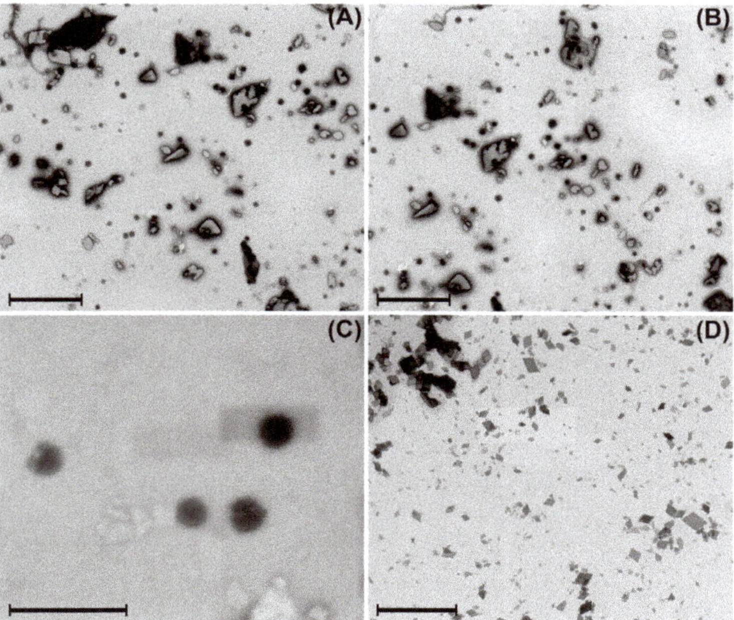

Figure A3. TEM micrographs of: (**A,B**) βCDNS with bar scales of 2000 nm; (**C**) βCDNS with bar scale of 400 nm; and (**D**) native βCD with bar scale of 4000 nm. Acceleration voltage used: 10.0 kV.

Appendix B

Appendix B.1. ^{1}H-NMR Full Spectra of Loading Drugs Process

Figures A4 and A5 show the full ^{1}H-NMR spectra of βCDNS, PhEA, AT, βCDNS-PhEA and βCDNS-AT, summarized in Figure 4.

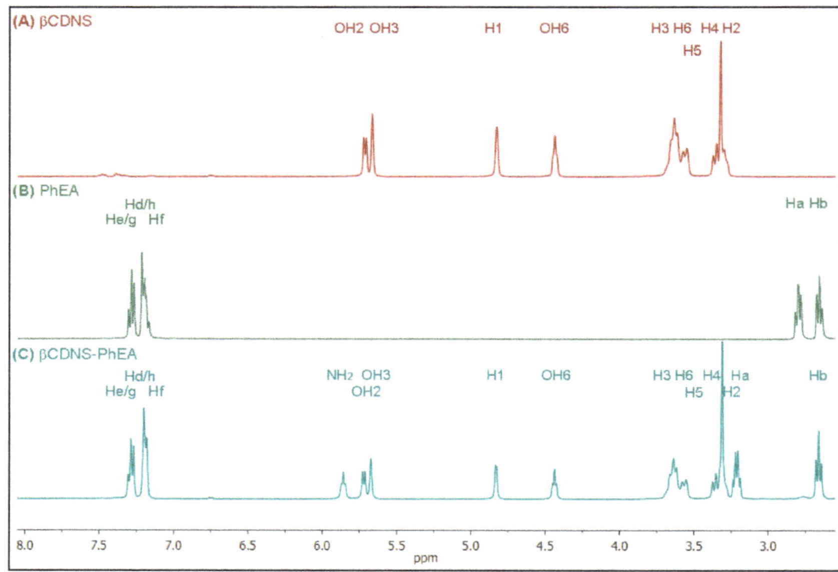

Figure A4. ^{1}H-NMR spectra of (**A**) βCDNS, (**B**) PhEA and (**C**) βCDNS-PhEA.

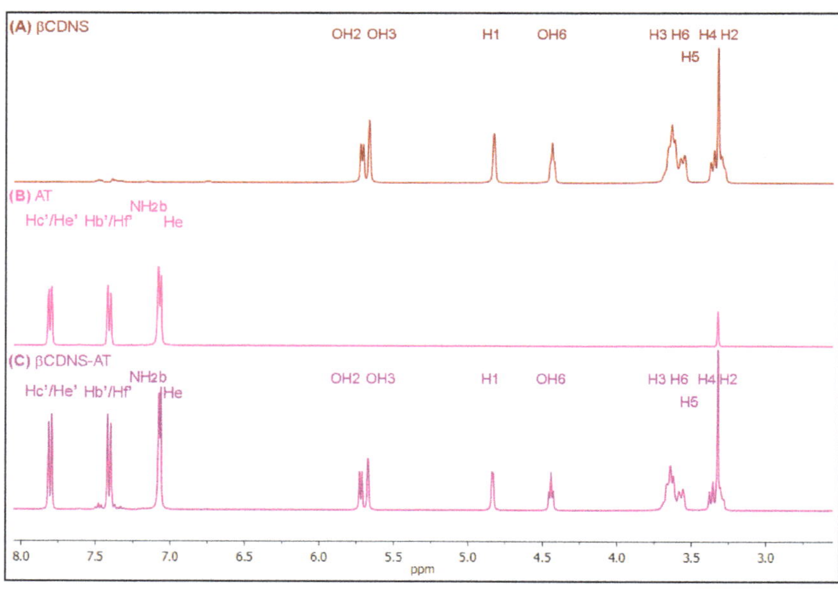

Figure A5. ^{1}H-NMR spectra of (**A**) βCDNS (**B**) AT and (**C**) βCDNS-AT.

Appendix B.2. Calculation of the Stoichiometric Ratio of the Drug Loading Process

The stoichiometric ratios were calculated in the ^{1}H-NMR spectra by comparing the integrals of the PhEA and AT protons with the integrals of the βCDNS protons from the βCDNS-PhEA and βCDNS-AT systems. First, the integrals of the PhEA signals (protons He/g, Hd/h, and Hf) were analysed using the H1 signal of βCDNS as reference, which integrated 7. In turn, the integrals of the AT signals (He, Hb'/f', Hc'/e' and NH$_2$b) were analysed using the H1 signal of βCDNS as reference, which integrated 7 (see the data in Tables A1 and A2). Finally, the stoichiometric ratios calculated are summarized in Table A3.

Table A1. Values of the integrated PhEA and βCD proton signals in the ^{1}H-NMR spectra of the βCDNS-PhEA system, with the integrated H1 proton signals of βCDNS as a reference.

Proton Signal	Reference	Integral	Counts	Ratios
H1	7	7.0	1	1
He/g	2	16.0	8.0	8
Hd/h	2	16.0	8.0	8
Hf	1	7.7	7.7	8

Table A2. Values of the integrated AT and βCD proton signals in the ^{1}H-NMR spectra of the βCDNS-AT system, with the integrated H1 proton signals of βCDNS as a reference.

Proton Signal	Reference	Integral	Counts	Ratios
H1	7	7	1	1
He	1	8.5	8.5	8
Hb'/f'	2	16.1	8.1	8
Hc'/e'	2	16.1	8.1	8
NH$_2$b	2	16.5	8.3	8

Table A3. Summary of the molar ratios of the drugs in the βCDNS for βCDNS-drug systems calculated from Tables A1 and A2.

	Molar Ratios	
System	βCDNS	Drug
βCDNS-PhEA	1	8 ($\pm$0.3)
βCDNS-AT	1	8 ($\pm$0.4)

Appendix B.3. Extinction Coefficient Determination of Drugs

For each drug, a calibration curve was made with stocks of aqueous solutions of known concentrations to obtain the value of ε. Table A4 presents the data used for this determination for PhEA and AT.

Table A4. Data on the concentrations of the drugs, PhEA and AT, and their absorbance maxima at 310 and 290 nm, respectively.

[PhEA] (mM)	Absorbance	[AT] (mM)	Absorbance
0.0441	0.2441	0.095	0.078
0.0882	0.3648	0.190	0.172
0.1323	0.5642	0.285	0.305
0.1764	0.7851	0.380	0.370
0.2205	1.0267	0.569	0.518
0.2646	1.3023	0.759	0.609
0.3087	1.4199	—	—

By plotting the PhEA concentration versus the maximum absorbance at 310 nm, the line shown in Figure A6A was obtained, and the value of ε was 4.7497 ± 0.2110 mM^{-1}cm^{-1}.

By plotting the AT concentration versus the maximum absorbance at 290 nm, the line presented in Figure A6B was obtained; the slope corresponds to the value of ε, which was $0.8031 \pm 0.0692\ \text{mM}^{-1}\text{cm}^{-1}$.

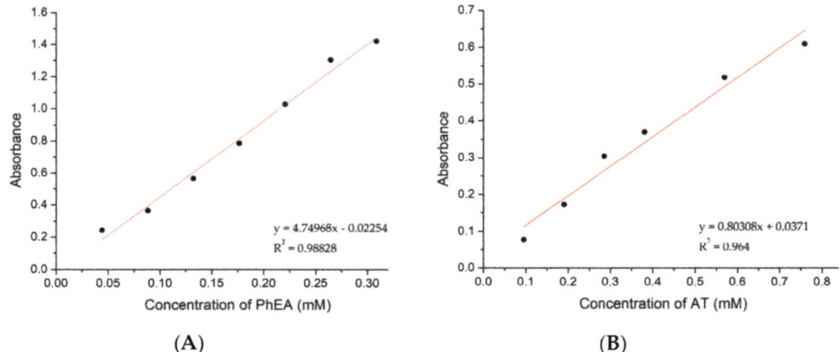

Figure A6. Linear plots of the (**A**) PhEA and (**B**) AT absorbance maxima at 310 and 290 nm, respectively, vs. concentrations.

Appendix B.4. Determinations of Association Constants of Drug Loading Systems

For the determination of K_a, a stock solution was prepared with 200 mg of βCDNS and water in a 25 mL measuring flask. Volumes of 0 to 2 mL of the stock were taken and diluted with water to produce a total volume of 2 mL with a fixed amount of each drug, 0.5 mL for PhEA and 5.0 mg for AT, added. All the data obtained are presented in Table A5 for PhEA and Table A6 for AT. By applying the extinction coefficient value ε to the Lambert-Beer equation, it was possible to determine the PhEA and AT concentrations in the different assays using the Higuchi-Connors method.

Table A5. Values of the different tests carried out to calculate the K_a and complexation efficiency of the βCDNS-PhEA system in water.

[βCDNS] (mM)	Absorbance	[PhEA] (mM)
5.920	1.126	0.237
5.328	1.006	0.212
4.736	0.896	0.189
4.144	0.775	0.163
3.552	0.654	0.138
2.960	0.564	0.119
2.368	0.478	0.101
1.776	0.404	0.085
0	0.132	0.028

Table A6. Values of the different tests carried out to calculate the K_a and complexation efficiency of the βCDNS-AT system in water.

[βCDNS] (mM)	Absorbance	[AT] (mM)
4.736	1.833	2.282
3.552	1.566	1.950
2.960	1.343	1.672
2.368	1.178	1.467
1.776	1.19	1.482
1.184	1.088	1.355
0.592	0.739	0.920
0	0.7	0.872

The linear relationship obtained from a plot of the solubilized PhEA concentration versus the added βCDNS concentration is shown in Figure A7. The value of the slope was 0.03534 (±0.00115). Using Equation (1), the association constant K_a was calculated, resulting in a value of 1318 M^{-1}. Finally, using Equation (2), the value of complexation efficiency was calculated, resulting in a value of 0.03663 for the βCDNS-PhEA system.

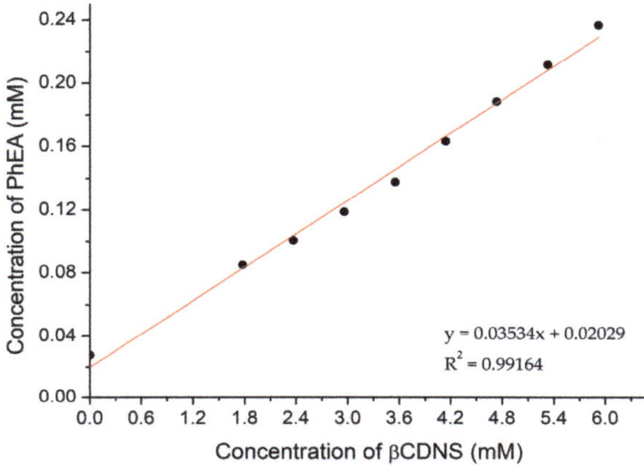

Figure A7. Graph of the concentration of solubilized PhEA versus the concentration of added βCDNS and the linear fit.

Figure A8 shows the linear relationship obtained from a plot of the solubilized AT concentrations versus the added βCDNS concentration. The value of the slope was 0.297 (±0.024). The K_a for the βCDNS-AT system was 484 M^{-1} and the complexation efficiency value was 0.422.

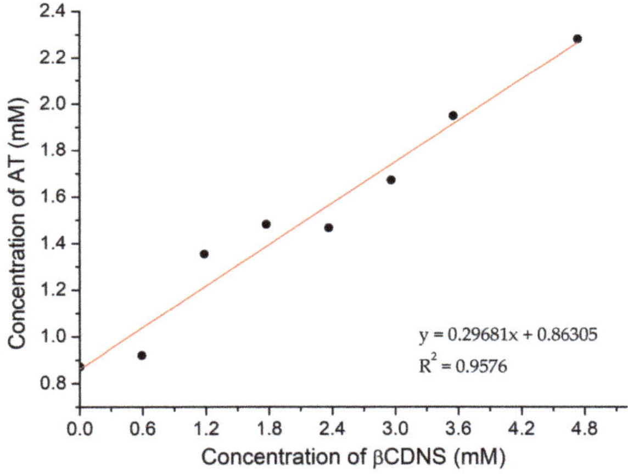

Figure A8. Graph of the concentration of solubilized AT versus the concentration of added βCDNS and the linear fit.

Appendix C

Appendix C.1. Size Histogram of Gold Nanoparticles

Figure A9 shows the size distribution histogram for synthesised AuNPs from representative TEM images. The observed diameter was 18 (±4) nm.

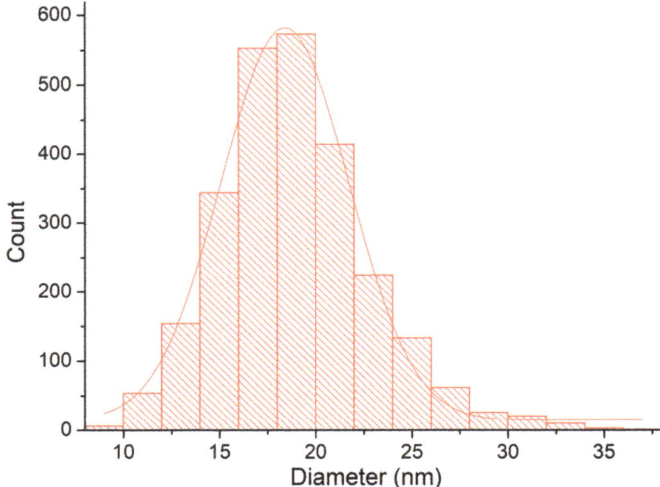

Figure A9. Histogram of size distribution of synthesised AuNPs.

Appendix C.2. Dynamic Light Scattering and ζ Potential Studies of Loading Systems and Gold Nanoparticles Interacting with Supramolecular Systems

The studies using DLS and the ζ potential are summarized in Table A7. βCDNS was dispersed in water (pH 8.8) and measured, while βCDNS-drug systems were sonicated and measured. AuNPs stabilized with citrate were filtered and then characterized. AuNPs with βCDNS and drugs were centrifuged and resuspended in water (pH 8.8) prior to characterization.

Table A7. Data obtained using dynamic light scattering (DLS) and the ζ potential of βCDNS and βCDNS-drugs, and AuNPs with citrate, βCDNS, βCDNS-PhEA and βCDNS-AT.

System	Hydrodynamic Diameter (nm)	Intensity Peak Area (%)	PDI	Surface Charge (mV)
βCDNS	133.9 ± 66.9	100.0	0.197	
βCDNS-PhEA	270.5 ± 48.0	87.0	0.398	
βCDNS-AT	335.5 ± 150.5	89.0	0.355	
AuNPs-citrate	33.9 ± 13.2	69.7	0.537	−51.40 ± 7.86
AuNPs-βCDNS	34.6 ± 12.9	70.0	0.542	−58.03 ± 7.01
βCDNS-PhEA-AuNP	51.2 ± 24.7	85.8	0.571	−33.03 ± 5.26
βCDNS-AT-AuNP	114.0 ± 42.2	86.3	0.663	−38.37 ± 6.90

References

1. Kurkov, S.V.; Loftsson, T. Cyclodextrins. *Int. J. Pharm.* **2013**, *453*, 167–180. [CrossRef] [PubMed]
2. Hartlieb, K.J.; Holcroft, J.M.; Moghadam, P.Z.; Vermeulen, N.A.; Algaradah, M.M.; Nassar, M.S.; Botros, Y.Y.; Snurr, R.Q.; Stoddart, J.F. CD-MOF: A versatile separation medium. *J. Am. Chem. Soc.* **2016**, *138*, 2292–2301. [CrossRef]
3. Ma, X.; Zhao, Y. Biomedical applications of supramolecular systems based on host-guest interactions. *Chem. Rev.* **2015**, *115*, 7794–7839. [CrossRef]
4. Mejia-Ariza, R.; Graña-Suárez, L.; Verboom, W.; Huskens, J. Cyclodextrin-based supramolecular nanoparticles for biomedical applications. *J. Mater. Chem. B* **2017**, *5*, 36–52. [CrossRef] [PubMed]
5. Wang, Y.; Liu, Y.; Liang, J.; Zou, M. A cyclodextrin-core star copolymer with Y-shaped ABC miktoarms and its unimolecular micelles. *RSC Adv.* **2017**, *7*, 11691–11700. [CrossRef]
6. Zhang, J.; Ma, P.X. Cyclodextrin-based supramolecular systems for drug delivery: Recent progress and future perspective. *Adv. Drug Deliv. Rev.* **2013**, *65*, 1215–1233. [CrossRef] [PubMed]
7. Sherje, A.P.; Dravyakar, B.R.; Kadam, D.; Jadhav, M. Cyclodextrin-based nanosponges: A critical review. *Carbohydr. Polym.* **2017**, *173*, 37–49. [CrossRef]

8.	Crini, G.; Fourmentin, S.; Fenyvesi, É.; Torri, G.; Fourmentin, M.; Morin-Crini, N. Cyclodextrins, from molecules to applications. *Environ. Chem. Lett.* **2018**, *16*, 1361–1375. [CrossRef]

9.	Singh, P.; Ren, X.; Guo, T.; Wu, L.; Shakya, S.; He, Y.; Wang, C.; Maharjan, A.; Singh, V.; Zhang, J. Biofunctionalization of β-cyclodextrin nanosponges using cholesterol. *Carbohydr. Polym.* **2018**, *190*, 23–30. [CrossRef]

10.	Caldera, F.; Tannous, M.; Cavalli, R.; Zanetti, M.; Trotta, F. Evolution of cyclodextrin nanosponges. *Int. J. Pharm.* **2017**, *531*, 470–479. [CrossRef]

11.	Chilajwar, S.V.; Pednekar, P.P.; Jadhav, K.R.; Gupta, G.J.C.; Kadam, V.J. Cyclodextrin-based nanosponges: A propitious platform for enhancing drug delivery. *Expert Opin. Drug Deliv.* **2014**, *11*, 111–120. [CrossRef] [PubMed]

12.	Trotta, F.; Cavalli, R. Characterization and applications of new hyper-cross-linked cyclodextrins. *Compos. Interfaces* **2009**, *16*, 39–48. [CrossRef]

13.	Alongi, J.; Poŝsković, M.; Frache, A.; Trotta, F. Novel flame retardants containing cyclodextrin nanosponges and phosphorus compounds to enhance EVA combustion properties. *Polym. Degrad. Stab.* **2010**, *95*, 2093–2100. [CrossRef]

14.	Ansari, K.A.; Torne, S.J.; Vavia, P.P.R.; Trotta, F.; Cavalli, R. Paclitaxel loaded nanosponges: In-vitro characterization and cytotoxicity study on MCF-7 cell line culture. *Curr. Drug Deliv.* **2011**, *8*, 194–202. [CrossRef]

15.	Gigliotti, C.L.; Minelli, R.; Cavalli, R.; Occhipinti, S.; Barrera, G.; Pizzimenti, S.; Cappellano, G.; Boggio, E.; Conti, L.; Fantozzi, R.; et al. In vitro and in vivo therapeutic evaluation of camptothecin-encapsulated β-cyclodextrin nanosponges in prostate cancer. *J. Biomed. Nanotechnol.* **2016**, *12*, 114–127. [CrossRef] [PubMed]

16.	Swaminathan, S.; Cavalli, R.; Trotta, F. Cyclodextrin-based nanosponges: A versatile platform for cancer nanotherapeutics development. *Wiley Interdiscip. Rev. Nanomed. Nanobiotechnol.* **2016**, *8*, 579–601. [CrossRef] [PubMed]

17.	Swaminathan, S.; Pastero, L.; Serpe, L.; Trotta, F.; Vavia, P.; Aquilano, D.; Trotta, M.; Zara, G.P.; Cavalli, R. Cyclodextrin-based nanosponges encapsulating camptothecin: Physicochemical characterization, stability and cytotoxicity. *Eur. J. Pharm. Biopharm.* **2010**, *74*, 193–201. [CrossRef]

18.	Venuti, V.; Rossi, B.; Mele, A.; Melone, L.; Punta, C.; Majolino, D.; Masciovecchio, C.; Caldera, F.; Trotta, F. Tuning structural parameters for the optimization of drug delivery performance of cyclodextrin-based nanosponges. *Expert Opin. Drug Deliv.* **2017**, *14*, 331–340. [CrossRef]

19.	Seglie, L.; Martina, K.; Devecchi, M.; Roggero, C.; Trotta, F.; Scariot, V. The effects of 1-MCP in cyclodextrin-based nanosponges to improve the vase life of *Dianthus caryophyllus* cut flowers. *Postharvest Biol. Technol.* **2011**, *59*, 200–205. [CrossRef]

20.	Irsfeld, M.; Spadafore, M.; Prüß, B.M. β-phenylethylamine, a small molecule with a large impact. *Webmedcentral* **2013**, *4*, 4409.

21.	Szabo, A.; Billett, E.; Turner, J. Phenylethylamine, a possible link to the antidepressant effects of exercise? *Br. J. Sports Med.* **2001**, *35*, 342–343. [CrossRef]

22.	Venkatachalam, T.K.; Sudbeck, E.A.; Mao, C.; Uckun, F.M. Anti-HIV activity of aromatic and heterocyclic Thiazolyl Thiourea compounds. *Bioorg. Med. Chem. Lett.* **2001**, *11*, 523–528. [CrossRef]

23.	Holla, B.S.; Malini, K.V.; Rao, B.S.; Sarojini, B.K.; Kumari, N.S. Synthesis of some new 2,4-disubstituted thiazoles as possible antibacterial and anti-inflammatory agents. *Eur. J. Med. Chem.* **2003**, *38*, 313–318. [CrossRef]

24.	Das, J.; Chen, P.; Norris, D.; Padmanabha, R.; Lin, J.; Moquin, R.V.; Shen, Z.; Cook, L.S.; Doweyko, A.M.; Pitt, S.; et al. 2-Aminothiazole as a novel kinase inhibitor template. Structure-activity relationship studies toward the discovery of N-(2-chloro-6-methylphenyl)-2-[[6-[4-(2-hydroxyethyl)-1-piperazinyl]-2-methyl-4-pyrimidinyl]amino]-1,3-thiazole-5-carboxamide (dasatinib, BMS-354825) as a potent *pan*-Src kinase inhibitor. *J. Med. Chem.* **2006**, *49*, 6819–6832. [CrossRef] [PubMed]

25.	Sierpe, R.; Lang, E.; Jara, P.; Guerrero, A.R.; Chornik, B.; Kogan, M.J.; Yutronic, N. Gold nanoparticles interacting with β-cyclodextrin-phenylethylamine inclusion complex: A ternary system for photothermal drug release. *ACS Appl. Mater. Interfaces* **2015**, *7*, 22–31. [CrossRef]

26.	Asela, I.; Noyong, M.; Simon, U.; Andrades-Lagos, J.; Campanini-Salinas, J.; Vásquez-Velásquez, D.; Kogan, M.; Yutronic, N.; Sierpe, R. Gold nanoparticles stabilized with βcyclodextrin-2-amino-4-(4-chlorophenyl) thiazole complex: A novel system for drug transport. *PLoS ONE* **2017**, *12*, e0185652. [CrossRef]

27.	Yeh, Y.C.; Creran, B.; Rotello, V.M. Gold nanoparticles: Preparation, properties, and applications in bionanotechnology. *Nanoscale* **2012**, *4*, 1871–1880. [CrossRef]

28.	Elahi, N.; Kamali, M.; Baghersad, M.H. Recent biomedical applications of gold nanoparticles: A review. *Talanta* **2018**, *184*, 537–556. [CrossRef] [PubMed]

29.	Zhang, Y.; Qian, J.; Wang, D.; Wang, Y.; He, S. Multifunctional gold nanorods with ultrahigh stability and tunability for in vivo fluorescence imaging, SERS detection, and photodynamic therapy. *Angewandte Chemie* **2013**, *52*, 1148–1151. [CrossRef]

30.	Shi, J.; Votruba, A.R.; Farokhzad, O.C.; Langer, R. Nanotechnology in drug delivery and tissue engineering: From discovery to applications. *Nano Lett.* **2010**, *10*, 3223–3230. [CrossRef]

31.	Guo, J.; Rahme, K.; He, Y.; Li, L.L.; Holmes, J.D.; O'Driscoll, C.M. Gold nanoparticles enlighten the future of cancer theranostics. *Int. J. Nanomed.* **2017**, *12*, 6131–6152. [CrossRef]

32.	Murphy, C.J.; Gole, A.M.; Stone, J.W.; Sisco, P.N.; Alkilany, A.M.; Goldsmith, E.C.; Baxter, S.C. Gold nanoparticles in biology: Beyond toxicity to cellular imaging. *Acc. Chem. Res.* **2008**, *41*, 1721–1730. [CrossRef]

33.	Saha, K.; Agasti, S.S.; Kim, C.; Li, X.; Rotello, V.M. Gold nanoparticles in chemical and biological sensing. *Chem. Rev.* **2012**, *112*, 2739–2779. [CrossRef]

34. Duncan, B.; Kim, C.; Rotello, V.M. Gold nanoparticle platforms as drug and biomacromolecule delivery systems. *J. Control. Release* **2010**, *148*, 122–127. [CrossRef] [PubMed]
35. Tapia-Arellano, A.; Gallardo-Toledo, E.; Ortiz, C.; Henríquez, J.; Feijóo, C.G.; Araya, E.; Sierpe, R.; Kogan, M.J. Functionalization with PEG/Angiopep-2 peptide to improve the delivery of gold nanoprisms to central nervous system: In vitro and in vivo studies. *Mater. Sci. Eng. C* **2021**, *121*, 111785. [CrossRef]
36. Gao, N.; Sun, H.; Dong, K.; Ren, J.; Qu, X. Gold-nanoparticle-based multifunctional amyloid-β inhibitor against Alzheimer's disease. *Chem. A Eur. J.* **2015**, *21*, 829–835. [CrossRef] [PubMed]
37. Hainfeld, J.F.; Smilowitz, H.M.; O'Connor, M.J.; Dilmanian, F.A.; Slatkin, D.N. Gold nanoparticle imaging and radiotherapy of brain tumors in mice. *Nanomedicine* **2013**, *8*, 1601–1609. [CrossRef] [PubMed]
38. Riley, R.S.; Day, E.S. Gold nanoparticle-mediated photothermal therapy: Applications and opportunities for multimodal cancer treatment. *Wiley Interdiscip. Rev. Nanomed. Nanobiotechnol.* **2017**, *9*. [CrossRef] [PubMed]
39. Morales-Zavala, F.; Arriagada, H.; Hassan, N.; Velasco, C.; Riveros, A.; Álvarez, A.R.; Minniti, A.N.; Rojas-Silva, X.; Muñoz, L.L.; Vasquez, R.; et al. Peptide multifunctionalized gold nanorods decrease toxicity of β-amyloid peptide in a *Caenorhabditis elegans* model of Alzheimer's disease. *Nanomed. Nanotechnol. Biol. Med.* **2017**, *13*, 2341–2350. [CrossRef] [PubMed]
40. Hassan, N.; Cordero, M.L.; Sierpe, R.; Almada, M.; Juárez, J.; Valdez, M.; Riveros, A.; Vargas, E.; Abou-Hassan, A.; Ruso, J.M.; et al. Peptide functionalized magneto-plasmonic nanoparticles obtained by microfluidics for inhibition of β-amyloid aggregation. *J. Mater. Chem. B* **2018**, *6*, 5091–5099. [CrossRef]
41. Guerrero, A.R.; Hassan, N.; Escobar, C.A.; Albericio, F.; Kogan, M.J.; Araya, E. Gold nanoparticles for photothermally controlled drug release. *Nanomedicine* **2014**, *9*, 2023–2039. [CrossRef] [PubMed]
42. Wust, P.; Hildebrandt, B.; Sreenivasa, G.; Rau, B.; Gellermann, J.; Riess, H.; Felix, R.; Schlag, P. Hyperthermia in combined treatment of cancer. *Lancet Oncol.* **2002**, *3*, 487–497. [CrossRef]
43. Liu, Y.; Sun, D.; Fan, Q.; Ma, Q.; Dong, Z.; Tao, W.; Tao, H.; Liu, Z.; Wang, C. The enhanced permeability and retention effect based nanomedicine at the site of injury. *Nano Res.* **2020**, *13*, 564–569. [CrossRef]
44. Jain, S.; Hirst, D.G.; O'Sullivan, J.M. Gold nanoparticles as novel agents for cancer therapy. *Br. J. Radiol.* **2012**, *85*, 101–113. [CrossRef]
45. Maeda, H. The enhanced permeability and retention (EPR) effect in tumor vasculature: The key role of tumor-selective macromolecular drug targeting. *Adv. Enzyme Regul.* **2001**, *41*, 189–207. [CrossRef]
46. Madan, J.; Dhiman, N.; Sardana, S.; Aneja, R.; Chandra, R.; Katyal, A. Long-circulating poly(ethylene glycol)-grafted gelatin nanoparticles customized for intracellular delivery of noscapine: Preparation, in-vitro characterization, structure elucidation, pharmacokinetics, and cytotoxicity analyses. *Anticancer Drugs* **2011**, *19*, 1061–1067. [CrossRef]
47. Silva, N.; Riveros, A.; Yutronic, N.; Lang, E.; Chornik, B.; Guerrero, S.; Samitier, J.; Jara, P.; Kogan, M. Photothermally controlled methotrexate release system using β-cyclodextrin and gold nanoparticles. *Nanomaterials* **2018**, *8*, 985. [CrossRef] [PubMed]
48. Xue, Q.; Liu, Z.; Guo, Y.; Guo, S. Cyclodextrin functionalized graphene-gold nanoparticle hybrids with strong supramolecular capability for electrochemical thrombin aptasensor. *Biosens. Bioelectron.* **2015**, *68*, 429–436. [CrossRef]
49. Park, C.; Youn, H.; Kim, H.; Noh, T.; Kook, Y.H.; Oh, E.T.; Park, H.J.; Kim, C. Cyclodextrin-covered gold nanoparticles for targeted delivery of an anti-cancer drug. *J. Mater. Chem.* **2009**, *19*, 2310–2315. [CrossRef]
50. Barlas, F.B.; Aydindogan, E.; Arslan, M.; Timur, S.; Yagci, Y. Gold nanoparticle conjugated poly(p-phenylene-β-cyclodextrin)-graft-poly(ethylene glycol) for theranostic applications. *J. Appl. Polym. Sci.* **2019**, *136*, 47250. [CrossRef]
51. Wang, H.; Chen, Y.; Li, X.Y.; Liu, Y. Synthesis of oligo(ethylenediamino)-β-cyclodextrin modified gold nanoparticle as a DNA concentrator. *Mol. Pharm.* **2007**, *4*, 189–198. [CrossRef] [PubMed]
52. Liu, Y.; Zhao, Y.L.; Chen, Y.; Wang, M. Supramolecular assembly of gold nanoparticles mediated by polypseudorotaxane with thiolated β-cyclodextrin. *Macromol. Rapid Commun.* **2005**, *26*, 401–406. [CrossRef]
53. Donoso-González, O.; Lodeiro, L.; Aliaga, Á.E.; Laguna-Bercero, M.A.; Bollo, S.; Kogan, M.J.; Yutronic, N.; Sierpe, R. Functionalization of gold nanostars with cationic β-cyclodextrin-based polymer for drug co-loading and SERS monitoring. *Pharmaceutics* **2021**, *13*, 261. [CrossRef]
54. Patel, P.; Deshpande, A. Patent review on cyclodextrin based nanosponges prepared by different methods: Physicochemical characterization, factors influencing formation and applications. *World J. Pharm. Sci.* **2014**, *2*, 380–385.
55. Coleman, A.W.; Nicolis, I.; Keller, N.; Dalbiez, J.P. Aggregation of cyclodextrins: An explanation of the abnormal solubility of β-cyclodextrin. *J. Incl. Phenom. Mol. Recognit. Chem.* **1992**, *13*, 139–143. [CrossRef]
56. Rita, L.; Amit, T.; Chandrashekhar, G. Current trends in B-cyclodextrin based drug delivery systems. *Int. J. Res. Ayurveda Pharm.* **2011**, *2*, 1520–1526.
57. Bolmal, U.B.; Manvi, F.V.; Kotha, R.; Palla, S.S.; Paladugu, A.; Reddy, K.R. Recent advances in nanosponges as drug delivery system. *Int. J. Pharm. Sci. Nanotechnol.* **2013**, *6*, 1934–1944. [CrossRef]
58. Omar, S.M.; Ibrahim, F.; Ismail, A. Formulation and evaluation of cyclodextrin-based nanosponges of griseofulvin as pediatric oral liquid dosage form for enhancing bioavailability and masking bitter taste. *Saudi Pharm. J.* **2020**, *28*, 349–361. [CrossRef] [PubMed]
59. Higuchi, T.; Connors, K.A. Phase solubility techniques. In *Advances in Analytical Chemistry and Instrumentation*; Wiley-Interscience: New York, NY, USA, 1965; Volume 4, pp. 117–212.

60. Turkevich, J.; Stevenson, P.C.; Hillier, J. A study of the nucleation and growth processes in the synthesis of colloidal gold. *Discuss. Faraday Soc.* **1951**, *11*, 55–75. [CrossRef]
61. Liu, X.; Atwater, M.; Wang, J.; Huo, Q. Extinction coefficient of gold nanoparticles with different sizes and different capping ligands. *Colloids Surf. B Biointerfaces* **2007**, *58*, 3–7. [CrossRef]
62. Near, R.D.; Hayden, S.C.; Hunter, R.E.; Thackston, D.; El-Sayed, M.A. Rapid and efficient prediction of optical extinction coefficients for gold nanospheres and gold nanorods. *J. Phys. Chem. C* **2013**, *117*, 23950–23955. [CrossRef]
63. Darandale, S.S.; Vavia, P.R. Cyclodextrin-based nanosponges of curcumin: Formulation and physicochemical characterization. *J. Incl. Phenom. Macrocycl. Chem.* **2013**, *75*, 315–322. [CrossRef]
64. Sambasevam, K.P.; Mohamad, S.; Sarih, N.M.; Ismail, N.A. Synthesis and characterization of the inclusion complex of β-cyclodextrin and azomethine. *Int. J. Mol. Sci.* **2013**, *14*, 3671–3682. [CrossRef]
65. Yaşayan, G.; Şatıroğlu Sert, B.; Tatar, E.; Küçükgüzel, İ. Fabrication and characterisation studies of cyclodextrin-based nanosponges for sulfamethoxazole delivery. *J. Incl. Phenom. Macrocycl. Chem.* **2020**, *97*, 175–186. [CrossRef]
66. Dobkowski, Z. Thermal analysis techniques for characterization of polymer materials. *Polym. Degrad. Stab.* **2006**, *91*, 488–493. [CrossRef]
67. Trotta, F.; Zanetti, M.; Camino, G. Thermal degradation of cyclodextrins. *Polym. Degrad. Stab.* **2000**, *69*, 373–379. [CrossRef]
68. Jiang, H.L.; Lin, J.C.; Hai, W.; Tan, H.W.; Luo, Y.W.; Xie, X.L.; Cao, Y.; He, F.A. A novel crosslinked β-cyclodextrin-based polymer for removing methylene blue from water with high efficiency. *Colloids Surf. A Physicochem. Eng. Asp.* **2019**, *560*, 59–68. [CrossRef]
69. Matencio, A.; Dhakar, N.K.; Bessone, F.; Musso, G.; Cavalli, R.; Dianzani, C.; García-Carmona, F.; López-Nicolás, J.M.; Trotta, F. Study of oxyresveratrol complexes with insoluble cyclodextrin based nanosponges: Developing a novel way to obtain their complexation constants and application in an anticancer study. *Carbohydr. Polym.* **2020**, *231*, 115763. [CrossRef] [PubMed]
70. Han, L.; Ma, G.; Xie, S.; Sun, J.; Jia, Y.; Jing, Y. Thermal properties and stabilities of the eutectic mixture: 1,6-hexanediol/lauric acid as a phase change material for thermal energy storage. *Appl. Therm. Eng.* **2017**, *116*, 153–159. [CrossRef]
71. Herrera, B.A.; Bruna, T.C.; Sierpe, R.A.; Lang, E.P.; Urzúa, M.; Flores, M.I.; Jara, P.S.; Yutronic, N.I. A surface functionalized with per-(6-amino-6-deoxy)-β-cyclodextrin for potential organic pollutant removal from water. *Carbohydr. Polym.* **2020**, *233*, 115865. [CrossRef]
72. Rao, V.M.; Stella, V.J. When can cyclodextrins be considered for solubilization purposes? *J. Pharm. Sci.* **2003**, *92*, 927–932. [CrossRef] [PubMed]
73. Connors, K.A. The stability of cyclodextrin complexes in solution. *Chem. Rev.* **1997**, *97*, 1325–1358. [CrossRef]
74. Sierpe, R.; Noyong, M.; Simon, U.; Aguayo, D.; Huerta, J.; Kogan, M.J.; Yutronic, N. Construction of 6-thioguanine and 6-mercaptopurine carriers based on βcyclodextrins and gold nanoparticles. *Carbohydr. Polym.* **2017**, *177*, 22–31. [CrossRef] [PubMed]
75. Kim, H.S.; Lee, D.Y. Photothermal therapy with gold nanoparticles as an anticancer medication. *J. Pharm. Investig.* **2017**, *47*, 19–26. [CrossRef]
76. Tao, Y.; Chan, H.F.; Shi, B.; Li, M.; Leong, K.W. Light: A magical tool for controlled drug delivery. *Adv. Funct. Mater.* **2020**, *30*, 2005029. [CrossRef]
77. Yen, H.J.; Hsu, S.H.; Tsai, C.L. Cytotoxicity and immunological response of gold and silver nanoparticles of different sizes. *Small* **2009**, *5*, 1553–1561. [CrossRef]

Article

Molecular Recognition by Pillar[5]arenes: Evidence for Simultaneous Electrostatic and Hydrophobic Interactions

Borja Gómez-González [1], Luis García-Río [1,*], Nuno Basílio [2], Juan C. Mejuto [3,*] and Jesus Simal-Gandara [4]

[1] Departamento de Química Física, Facultade de Química, Universidade de Santiago de Compostela, 15782 Santiago, Spain; gomezgonzalezborja@gmail.com
[2] Laboratório Associado Para a Química Verde (LAQV), Rede de Química e Tecnologia (REQUIMTE), Departamento de Química, Faculdade de Ciências e Tecnologia, Universidade NOVA de Lisboa, 2829-516 Caparica, Portugal; nuno.basilio@fct.unl.pt
[3] Department of Physical Chemistry, Faculty of Science, University of Vigo, 32004 Ourense, Spain
[4] Nutrition and Bromatology Group, Analytical and Food Chemistry Department, Faculty of Food Science and Technology, University of Vigo, 32004 Ourense, Spain; jsimal@uvigo.es
* Correspondence: luis.garcia@usc.es (L.G.-R.); xmejuto@uvigo.es (J.C.M.)

Abstract: The formation of inclusion complexes between alkylsulfonate guests and a cationic pillar[5]arene receptor in water was investigated by NMR and ITC techniques. The results show the formation of host-guest complexes stabilized by electrostatic interactions and hydrophobic effects with binding constants of up to 10^7 M^{-1} for the guest with higher hydrophobic character. Structurally, the alkyl chain of the guest is included in the hydrophobic aromatic cavity of the macrocycle while the sulfonate groups are held in the multicationic portal by ionic interactions.

Keywords: pillararene; host:guest; supramolecular; hydrophobic; ITC; NMR

Citation: Gómez-González, B.; García-Río, L.; Basílio, N.; Mejuto, J.C.; Simal-Gandara, J. Molecular Recognition by Pillar[5]arenes: Evidence for Simultaneous Electrostatic and Hydrophobic Interactions. *Pharmaceutics* **2022**, *14*, 60. https://doi.org/10.3390/pharmaceutics14010060

Academic Editors: Francisco José Ostos, José Antonio Lebrón and Pilar López-Cornejo

Received: 16 November 2021
Accepted: 22 December 2021
Published: 28 December 2021

Publisher's Note: MDPI stays neutral with regard to jurisdictional claims in published maps and institutional affiliations.

1. Introduction

Supramolecular chemistry is a topic of great interest to the scientific community that wants to take advantage of non-covalent interactions, such as van der Waals forces, hydrogen bonds, π-π stacking interaction, electrostatic interactions, or hydrophobic/hydrophilic interactions, with the aim of implementing and explaining increasing complexity systems (bottom-up approach) [1–3]. During the last decades, numerous supramolecular systems have been successfully developed and in the literature, there are numerous investigations regarding their applications as functional materials, in catalytic processes, electronic devices, sensors, or drug carriers, etc., [4–6]. Among these applications, nanomedicine presents a promising potential for modernizing traditional biomedical practices, and in this context, the design of new supramolecular systems in the nanometric range is one of the new frontiers that will offer new diagnostic and therapeutic applications in the field of nanomedicine (drug delivery, gene delivery, drug/gene co-delivery, bioimaging or photodynamic therapy) [7,8].

Noncovalent interactions present several advantages in comparison to covalent ones:
(i) An easy and facile approach for building supramolecular structures avoiding synthetic processes [9].
(ii) Supramolecular methods are cost-effective and environmentally friendly.
(iii) Supramolecular materials consist of components connected by non-covalent interactions and experiencing spontaneous assembly and disassembly processes [10–12].
(iv) The formation of supramolecular materials is reversible and capable of being recycled and self-repaired from external mechanical damage.
(v) Supramolecular materials have the ability to respond to external stimuli being able to rearrange their structures or morphologies toward more stable states. This adaptive capability can be utilized for the development of stimuli-responsive functional materials [13–16].

(vi)   In addition, it allows the manipulation of supramolecular molecules or building blocks at the molecular level, modulating sizes and morphologies using the "bottom-up" method, providing a variety of novel diagnostic and therapeutic platforms toward applications in nanomedicine.

Within the different non-covalent interactions considered as supramolecular phenomena, the host:guest between different substrates and macrocycles have been studied extensively in recent decades. By including host:guest, two or more molecules can be integrated in a simple and reversible way. This offers us multiple possibilities for new supramolecular structure design. Molecular recognition that involves host:guest interactions play a vital role in life-sustaining biological processes [17,18]. Macrocyclic compounds have been extensively used and intensively investigated as prime host receptors with high affinity and selectivity for complementary small guest molecules or ions. Examples of macrocycles include cryptands [19], crown ethers [20–22], cyclophanes [23], cyclopeptides [24–26], cyclodextrins [27–29], resorcinarenes [30], cucurbit[n]urils [31–34], calix[n]arenes [35–38], and pillar[n]arenes [39]. These macrocycles (hosts) have cavities that allow the encapsulation of substrates of interest (guest). The external properties exhibited by the host molecules favor interaction with the surrounding solvent. On the other hand, the characteristics of the cavities allow the inclusion of the guest. This inclusion occurs through different causes (hydrophobic and/or electrostatic interactions, formation of hydrogen bonds, suitable molecular shape and/or size, etc. In fact, encapsulation in an aqueous solution of hydrophobic guest molecules in macrocyclic hydrophobic cavities is one of the most common cases. The host:guest complex will exhibit high stabilities, providing robust and reliable structures for obtaining supramolecular systems in aqueous media.

Pillar[n]arenes are one of the most recent families of macrocyclic hosts used in supramolecular chemistry [40]. Pillararenes bring together some interesting characteristics of other host systems in a single molecular structure, such as a highly symmetrical pillar-shaped structure which is similar in many respects to that of highly symmetrical cucurbiturils, a π-rich aromatic cavity, also found in calixarenes, and several hydroxyl moieties on both rims, a feature shared with the highly functionalized cyclodextrins. Substituents on both rims of pillararenes affect their physical properties, such as solubility, conformational and host:guest properties. Pillar[n]arenes are very useful structures useful for the design of different supramolecular systems [41–51]. In particular, these systems have interrelated applications in the biomedical and pharmacological fields as drug carriers [46], transmembrane channels [47], or cellular glue [48].

Another aspect to consider is that the presence of charged groups on the pillars[n]arenes convert them into water-soluble substrates (see Scheme 1). Furthermore, its ability to incorporate a wide variety of hosts into its cavity [52,53], its applications as catalysts [54–56], detection [57,58] and gene nanocarriers [59] in aqueous media, has caused that numerous investigations focus on them. When oppositely charged molecules were evaluated as hosts, electrostatic interactions contribute significantly to stabilizing the resulting supramolecular system.

Pillararenes-based host-guest systems comprising amphiphilic guests offer interesting strategies for the development of novel stimuli-responsive drug-delivery systems which improve precision and efficiency in drug delivery [60–66].

In this sense, knowledge of the driving forces behind the host:guest complex formation process is important. We know about the electrostatic interactions between the charged groups of the pillar[5]arene and the ionic substrates, however, the role played by hydrophobic/hydrophilic interactions must be explored in detail. In this context, fundamental studies on the interaction between charged pillararene receptors and model amphiphilic compounds are of utmost importance for the intended pharmaceutical applications in relation to these macrocycles.

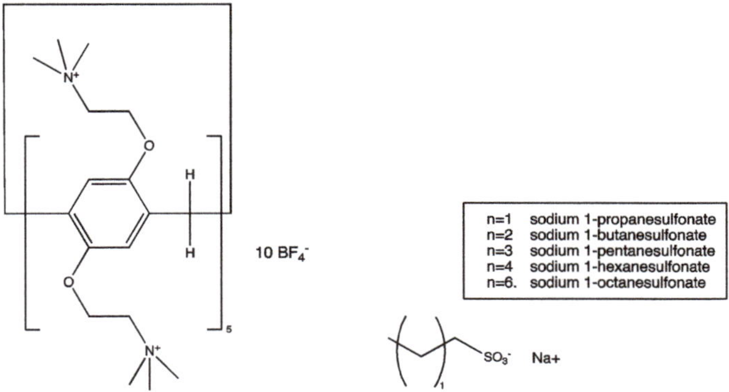

Scheme 1. Cationic pillar[5]arene and alkylsulfonates.

In this article, we present a structural and thermodynamic study on the host-host complexes between a cationic pillar[5]arene and charged amphiphilic compounds, which by keeping the head group constant, its hydrophobicity can be modulated by modifying the length of the chain hydrocarbon. The family of amphiphiles chosen was the alkylsufonates (see Scheme 1).

2. Materials and Methods

2.1. Materials

The highest purity commercially available reagents were supplied by Sigma-Aldrich (Madrid, Spain) and were used without further purification. The water-soluble cationic pillar[5]arene was obtained by a synthetic procedure described elsewhere [67]. Br- exchange by BF4- was carried out as follows: To a solution with Br- (1.17 g, 0.514 mmol) in Milli-Q water at room temperature and with stirring, AgBF4 was added slowly in little portions. A grayish precipitate was obtained. The suspension was centrifuged, and supernatant was collected and filtered (0.45 μm). A yellowish solid was obtained after removing the solvent (1.15 g, 96%). ^{1}H NMR (D$_2$O, 300 MHz): δ = 6.89 (s, 10H); δ = 4.36 (s, 20H); δ = 3.91 (s, 10H); δ = 3.72 (s, 20H); δ = 3.19 (s, 90H); ^{13}C NMR (D$_2$O, 75 MHz): δ = 149.2 (C, 10C); δ = 129.8 (C, 10C); δ = 115.9 (CH, 10C); δ = 64.8 (CH$_2$, 10C); δ = 62.3 (CH$_2$, 10C); δ = 53.7 (CH$_3$, 30C); δ = 29.3 (CH$_2$, 5C); MS (ESI): m/z calcd for [TMAP5^{10+}.9BF$_4$$^{-}$]$^{2+}$ 2253.4; found 2253.2; calcd for [TMAP5^{10+}.8BF$_4$$^{-}$]$^{2+}$ 1083.3; found 1083.1. The final product was analyzed by thermal gravimetric analysis to assess volatile content.

2.2. Microcalorimetry

An isothermal titration microcalorimeter (VP-ITC) supplied by Microcal Co. (Northamptoh, MA, USA) at 1 atm and 25 °C to carry out the microcalorimetric titrations. The procedure used for each titration consisted in sequentially injecting a guest solution in a syringe (0.270 mL) with shaking (459 rpm) into a host solution in the sample cell (1.459 mL). Before each titration, the samples were degassed and thermostatted using an accessory supplied by ThermoVac (Leybold Hispánica, Barcelona, SPAIN). For the reference cell, the same sample was used as in the sample cell. The first injection was discarded in all the experiments carried out in order to suppress the diffusion effects in the calorimetric cell of the syringe material. The number of injections, their volume and the spacing time between each one were varied according to the experiment. Binding constants were calculated from the titration curve by using the AFFINImeter software (S4SD, Santiago de Comostela, SPAIN).

2.3. NMR Spectrometry

NMR experiments were conducted at 25 °C on a spectrometer supplied by Bruker (Bruker NEO 17.6 T) (Billerica, MA, USA) with 750 MHz proton resonance, equipped with a ^{1}H/^{13}C/^{15}N triple resonance PA-TXI probe with deuterium lock channel and shielded PFG z-gradient. The control software was TopSpin 4.0. Chemical shifts were referenced to the lock deuterium solvent. Spectra have been processed and analyzed using Mestrenova software v14.0 supplied by Mestrelab Inc (Santiago de Compostela, SPAIN). The 1D ^{1}H spectrum has been measured with 128 scans, d1 2s relaxation delay and 2.75 s FID acquisition time (aq). The FID has been acquired with 64k complex data points. It has been processed using Fourier Transformation (FT) and zero-filling. 131 k data points spectra have been obtained. The total measurement time was ~10 min.

A two-dimensional 2D COSY spectrum magnitude mode was measured (pulse sequence "cosygpppqf" of Bruker library). The relaxation delay (d_1) and the FID acquisition time (at) were 2 and 0.172 s, respectively. The spectrum was measured with eight scans. The number of points in the direct and indirect dimensions was 4 k and 160, respectively. The spectrum was processed with apodization with a sine-bell function in both dimensions and represented in the magnitude mode. The total measurement time was ~48 min.

A two-dimensional 2D HSQC multiplicity edited 1H-13C spectrum was measured (pulse sequence "hsqcedetgpsisp 2.4" of the Bruker library). The spectrum includes adiabatic inversion pulses in ^{13}C and suppression of COSY type artifacts. The INEPTs transfers were optimized for a nominal value of ^{1}J$_{CH}$ of 145 Hz. The delay for multiplicity selection was set to $1/(2 \cdot {}^1J_{CH})$ to detect with the same sign signals of CH$_3$ and CH groups and with opposite phase CH$_2$ groups. The relaxation delay (d_1) and the FID acquisition time (at) were 1.6 and 0.112s, respectively; 2048 and 160 complex points in the t2 and t1 dimensions spectrum were acquired. Scans number per t1 increment was 8. The total measurement time was ~1 h 15 min.

3. Results

The hydrophobic cavity of the pillar[5]arene, together with the presence of five positive charges in each rim makes this macrocycle an excellent receptor for amphiphilic anionic guests. The complexation of the different alkylsulfonates, whose hydrophilic head exhibits a negative charge, (G) by the pillar[5]arene (H) was studied by different experimental techniques.

3.1. NMR Evidence of N-Octylsulfonate Complexation by Pillararene

NMR spectroscopy has been used to determine the structures of macrocycles complexes. The ^{1}H NMR spectra of octylsulfonate upon mixing in different proportions with pillararene can be observed in Figure 1.

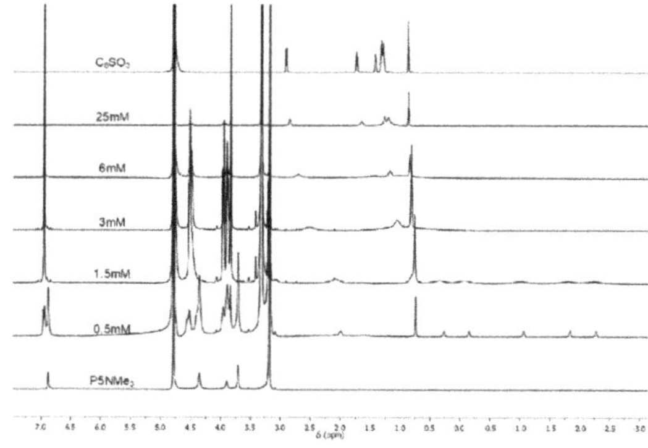

Figure 1. ^{1}H NMR spectra in D$_2$O at 25 °C for pillararene (1.5 mM); octylsulfonate (1.5 mM) and mixtures of both with a constant concentration of pillararene (1.5 mM) and different concentrations of octylsulfonate.

All protons of octylsulfonate appear upfield-shifted with respect to the free guest upon addition of pillararene, indicating that an inclusion complex was formed. These results indicate that octylsulfonate is incorporated into the magnetic shielding region of the pillararene aromatic cavity with the sulfonate group pointing towards the trimethylammonium groups of the host. Moreover, the host proton signals are also affected by complexation due to the asymmetric structure of the guest and the manner in which it is inserted into the host cavity [52]. To determine the binding stoichiometry of the host:guest complex, considering that fast exchange on the NMR chemical shift timescale was observed for this complex, an NMR titration at constant host concentration, was carried out. Figure 2 shows that the magnitude of the upfield shift for guest hydrogen atoms increases upon a gradual increase of the [host]/[guest] ratio, reaching a plateau for values higher than 1, indicating a 1:1 stoichiometry for the inclusion complex.

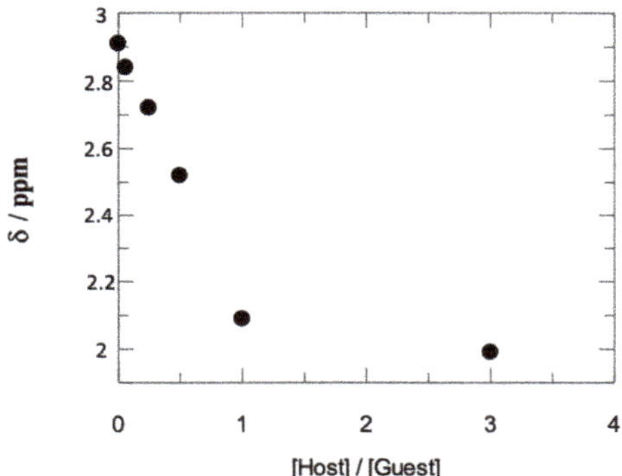

Figure 2. Chemical shifts for hydrogen atoms in alpha position to the sulfonate group in octylsulfonate in the presence of increasing concentrations of host.

Detailed analysis of spectrum for [Pillararene] = 1.5 mM and [octylsulfonate] = 0.50 mM (Figure 1) reveals that the signal corresponding to the methylene groups in positions C4–C7 of octylsulfonate splits into different signals, allowing a clear characterization of the inclusion complex. Figures 3 and 4 show the HSQC and COSY spectra respectively allowing the assignment of all signals in the NMR spectrum.

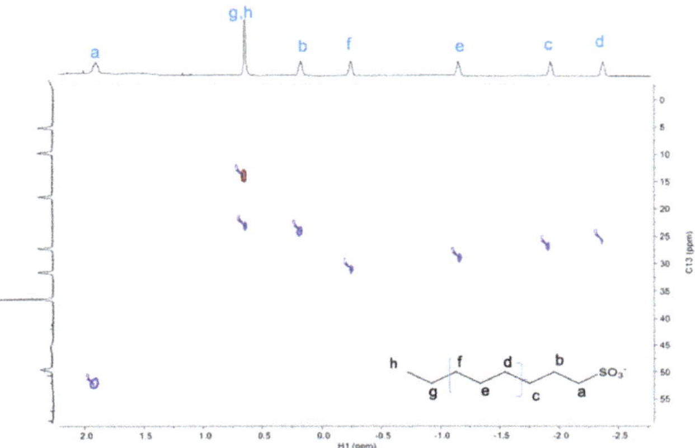

Figure 3. HSQC spectra for a mixture of 1.5 mM of Pillararene and 0.5 mM of octylsulfonate in D$_2$O a 25 °C. Labels for octylsulfonate hydrogen atoms are according to the picture shown in the figure.

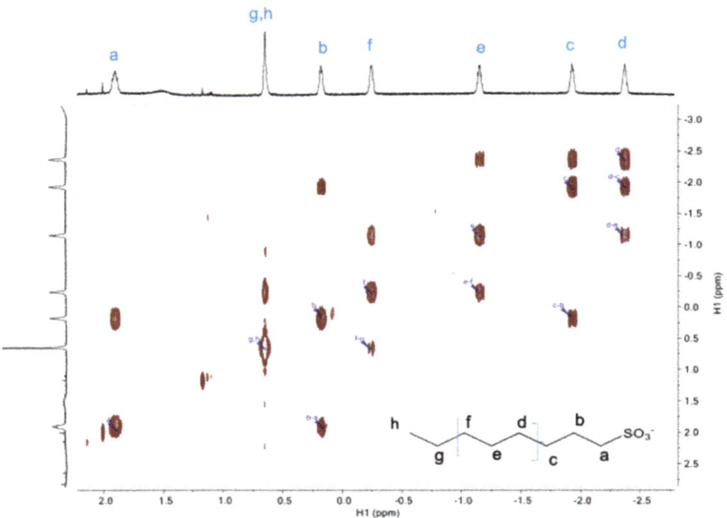

Figure 4. COSY spectra for a mixture of 1.5 mM of Pillararene and 0.5 mM of octylsulfonate in D_2O a 25 °C. Labels for octylsulfonate hydrogen atoms are according to the picture shown in the figure.

Assignment of NMR signals allows us to quantify the magnitude of the complexation-induced upfield effect for each hydrogen atom in octylsulfonate (results showed in Table 1). We refer to a complexation-induced chemical shift as the difference between the chemical shift observed for the guest free and complexed, $\Delta\delta = \delta_{free} - \delta_{bound}$. The magnitude of $\Delta\delta$ is dependent on the hydrogen atom position along the alkyl chain of octylsulfonate. It is remarkable the very large magnitude of the upfield effects with values larger than $\Delta\delta = 3$ ppm for some central chain nuclei. Hydrogen atoms Hc and Hd show the large $\Delta\delta$ values allowing to propose a structure for the host:guest complex as shown in Figure 5. Hydrogen atoms in positions c and d are located inside the aromatic region of the pillararene allowing the large $\Delta\delta$ values, $\Delta\delta > 3$ ppm. Hydrogens at position e should be just below this region but close to the aromatic groups ($\Delta\delta = 2.3$ ppm). It is remarkable that hydrogen atoms at positions g and h ($\Delta\delta < 1$ ppm), as well as in the alpha position to the sulfonate group, are clearly located outside the aromatic region.

Table 1. Magnitude of the complexation induced chemical shifts (ppm) for host:guest complexes between pillararene and different alkylsulfonates.

	Ha	Hb	Hc	Hd	He	Hf	Hg	Hh
$C_8SO_3^-$	0.91	1.48	3.27	3.58	2.37	1.46	0.56	0.14
$C_6SO_3^-$	1.11	1.72	3.7	3.8	2.02	0.98		
$C_5SO_3^-$	1.24	1.78	3.84	3.85	1.76			
$C_4SO_3^-$	1.24	1.84	3.33	2.81				
$C_3SO_3^-$	0.94	1.2	2.16					

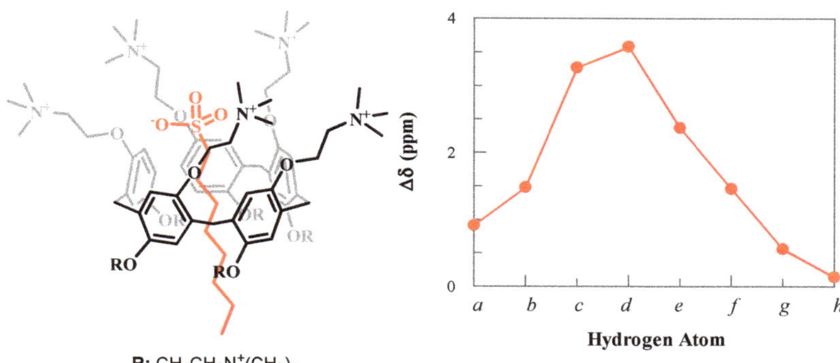

Figure 5. (**Left**) Plot of the magnitude of complexation induced chemical shift, $\Delta\delta$, as a function of the hydrogen atom position (starting at the sulfonate group). (**Right**) Schematic picture of the host:guest complex showing hydrogens Hc and Hd fully incorporated into the aromatic region of the host. For simplicity only two trimethylammonium groups of pillararene are shown.

Similar experiments were conducted for shorter chain alkylsulfonates with three to six carbon atoms (see Table 1) revealing that hydrogen atoms in positions Hc and Hd show the large upfield effects confirming that these atoms are clearly included inside the pillararene cavity. The complexation picture shows the sulfonate aligned with the trimethylammonium head groups of the receptor in such a way that electrostatic interaction should be the major driving force for complexation. It is remarkable that $\Delta\delta$ values are also dependent on the nature of the alkylsulfonate (see Figure 6). In fact, Ha hydrogen atoms show the large upfield effect for alkylsulfonates with four and five carbon atoms, meanwhile, alkyl sulfonates with three and eight carbon atoms present smaller values. On the other hand, hydrogens Hc show the large upfield effect for $C_5SO_3^-$ and $C_6SO_3^-$, and hydrogens Hb show the large $\Delta\delta$ for $C_4SO_3^-$ and $C_5SO_3^-$. More clearly, Figure 6-left shows that the magnitude of $\Delta\delta$ is strongly dependent on the number of carbon atoms in the alkylsulfonate for hydrogens Hc > Hb > Ha, being indicative of a different degree of penetration into the pillararene cavity. Figure 6-right represents normalized $\Delta\delta_{corr}$ by subtracting the values corresponding to hydrogens Ha. The normalized values are directly comparable and indicate that Hc hydrogens are much closer to the cavity than Hb and that an optimal degree of penetration is reached for five atoms of carbon. Alkylsulfonates with three and four carbon atoms can form external complexes where the carbon atoms do not fit neatly together. This causes that the magnitude of $\Delta\delta_{corr}$ does not reach an optimal value. Likewise, it is observed that for octylsulfonate, the Hc hydrogens present a lower inclusion than for the 5 carbon atom homolog. This behavior may be due to a hydrophobic pushup effect that compels the sulfonate group towards a plane superior to the portal of the pillararene in order to accommodate more methylene groups inside the cavity. At the same time, the possibility that the hydrophobic effect induces a greater degree of folding of the alkyl chain in order to maximize the number of carbon atoms that can be included in the cavity should be considered.

These results indicate that the location of the sulfonate group should be dependent on the number of carbon atoms, being closer to the positive portal of the pillararene for $C_5SO_3^-$ and $C_4SO_3^-$. This behavior can be observed for hydrogen atoms in positions Hb and Hc, being clear evidence of a different degree of guest penetration into the host cavity and, consequently, ruling out the electrostatic attraction as the only interaction stabilizing the host:guest complex.

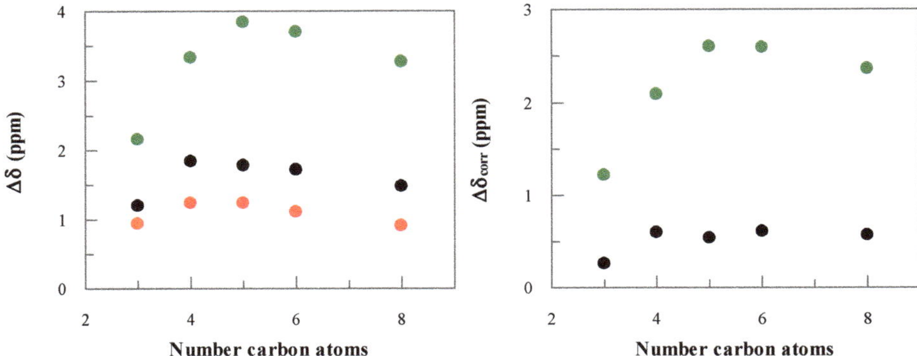

Figure 6. (**Left**) Influence of the number of carbon atoms in the alkyl chain of alkylsulfonate on the chemical induced upfield effect, $\Delta\delta$: (•) Ha; (•) Hb and (•) Hc. (**Right**) Values of upfield effect for hydrogen atoms in positions (•) Hb and (•) Hc after correction by upfield effect of hydrogens Ha.

3.2. Calorimetric Titrations for Alikylsulfonate Recognition by Pillararene

In order to quantitatively evaluate the complexation of pillar[5]arene with each guest and the stoichiometry of the complex formed, an isothermal calorimetry titration was carried out at 25 °C under neutral conditions. Each titration was done by consecutively adding the guest to the host in the sample cell. As an example, each butylsulfonate titration in the sample cell containing the pillar[5]arene is shown in Figure 7 (see Supplementary Materials for other alkylsulfonates). The experimental data were satisfactorily fitted to a model of "a set of binding sites", obtaining the binding constant (K) and the thermodynamic parameters (Table 2).

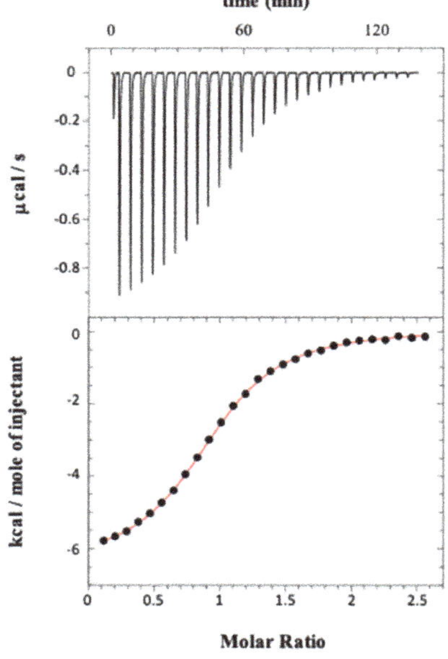

Figure 7. Microcalorimetric titration of butylsulfonate (G) with pillar[5]arene (H) in water at 25 °C. (**Top**): Raw data for the 28 sequential injections (10 µL per injection) of a solution of G (0.5 mM) into a solution of H (0.04 mM). (**Bottom**): "Net" heat effects fitted using the "one set of sites" binding model.

Table 2. Thermodynamic parameters obtained for host:guest complexes between pillararene and different alkylsulfonates.

	K (M^{-1})	ΔG^0 (kcal mol^{-1})	ΔH^0 (kcal mol^{-1})	TS^0 (kcal mol^{-1})
$C_8SO_3{}^-$	$(4.79 \pm 0.18) \times 10^7$	-10.43	-8.90 ± 0.01	1.5
$C_6SO_3{}^-$	$(1.67 \pm 0.02) \times 10^7$	-9.91	-8.86 ± 0.01	1.05
$C_5SO_3{}^-$	$(5.88 \pm 0.04) \times 10^6$	-9.19	-8.70 ± 0.01	0.49
$C_4SO_3{}^-$	$(2.63 \pm 0.01) \times 10^6$	-8.72	-6.60 ± 0.01	2.12
$C_3SO_3{}^-$	$(7.22 \pm 0.02) \times 10^3$	-5.24	-4.42 ± 0.01	0.82

The results indicate that complexation is mainly enthalpy-driven ($\Delta H^0 = -(6.60 \pm 0.01)$ kcal/mol) accompanied by favorable entropic changes ($T\Delta S^0 = 2.12$ kcal/mol), this balance is more favorable to the enthalpic term with the other alkylsufonates.

From the results of the experiments obtained for guests and other macrocyclic compounds, it has been shown that non-covalent interactions contribute to enthalpic changes, while entropy changes can be attributed to conformational changes and/or effects associated with desolvation processes [68]. Thus, hydrophobic or electrostatic interactions, together with dehydration processes, have a positive contribution to entropy. The negative contribution would be produced by the loss of conformational freedom degrees (both on the guest and on the host). Thus, the values obtained for the thermodynamic parameters would indicate that the electrostatic interactions, π-π, and C-H$\cdots\pi$ interactions between the aromatic ring and the methyl group of the alkylsulfonate and the electron-rich pillararene cavity would give rise to a favorable contribution on enthalpy. At the same time, the solvent molecules (water) that surround both the host and the guest are released into the bulk water and would be the cause of the entropic increase. The binding constant obtained, K = $(2.63 \pm 0.01) \times 10^6$ M^{-1}, is comparable with those reported for negatively charged pillararenos [52,53,68,69] or calixarenes [70–72].

Experimental results reported in Table 2 show alkylsulfonate binding constants to be very sensitive to alkylsulfonate chain length with an increase of almost 10^4 fold on-going from propane to octanesulfonate. Quantitative analysis of these binding constants requires correction of binding constant for propanesulfonate. Because of its smaller value, experimental results were obtained in the presence of [Pillararene] = 0.25 mM instead of [Pillararene] = 0.04 mM used for other alkylsulfonates. Previous results from our group have shown that toluenesulfonate binding constant to pillararene decreases from 1.37×10^6M^{-1} to 3.18×10^4M^{-1} by increasing the host concentration from 0.01 to 0.1 mM [53]. This behavior is due to BF$_4{}^-$ complexation by the pillararene, which difficult the entrance of the guest. Extrapolation to alkylsulfonates implies that propanesulfonate binding constant of 1.86×10^5 M^{-1} should be used for comparative proposes.

Figure 8 plots the dependence of the binding constant with the alkyl chain length and includes similar results using β-cyclodextrin as a receptor [73]. Quantitative analysis of the thermodynamic parameters involved in the complex formation between surfactant molecules and cyclodextrin can be simplified by considering the process divided into three stages:

(i) Dehydration of surfactants and cyclodextrin This process is entropically favored due to a strong water structuring that hydrates the exposed hydrophobic residue of the surfactant and to geometric constraints within the CD cavity. Water is structured around the surfactant hydrophobic chain, giving rise to a strong network of hydrogen bonds. The amount of water molecules involved in hydration scales linearly with the alkyl chain length, therefore, the linear relationship between the number of carbons present in the surfactant hydrocarbon chain and the micellization free energy, and similar phenomena involving removal of the surfactant chain from the aqueous medium.

(ii)   Inclusion of the surfactants in the CD's cavity. Inclusion takes place with the entry of the surfactant hydrocarbon chain inside the cavity, which is stabilized by Van der Waals interactions. The internal diameter of β-CD allows the loose accommodation of a methylene group.

(iii)   Hydration of the inclusion complex. In the last stage, water from the exposed part of the guest is restructured and integrated into the hydration shell of the host:guest complex [74].

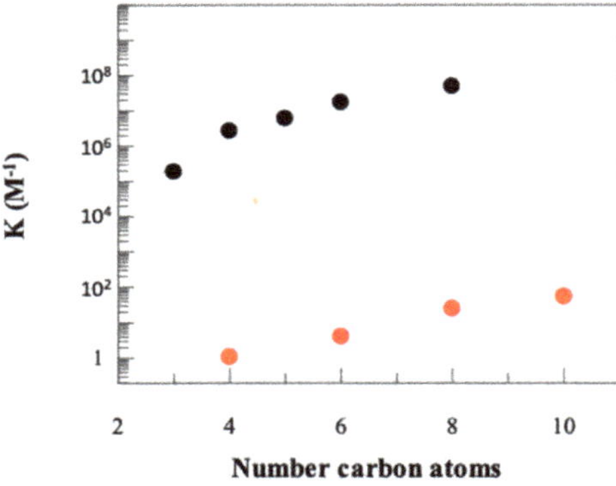

Figure 8. (●)Influence of alkyl chain length of alkylsulfonates on their binding constants to pillar[5]arene using a [pillararene] = 0.04 mM at 25 °C. Value for propanesulfonate was extrapolated from [host] = 0.25 mM (see text). (●) Binding constants for alkylsulfonates to β-cyclodextrin taken from ref. [73].

Alkylsulfonate binding constants to β-CD increase with the number of methylene groups into the alkyl chain in a non-linear way. The binding constants found for short and very large alkyl chains present lower values than expected due to the fact that the cavity occupation is not complete. This implies that a small amount of water molecules is expelled into the bulk. On the other hand, in the case of large chains, the fact that the binding constants present values lower than those expected would be due to the tolerance of the cyclodextrin cavity to accommodate 6–8 methylene groups.

Figure 8 shows that pillararene is a much more effective receptor for alkylsulfonates than β-CD by a factor of 10^6. This effect should be ascribed to electrostatic interactions between the negative charge of the guest and the positive ones on the upper and lower rim of pillararene. Note that this interaction is not possible in the case of β-CD as a receptor. The influence of the alkyl chain length on the binding constants to pillararene parallels that observed with β-CD indicating that hydrophobic interactions are playing an important role in the recognition ability of pillararene.

Hydrophobic effects in pillararene recognition are responsible for the different locations of the sulfonate group with respect to the positive upper or lower rim of the host. This different location is reflected by the complexation-induced upfield effect observed in Figure 6-left for hydrogens in alpha position (Ha) to the sulfonate group. Electrostatic interaction in the host:guest complex will compel the sulfonate group close to the trimethylammonium ones in such a way that the distance between the hydrogens Ha of the guest and the aromatic ring of the host keeps constant. However, experimental results indicate that this distance decrease for the following alkylsulfonates: $C_8SO_3^- > C_3SO_3^- > C_6SO_3^- > C_5SO_3^- \approx C_4SO_3^-$. X-ray crystal structure of 1,4-dipropoxypillar[5]arene confirmed that it is a pentagon from the upper view and a pillar structure from the side view. The diameter of the internal cavity was 4.7 Å, which is similar to that of cyclodextrin, allowing the perfect inclusion of methylene chain [75]. The height of pillararene cavity, taken as the distance between the oxygen atoms in the upper and lower rims, is 5.5 Å, allowing accommodation of 4–5 methylene groups. It means that $C_5SO_3^-$ and $C_4SO_3^-$ are deeply included in the pillararene cavity in comparison to $C_8SO_3^-$ and $C_3SO_3^-$. The smaller alkylsulfonate does not displace a large amount of water from the host cavity resulting in a small hydrophobic effect. On the other hand, three methylene groups of $C_8SO_3^-$ will be outside the cavity. Their hydration in the host:guest complex will contribute unfavorably to its stability.

4. Conclusions

To sum up, we have demonstrated that alkylsulfonates with different chain lengths are effectively bound by a decacationic pillar[5]arene receptor in an aqueous solution with binding constants in the micro/submicromolar range. The formation of the complexes is enthalpy and entropy driven suggesting that ionic, C-H$\cdots\pi$, van der Waals interaction along with hydrophobic effects contribute to the binding stability. The observed increase in the binding constants as the guest alkyl chain length increases provides strong evidence for the contribution of the hydrophobic effect for the recognition process. This view is supported by the structural NMR studies showing that hydrophobic alkyl chains are deeply included in the aromatic cavity of the macrocyclic receptor. The results obtained herein suggest that cationic pillararene receptors are potentially strong binders for anionic and eventually zwitterionic lipids, and therefore, further studies addressing this class of natural molecules as a guest should be considered due to the potential pharmaceutical applications of these macrocycles.

Supplementary Materials: The following are available online at https://www.mdpi.com/article/10.3390/pharmaceutics14010060/s1, Figures S1–S4 show each titration of alkylsulfonates fitted by the "one set of binding sites" model.

Author Contributions: Conceptualization, L.G.-R. and N.B.; methodology, L.G.-R.; software, J.C.M.; validation, L.G.-R. and N.B.; formal analysis, N.B.; investigation, B.G.-G.; resources, B.G.-G.; data curation, L.G.-R. and N.B.; writing—original draft preparation, L.G.-R.; writing—review and editing, N.B., J.C.M. and J.S.-G.; visualization, J.C.M.; supervision, L.G.-R.; project administration, L.G.-R. and J.S.-G.; funding acquisition, L.G.-R. and J.S.-G. All authors have read and agreed to the published version of the manuscript.

Funding: Financial support from the Ministerio de Economia y Competitividad of Spain (project CTQ2017-84354-P), Xunta de Galicia (GR 2007/085 and ED431C2018/42-GRC) and the European Regional Development Fund (ERDF) is gratefully acknowledged. This work was also supported by the Associate Laboratory for Green Chemistry-LAQV which is financed by national funds from FCT/MCTES (UIDB/50006/2020). N.B. acknowledges the FCT/MCTES for the research contract CEECIND/00466/2017.

Institutional Review Board Statement: Not applicable.

Informed Consent Statement: Not applicable.

Conflicts of Interest: The authors declare no conflict of interest. The funders had no role in the design of the study; in the collection, analyses, or interpretation of data; in the writing of the manuscript, or in the decision to publish the results.

References

1. Sijbesma, R.P.; Beijer, F.H.; Brunsveld, L.; Folmer, B.J.; Hirschberg, J.H.; Lange, R.F.; Lowe, J.K.L.; Meijer, E.W. Reversible polymers formed from self-complementary monomers using quadruple hydrogen bonding. *Science* **1997**, *278*, 1601–1604. [CrossRef] [PubMed]
2. Hartgerink, J.D.; Beniash, E.; Stupp, S.I. Self-Assembly and Mineralization of Peptide-Amphiphile Nanofibers. *Science* **2001**, *294*, 1684–1688. [CrossRef] [PubMed]
3. Zhang, W.; Jin, W.; Fukushima, T.; Saeki, A.; Seki, S.; Aida, T. Supramolecular Linear Heterojunction Composed of Graphite-Like Semiconducting Nanotubular Segments. *Science* **2011**, *334*, 340–343. [CrossRef] [PubMed]
4. Wang, F.; Zhang, J.; Ding, X.; Dong, S.; Liu, M.; Zheng, B.; Li, S.; Wu, L.; Yu, Y.; Gibson, H.W.; et al. Metal Coordination Mediated Reversible Conversion between Linear and Cross-Linked Supramolecular Polymers. *Angew. Chem. Int. Ed.* **2010**, *49*, 1090–1094. [CrossRef] [PubMed]
5. Zhang, Z.; Luo, Y.; Chen, J.; Dong, S.; Yu, Y.; Ma, Z.; Huang, F. Formation of Linear Supramolecular Polymers That Is Driven by C-H$\cdots\pi$ Interactions in Solution and in the Solid State. *Angew. Chem. Int. Ed.* **2011**, *50*, 1397–1401. [CrossRef]

6.	Wang, F.; Han, C.; He, C.; Zhou, Q.; Zhang, J.; Wang, C.; Li, N.; Huang, F. Self-Sorting Organization of Two Heteroditopic Monomers to Supramolecular Alternating Copolymers. *J. Am. Chem. Soc.* **2008**, *130*, 11254–11255. [CrossRef]
7.	Chen, K.-J.; Garcia, M.A.; Wang, H.; Tseng, H.-R. *Supramolecular Nanoparticles for Molecular Diagnostics and Therapeutics*; Supramolecular Chemistry; John Wiley & Sons Ltd.: Hoboken, NJ, USA, 2012.
8.	Ma, X.; Zhao, Y. Biomedical applications of supramolecular systems based on host-guest interactions. *Chem. Rev.* **2015**, *115*, 7794–7839. [CrossRef]
9.	Rybtchinski, B. Adaptive Supramolecular Nanomaterials Based on Strong Noncovalent Interactions. *ACS Nano* **2011**, *5*, 6791–6818. [CrossRef]
10.	Lehn, J.M. Supramolecular Shemistry. *Science* **1993**, *260*, 1762–1763. [CrossRef]
11.	Stupp, S.I.; Lebonheur, V.; Walker, K.; Li, L.S.; Huggins, K.E.; Keser, M.; Amstutz, A. Supramolecular Materials: Self-Organized Nanostructures. *Science* **1997**, *276*, 384–389. [CrossRef]
12.	Oshovsky, G.V.; Reinhoudt, D.N.; Verboom, W. Supramolecular Chemistry in Water. *Angew. Chem. Int. Ed.* **2007**, *46*, 2366–2393. [CrossRef]
13.	Yan, X.; Wang, F.; Zheng, B.; Huang, F. Stimuli-responsive supramolecular polymeric materials. *Chem. Soc. Rev.* **2012**, *41*, 6042–6065. [CrossRef]
14.	Sambe, L.; Belal, K.; Stoffelbach, F.; Lyskawa, J.; Delattre, F.; Bria, M.; Sauvage, F.X.; Sliwa, M.; Humblot, V.; Charleux, B.; et al. Multi-stimuli responsive supramolecular diblock copolymers. *Polym. Chem.* **2014**, *5*, 1031–1036. [CrossRef]
15.	Choudhury, S.D.; Barooah, N.; Aswal, V.K.; Pal, H.; Bhasikuttan, A.C.; Mohanty, J. Stimuli-responsive supramolecular micellar assemblies of cetylpyridinium chloride with cucurbit[5/7]urils. *Soft Matter* **2014**, *10*, 3485–3493. [CrossRef] [PubMed]
16.	Schaeffer, G.; Fuhr, O.; Fenske, D.; Lehn, J.-M. Self-Assembly of a Highly Organized, Hexameric Supramolecular Architecture: Formation, Structure and Properties. *Chem. Eur. J.* **2014**, *20*, 179–186. [CrossRef]
17.	Bissantz, C.; Kuhn, B.; Stahl, M. A Medicinal Chemist's Guide to Molecular Interactions. *J. Med. Chem.* **2010**, *53*, 5061–5084. [CrossRef]
18.	Morra, G.; Genoni, A.; Neves, M.A.C.; Merz, K.M.; Colombo, G. Molecular Recognition and Drug-Lead Identification: What Can Molecular Simulations Tell Us? *Curr. Med. Chem.* **2010**, *17*, 25–41. [CrossRef]
19.	Zhang, M.; Yan, X.; Huang, F.; Gibson, H.W. Stimuli-Responsive Host–Guest Systems Based on the Recognition of Cryptands by Organic Guests. *Acc. Chem. Res.* **2014**, *47*, 1995–2005. [CrossRef] [PubMed]
20.	Kralij, M.; Tusek-Bozic, L.; Frkanec, L. Biomedical potentials of crown ethers: Prospective antitumor agents. *ChemMedChem* **2008**, *3*, 1478–1492. [CrossRef] [PubMed]
21.	Goek, G.W.; Leevy, W.M.; Weber, M.E. Crown Ethers: Sensors for Ions and Molecular Scaffolds for Materials and Biological Models. *Chem. Rev.* **2004**, *104*, 2723–2750.
22.	Hof, F. Host–guest chemistry that directly targets lysine methylation: Synthetic host molecules as alternatives to bio-reagents. *Chem. Commun.* **2016**, *52*, 10093–10108. [CrossRef] [PubMed]
23.	Ramaiah, D.; Neelakandan, P.P.; Naira, A.K.; Aviraha, R.R. Functional cyclophanes: Promising hosts for optical biomolecular recognition. *Chem. Soc. Rev.* **2010**, *39*, 4158–4168. [CrossRef] [PubMed]
24.	Tan, N.-H.; Zhou, J. Plant Cyclopeptides. *Chem. Rev.* **2006**, *106*, 840–895. [CrossRef] [PubMed]
25.	Galan, M.C.; Dumy, P.; Renaudet, O. Multivalent glyco(cyclo)peptides. *Chem. Soc. Rev.* **2013**, *42*, 4599–4612. [CrossRef] [PubMed]
26.	Russo, A.; Aiello, C.; Grieco, P.; Marasco, D. Targeting "Undruggable" Proteins: Design of Synthetic Cyclopeptides. *Curr. Med. Chem.* **2016**, *23*, 748–762. [CrossRef]
27.	Davis, M.E.; Brester, M.E. Cyclodextrin-based pharmaceutics: Past, present and future. *Nat. Rev. Drug Discov.* **2004**, *3*, 1023–1035. [CrossRef]
28.	Mellet, C.O.; Fernandez, J.M.G.; Benito, J.M. Cyclodextrin-based gene delivery systems. *Chem. Soc. Rev.* **2011**, *40*, 1586–1608. [CrossRef]
29.	Crini, G. Review: A History of Cyclodextrins. *Chem. Rev.* **2014**, *114*, 10940–10975. [CrossRef]
30.	Timmerman, P.; Verboom, W.; Reinhoudt, D.N. Resorcinarenes. *Tetrahedron* **1996**, *52*, 2663–2704. [CrossRef]
31.	Assaf, K.I.; Nau, W.M. Cucurbiturils: From synthesis to high-affinity binding and catalysis. *Chem. Soc. Rev.* **2015**, *44*, 394–418. [CrossRef]
32.	Lagona, J.; Mukhopadhyay, P.; Chakrabarti, S.; Isaacs, L. The Cucurbit[n]uril Family. *Angew. Chem. Int. Ed.* **2005**, *44*, 4844–4870. [CrossRef]
33.	Barrow, S.J.; Kasera, S.; Rowland, M.J.; Barrio, J.; Scherman, O.A. Cucurbituril-Based Molecular Recognition. *Chem. Rev.* **2015**, *115*, 12320–12406. [CrossRef]
34.	Kim, K.; Selvapalam, N.; Ko, Y.H.; Park, K.M.; Kim, D.; Kim, J. Functionalized cucurbiturils and their applications. *Chem. Soc. Rev.* **2007**, *36*, 267–279. [CrossRef]
35.	Diamond, D.; Mckervey, M.A. Calixarene-based sensing agents. *Chem. Soc. Rev.* **1996**, *25*, 15–24. [CrossRef]
36.	Ikeda, A.; Shinkai, S. Novel Cavity Design Using Calix[n]arene Skeletons: Toward Molecular Recognition and Metal Binding. *Chem. Rev.* **1997**, *97*, 1713–1734. [CrossRef]
37.	Böhmer, V. Calixarenes, Macrocycles with (Almost) Unlimited Possibilities. *Angew. Chem. Int. Ed. Engl.* **1995**, *34*, 713–745. [CrossRef]
38.	Nimse, S.B.; Kim, T. Biological applications of functionalized calixarenes. *Chem. Soc. Rev.* **2013**, *42*, 366–386. [CrossRef] [PubMed]

39. Sathiyajith, C.W.; Shaikh, R.R.; Han, Q.; Zhang, Y.; Meguellati, K.; Yang, Y.-W. Biological and related applications of pillar[n]arenes. *Chem. Commun.* **2017**, *53*, 677–696. [CrossRef]

40. Ogoshi, T.; Kanai, S.; Fujinami, S.; Yamagishi, T.; Nakamoto, Y. para-Bridged Symmetrical Pillar[5]arenes: Their Lewis Acid Catalyzed Synthesis and Host-Guest Property. *J. Am. Chem. Soc.* **2008**, *130*, 5022–5023. [CrossRef] [PubMed]

41. Nierengarten, I.; Guerra, S.; Holler, M.; Karmazin-Brelot, L.; Barbera, J.; Deschenaux, R.; Nierengarten, J.-F. Macrocyclic Effects in the Mesomorphic Properties of Liquid-Crystalline Pillar[5]- and Pillar[6]arenes. *Eur. J. Org. Chem.* **2013**, 3675–3684. [CrossRef]

42. Li, C. Pillararene-based supramolecular polymers: From molecular recognition to polymeric aggregates. *Chem. Commun.* **2014**, *50*, 12420–12433. [CrossRef] [PubMed]

43. Wang, K.; Wang, C.Y.; Zhang, Y.; Zhang, S.X.A.; Yang, B.; Yang, Y.W. Ditopic pillar[5]arene-based fluorescence enhancement material mediated by [c2]daisy chain formation. *Chem. Commun.* **2014**, *50*, 9458–9461. [CrossRef] [PubMed]

44. Hua, B.; Shao, L.; Yu, G.; Huang, F. Fluorescence indicator displacement detection based on pillar[5]arene-assisted dye deprotonation. *Chem. Commun.* **2016**, *52*, 10016–10019. [CrossRef]

45. Ogoshi, T.; Kayama, H.; Yamafuji, D.; Aoki, T.; Yamagishi, T.-A. Supramolecular polymers with alternating pillar[5]arene and pillar[6]arene units from a highly selective multiple host–guest complexation system and monofunctionalized pillar[6]arene. *Chem. Sci.* **2012**, *3*, 3221–3226. [CrossRef]

46. Duan, Q.; Cao, Y.; Li, Y.; Hu, X.; Xiao, T.; Lin, C.; Pan, Y.; Wang, L. pH-Responsive Supramolecular Vesicles Based on Water-Soluble Pillar[6]arene and Ferrocene Derivative for Drug Delivery. *J. Am. Chem. Soc.* **2013**, *135*, 10542–10549. [CrossRef]

47. Si, W.; Xin, P.; Li, Z.-T.; Hou, J.-L. Tubular Unimolecular Transmembrane Channels: Construction Strategy and Transport Activities. *Acc. Chem. Res.* **2015**, *48*, 1612–1619. [CrossRef]

48. Yu, G.; Ma, Y.; Han, C.; Yao, Y.; Tang, G.; Mao, Z.; Gao, C.; Huang, F. A Sugar-Functionalized Amphiphilic Pillar[5]arene: Synthesis, Self-Assembly in Water, and Application in Bacterial Cell Agglutination. *J. Am. Chem. Soc.* **2013**, *135*, 10310–10313. [CrossRef]

49. Zhang, Z.; Zhao, Q.; Yuan, J.; Antonietti, M.; Huang, F. A hybrid porous material from a pillar[5]arene and a poly(ionic liquid): Selective adsorption of n-alkylene diols. *Chem. Commun.* **2014**, *50*, 2595–2597. [CrossRef]

50. Ogoshi, T.; Yamagishi, T.; Nakamoto, Y. Pillar-shaped macrocyclic hosts pillar[n]arenes: New key players for supramolecular chemistry. *Chem. Rev.* **2016**, *116*, 7937–8002. [CrossRef]

51. Strutt, N.L.; Zhang, H.; Schneebeli, S.T.; Stoddart, J.F. Amino-Functionalized Pillar[5]arene. *Chem. Eur. J.* **2014**, *20*, 10996–11004. [CrossRef]

52. Gómez, B.; Francisco, V.; Fernández-Nieto, F.; Garcia-Rio, L.; Martín-Pastor, M.; Paleo, M.R.; Sardina, F.J. Host–Guest Chemistry of a Water-Soluble Pillar[5]arene: Evidence for an Ionic-Exchange Recognition Process and Different Complexation Modes. *Chem. Eur. J.* **2014**, *20*, 12123–12132. [CrossRef]

53. Gómez, B.; Francisco, V.; Montecinos, R.; Garcia-Rio, L. Investigation of the binding modes of a positively charged pillar[5]arene: Internal and external guest complexation. *Org. Biomol. Chem.* **2017**, *15*, 911–919. [CrossRef] [PubMed]

54. Liz, D.G.; Manfredi, A.M.; Medeiros, M.; Montecinos, R.; Gómez, B.; Garcia-Rio, L.; Nome, F. Supramolecular phosphate transfer catalysis by pillar[5]arene. *Chem. Commun.* **2016**, *52*, 3167–3170. [CrossRef]

55. Wanderlind, E.H.; Liz, D.G.; Gerola, A.P.; Affeldt, R.F.; Nascimento, V.; Bretanha, L.C.; Montecinos, R.; Garcia-Rio, L.; Fiedler, H.D.; Nome, F. Imidazole-functionalized pillar[5]arenes: Highly reactive and selective supramolecular artificial enzymes. *ACS Catal.* **2018**, *8*, 3343–3347. [CrossRef]

56. Silveira, E.V.; Nascimento, V.; Wanderlind, E.H.; Affeldt, R.F.; Micke, G.A.; Garcia-Rio, L.; Nome, F. Inhibitory and cooperative effects regulated by pH in host-guest complexation between cationic pillar[5]arene and reactive 2-carboxyphthalanilic acid. *J. Org. Chem.* **2019**, *84*, 9684–9692. [CrossRef] [PubMed]

57. Montes-Garcia, V.; Fernández-López, C.; Gómez, B.; Pérez-Juste, I.; Garcia-Rio, L.; Liz-Marzán, L.M.; Pérez-Juste, J.; Pastoriza-Santos, I. Pillar[5]arene-mediated synthesis of gold nanoparticles: Size control and sensing capabilities. *Chem. Eur. J.* **2014**, *20*, 8404–8409. [CrossRef]

58. Montes-Garcia, V.; Gómez, B.; Martínez-Solís, D.; Taboada, J.M.; Jiménez-Otero, N.; Uña-Alvarez, J.; Obelleiro, F.; Garcia-Rio, L.; Pérez-Juste, J.; Pastoriza-Santos, I. Pillar[5]arene-based supramolecular plasmonic thin films for label-free, quantitative and multiplex SERS detection. *ACS Appl. Mater. Interfaces* **2017**, *9*, 26372–26382. [CrossRef]

59. Barrán-Berdón, A.L.; Martínez-Negro, M.; Garcia-Rio, L.; Domènech, O.; Tros de Ilarduya, C.; Aicart, E.; Junquera, E. Biophysical study of gene nanocarriers formed by anionic/zwitterionic mixed lipids and pillar[5]arene policationic macrocycles. *J. Mat. Chem. B* **2017**, *5*, 3122–3131. [CrossRef]

60. Zhang, H.; Liu, Z.; Zhao, Y. Pillararene-based self-assembled amphiphiles. *Chem. Soc. Rev.* **2018**, *47*, 5491–5528. [CrossRef]

61. Feng, W.; Ji, M.; Yang, K.; Pei, Y.; Pei, Z. Supramolecular delivery systems based on pillararenes. *Chem. Commun.* **2018**, *54*, 13626–13640. [CrossRef]

62. Wu, X.; Li, Y.; Lin, C.; Hu, X.-Y.; Wang, L. GSH- and pH- responsive drug delivery system constructed by water-soluble pillar[5]arene and lysine derivative for controllable drug release. *Chem. Commun.* **2015**, *51*, 6832–6835. [CrossRef]

63. Braegelman, A.S.; Webber, M.J. integrating stimuli-responsive properties in host-guest supramolecular drug delivery systems. *Theranostics* **2019**, *9*, 3017–3040. [CrossRef]

64. Peng, H.; Xie, B.; Yang, X.; Dai, J.; Wei, G.; He, Y. Pillar[5]arene-based, dual pH and enzyme responsibe supramolecular vesicles for targeted antibiotic delivery against intracellular MRSA. *Chem. Commun.* **2020**, *56*, 8115–8118.

65. Shurpik, D.N.; Mostavaya, O.A.; Sevastyanov, D.A.; Lenina, O.A.; Sapunova, A.S.; Voloshina, A.D.; Petrov, K.A.; Kovyazina, I.V.; Cragg, P.J.; Stoikov, I.I. Supramolecular neuromuscular blocker inhibition by a pillar[5]arene through aqueous inclusion of rocuronium bromide. *Org. Biomol. Chem.* **2019**, *17*, 9951–9959. [CrossRef] [PubMed]
66. Shurpik, D.N.; Sevastyanov, D.A.; Zelenikhin, P.V.; Subakaeva, E.V.; Evtugyn, V.G.; Osin, Y.N.; Cragg, P.J.; Stoikov, I.I. Hydrazides of glycine-containing decasubstituted pillar[5]arene: Synthesis and ecapsulation of fluxoridine. *Tetrahedron Lett.* **2018**, *59*, 4410–4415. [CrossRef]
67. Ma, Y.; Ji, X.; Xiang, F.; Chi, X.; Han, C.; He, J.; Abliz, Z.; Chen, W.; Huang, F. A cationic water-soluble pillar[5]arene: Synthesis and host–guest complexation with sodium 1-octanesulfonate. *Chem. Commun.* **2011**, *47*, 12340–12342. [CrossRef]
68. Douteau-Guével, N.; Coleman, A.W.; Morel, J.-P.; Morel-Desrosiers, N. Complexation of the basic amino acids lysine and arginine by three sulfonatocalix[n]arenes (n = 4, 6 and 8) in water: Microcalorimetric determination of the Gibbs energies, enthalpies and entropies of complexation. *J. Chem. Soc. Perkin Trans.* **1999**, *2*, 629–633. [CrossRef]
69. Yu, G.; Zhou, X.; Zhang, Z.; Han, C.; Mao, Z.; Gao, C.; Huang, F. Pillar[6]arene/Paraquat Molecular Recognition in Water: High Binding Strength, pH-Responsiveness, and Application in Controllable Self-Assembly, Controlled Release, and Treatment of Paraquat Poisoning. *J. Am. Chem. Soc.* **2012**, *134*, 19489–19497. [CrossRef] [PubMed]
70. Francisco, V.; Basilio, N.; Garcia-Rio, L. Counterion Exchange as a Decisive Factor in the Formation of Host:Guest Complexes by p-Sulfonatocalix[4]arene. *J. Phys. Chem. B* **2012**, *116*, 5308–5315. [CrossRef]
71. Arena, G.; Gentile, S.; Gulino, F.G.; Sciotto, D.; Sgarlata, C. Water-soluble pentasulfonatocalix[5]arene: Selective recognition of ditopic trimethylammonium cations by a triple non-covalent interaction. *Tetrahedron Lett.* **2004**, *45*, 7091–7094. [CrossRef]
72. Guo, D.-S.; Wang, L.-H.; Liu, Y. Highly Effective Binding of Methyl Viologen Dication and Its Radical Cation by p-Sulfonatocalix[4,5]arenes. *J. Org. Chem.* **2007**, *72*, 7775–7778. [CrossRef] [PubMed]
73. Cepeda, M.; Daviña, R.; García-Río, L.; Parajó, M. Cyclodextrin-surfactant binding constant as driven force for uncomplexed cyclodextrin in equilibrium with micellar systems. *Chem. Phys. Lett.* **2010**, *499*, 70–74. [CrossRef]
74. Araujo, L.S.S.; Lazzara, G.; Chiappisi, L. Cyclodextrin/surfactant inclusion complexex: An integrated view of their thermodynamic and structural properties. *Adv. Coll. Int. Sci.* **2021**, *289*, 102375–102386. [CrossRef] [PubMed]
75. Xue, M.; Yang, Y.; Chi, X.; Zhang, Z.; Huang, F. Pillararenes, A New Class of Macrocycles for Supramolecular Chemistry. *Acc. Chem. Res.* **2012**, *45*, 1294–1308. [CrossRef]

pharmaceutics

MDPI

Article

Multivalent Calixarene-Based Liposomes as Platforms for Gene and Drug Delivery

José Antonio Lebrón [1], Manuel López-López [2], Clara B. García-Calderón [3], Ivan V. Rosado [3], Fernando R. Balestra [4,5], Pablo Huertas [4,5], Roman V. Rodik [6], Vitaly I. Kalchenko [6], Eva Bernal [1], María Luisa Moyá [1,*], Pilar López-Cornejo [1,*] and Francisco J. Ostos [1,*]

[1] Department of Physical Chemistry, Faculty of Chemistry, University of Seville, C/Profesor García González 1, 41012 Seville, Spain; jlebron@us.es (J.A.L.); evabernal@us.es (E.B.)
[2] Department of Chemical Engineering, Physical Chemistry and Materials Science, Faculty of Experimental Sciences, University of Huelva, Campus de El Carmen, Avda. de las Fuerzas Armadas s/n, 21071 Huelva, Spain; manuel.lopez@diq.uhu.es
[3] Institute of Biomedicine of Seville (IBiS), University Hospital Virgen del Rocío/CSIC/University of Seville, Avda. Manuel Siurot s/n, 41013 Seville, Spain; claragarcia@us.es (C.B.G.-C.); ivrosado@us.es (I.V.R.)
[4] Department of Genetics, Faculty of Biology, University of Seville, C/Profesor García González 1, 41012 Seville, Spain; fernando.balestra@cabimer.es (F.R.B.); phuertas@us.es (P.H.)
[5] Andalusian Center of Molecular Biology and Regenerative Medicine (CABIMER), University of Seville-CSIC-University Pablo de Olavide, Avda. Américo Vespucio 24, 41092 Seville, Spain
[6] Institute of Organic Chemistry, National Academy of Science of Ukraine, Murmanska Str. 5, 02660 Kiev, Ukraine; dmso@ukr.net (R.V.R.); vik@ioch.kiev.ua (V.I.K.)
*   Correspondence: moya@us.es (M.L.M.); pcornejo@us.es (P.L.-C.); fostos@us.es (F.J.O.); Tel.: +34-954-557-175 (M.L.M.)

Citation: Lebrón, J.A.; López-López, M.; García-Calderón, C.B.; V. Rosado, I.; Balestra, F.R.; Huertas, P.; Rodik, R.V.; Kalchenko, V.I.; Bernal, E.; Moyá, M.L.; et al. Multivalent Calixarene-Based Liposomes as Platforms for Gene and Drug Delivery. *Pharmaceutics* **2021**, *13*, 1250. https://doi.org/10.3390/pharmaceutics13081250

Academic Editors: Franco Dosio and Giovanna Della Porta

Received: 20 July 2021
Accepted: 8 August 2021
Published: 12 August 2021

Publisher's Note: MDPI stays neutral with regard to jurisdictional claims in published maps and institutional affiliations.

Abstract: The formation of calixarene-based liposomes was investigated, and the characterization of these nanostructures was carried out using several techniques. Four amphiphilic calixarenes were used. The length of the hydrophobic chains attached to the lower rim as well as the nature of the polar group present in the upper rim of the calixarenes were varied. The lipid bilayer was formed with one calixarene and with the phospholipid 1,2-dioleoyl-sn-glycero-3-phosphoethanolamine, DOPE. The cytotoxicity of the liposomes for various cell lines was also studied. From the results obtained, the liposomes formed with the least cytotoxic calixarene, $(TEAC_{12})_4$, were used as nanocarriers of both nucleic acids and the antineoplastic drug doxorubicin, DOX. Results showed that $(TEAC_{12})_4$/DOPE/p-EGFP-C1 lipoplexes, of a given composition, can transfect the genetic material, although the transfection efficiency substantially increases in the presence of an additional amount of DOPE as coadjuvant. On the other hand, the $(TEAC_{12})_4$/DOPE liposomes present a high doxorubicin encapsulation efficiency, and a slow controlled release, which could diminish the side effects of the drug.

Keywords: cationic calix[4]arenes; liposomes; nucleic acids; transfection efficiency; doxorubicin; encapsulation

1. Introduction

Liposomes are spherical structures, similar to vesicles, which have an inner aqueous polar region and a hydrophobic lipid bilayer [1]. Their spontaneous formation in aqueous solutions is due to interactions among water molecules, hydrophilic head groups and hydrophobic chains of the amphiphilic molecules forming the lipid bilayers [2–5]. Liposomes have been prepared by several techniques such as thin lipid film hydration, solvent injection, detergent dialysis or reverse phase evaporation [2,4,5]. They are usually characterized according to their size and number of bilayers, and their charge can be positive, negative or neutral [6,7].

Since the pioneering work by Bangham [8], liposomes have been broadly used as delivery systems for several diagnostic and therapeutic compounds including drugs, genes, imaging agents, proteins, or vaccines, among others [9–13]. Liposomes draw great interest for many researchers working on biomedical applications due to the numerous advantages they offer. It is easy to prepare biocompatible and biodegradable liposomes, with low toxicity and immunogenicity, high stability, and with the ability to host hydrophilic and hydrophobic compounds, whose release can be controlled [2,13,14]. A great variety of amphiphilic molecules have been used to prepare liposomes, making the formation of vesicles in aqueous solution favorable [15–17]. Among them, calixarenes have been frequently used [18–20].

Calixarenes are macrocycles prepared by the base-catalyzed condensation of formaldehyde and p-substituted phenols [21]. They are composed of phenolic units disposed in cyclic arrays and linked by methylene spacers. Their name comes from the word *calix crater* because their tridimensional structure resembles that of an ancient Greek vase. The most common calixarenes are calix[4]arenes, calix[6]arenes, and calix[8]arenes, where [n] indicates the number of phenolic units. The synthetic, easy-to-obtain multivalent ligands by introducing substituents of different nature at the upper and lower rim is one of their main advantages. The structural variety of calixarenes permits to have selective receptors for the inclusion of several neutral molecules, metal and ammonium anions, and cations [22]. Besides, calixarenes can be used in the self-assembly of nanoparticles [23], in the building of molecular machines and rotaxanes [24], for molecular encapsulation [25], and many other applications [22,26,27].

The ability of calixarenes to bind, condense, and transport DNA across cell membranes has been previously investigated [28]. In particular, calix[4]arenes in cone conformation have been found to facilitate cell transfection effectively. Among them, those with long alkyl chains usually lead to small aggregates with low polydispersity, promoting more efficiently gene transfection [29]. On the other hand, it was previously shown that calixarenes interact with the antineoplastic drug doxorubicin, DOX (Scheme 1) [30]. The interaction between calixarenes and DOX is mainly mediated by host–guest and π–π interactions. Works in the literature have shown that calixarenes can be used for the treatment of different types of cancer [31], the results obtained for distinct cancer lines showing that a higher therapeutic effect of the drug is achieved as well as a decrease in side effects.

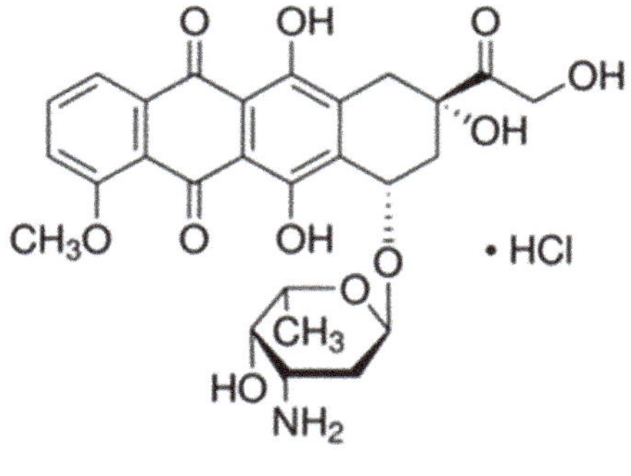

Scheme 1. Structure of doxorubicin.

The authors have been interested in the interaction of calixarenes with nucleic acids and antineoplastic drugs such as DOX [30,32–34]. In this work, the authors wanted to investigate if the use of calixarene-based liposomes improves the results of their use as nanodelivery systems as compared to the naked calixarenes. It is worth noting that thanks to their unique structural properties, calixarene-based liposomes could provide hybrid systems that will synergistically lead to non-viral vectors with enhanced cell transfection properties [35,36]. Moreover, the hydrophobic cavity of calixarenes provide the posibility that, within the calixarene-based liposomes, host–guest phenomena with different drugs can occur [37]. The more appropriate type of calixarenes for preparing calixarene-based liposomes are the amphiphilic ones. They can be obtained by introducing polar groups at one rim and hydrophobic chains at the other rim [30]. With the goal of preparing calixarene-based liposomes for delivering of genetic material and doxorubicin, in this work the cationic calix[4]arenes 5,11,17,23-tetratriethylammonium-methylene-25,26,27,28- tetradodecyloxycalix[4]arene tetrachloride, $(TEAC_{12})_4$; 5,11,17,23-tetra(3-methylimidazolium)-methylene-25,26,27,28-tetradodecyloxycalix[4]arene tetrachloride, $(Im_{12})_4$; 5,11,17,23-tetra(3-methylimidazolium)-methylene-25,27-dihexadecyloxy-26,28-dipropoxycalix[4]arene tetrachloride, $(Im_{16}Im_3)_2$; and 5,11,17,23-tetra(3-methylimidazolium)-methylene-25,26,27,28-tetrahexadecyloxycalix[4]arene tetrachloride, $(Im_{16})_4$, have been used to prepared liposomes (see Scheme 2). The abbreviations used tried to inform the reader about the structure of the calixarenes. TEAC stands for tetraethylammonium chloride and Im for imidazolinium, in order to distinguish between the two different charged heads present in the upper rim of the calixarenes. The subscripts 3, 12, and 16 inside the parentheses indicate the length of the hydrophobic chains attached to the lower rim. The subscripts 2 and 4 outside the parentheses indicate the number of units inside the parentheses present in the calixarene molecules. Throughout this work the abbreviation CAL means calixarene.

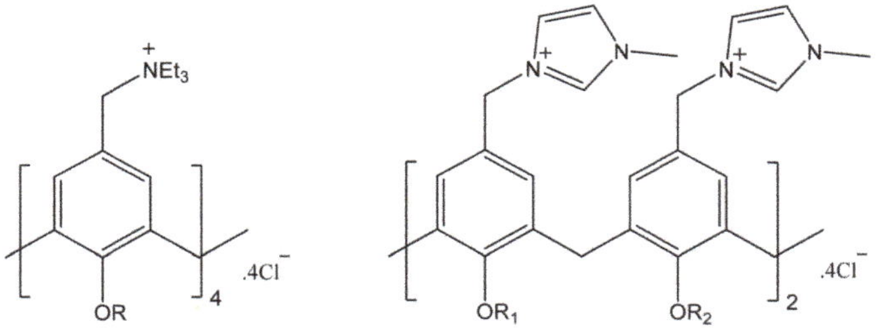

$R = (CH_2)_{11}CH_3 \, (TEAC_{12})_4$

i) $R_1 = (CH_2)_{11}CH_3 \quad R_2 = (CH_2)_{11}CH_3 \quad (Im_{12})_4$
ii) $R_1 = (CH_2)_{15}CH_3 \quad R_2 = (CH_2)_2CH_3 \quad (Im_{16}Im_3)_2$
iii) $R_1 = (CH_2)_{15}CH_3 \quad R_2 = (CH_2)_{15}CH_3 \quad (Im_{16})_4$

Scheme 2. Structure of the calix[4]arenes investigated in this work. (**A**) Calix[4]arene ammonium derivate; and (**B**) Calix[4]arene imidazolinium derivates.

The results obtained will be of interest for researchers working on the use of calixarenes in several biotechnological applications.

2. Materials and Methods

2.1. Materials

The lipid 1,2-dioleoyl-sn-glycero-3-phosphoethanolamine (DOPE) was purchased from Avanti Polar Lipids (Alabaster, AL, USA). Red Safe was from iNtRON (Biotechnologiy, Chicago, IL, USA). The rest of the materials, including doxorubicin, were from Sigma-Aldrich (Darmstadt, Germany) and used without further purification.

ctDNA concentration was estimated by UV–visible spectroscopy measuring at 260 nm (molar absorptivity 6600 mol^{-1} dm^3 cm^{-1} [38]). The average number of base pairs was estimated by agarose gel electrophoresis, using ethidium bromide, EB. The results indicate that there are above 10,000 bp [39]. Throughout the manuscript, the ctDNA concentration will be expressed per base-pairs. The pEGFP-C1 plasmid (Clontech, Biocientífica S.A., Buenos Aires, Argentina), pDNA, was extracted from competent *E. coli* bacteria previously transformed with pEGFP-C1; the extraction was done using a GenElute HP Select Plasmid Gigaprep kit (Sigma Aldrich, Darmstadt, Germany). A protocol previously described was used [40]. FuGENE 6 was from Promega Corporation (Madison, WI, USA).

The syntheses of the cationic calixarenes $(TEAC_{12})_4$ and $(Im_{16}Im_3)_2$ were previously described [30], and $(Im_{12})_4$ and $(Im_{16})_4$ were purchased from Life Chemicals Inc. (Niagara-on-the-Lake, ON, Canada). Their purity ($\geq$99%) was checked by ^{1}H and ^{13}C NMR, elemental analysis, and mass spectra.

Solutions were prepared with MilliQ water (resistivity > 18 MΩ $\times$ cm). The pH was kept constant at 7.4 by using 10 mM HEPES (4-(2-hydroxyethyl)piperazine-1-ethanesulfonic acid sodium salt) buffer.

2.2. Preparation of Liposomes

Liposomes were prepared using the lipid thin-film hydration method [41]. Briefly, adequate quantities of calixarenes, CAL, and DOPE were dissolved in chloroform. Different volumes of these solutions were mixed in order to obtain the desired cationic calixarene molar fraction, α, given by:

$$\alpha = \frac{n_{CAL}}{n_{CAL} + n_{DOPE}} \tag{1}$$

where n_{CAL} and n_{DOPE} are the mole number of the cationic calixarene and the zwitterionic DOPE, respectively, in the total volume of the organic solution.

A rotary evaporator was used to evaporate the organic solvent, at 303 K for 50 min. The resultant dry lipid film was stored at 193 K for at least 24 h. In this way, degradation is avoided [42]. Afterwards, 2 mL of HEPES 10 mM, pH = 7.4, was added for hydrating the lipid film, and the mixture was submitted to 10 cycles of vortex (3 min/1200 rpm) and sonication (2 min, JP Selecta Ultrasons system 200 W, 50 kHz, Abrera, Barcelona, Spain). In the final step the solution was vortexed for 2 h at room temperature. The liposome solution had a high polydispersity, with multilamellar liposomes. In order to obtain a homogeneous size distribution solution with unilamellar liposomes, 1 mL of liposome solution was extruded 10 times with a manual mini extruder from Avanti Polar Lipids (Alabaster, AL, USA), using polycarbonate membranes of 100 and 200 nm (Whatman, Maidstone, UK). After extrusion, the solutions were maintained in the dark at 277 K for 24 h for a complete stabilization. In this work the calixarene/DOPE liposomes will be named CAL/DOPE liposomes.

Only in the case of $(Im_{16})_4$ liposomes was a mixture of ethyl acetate:ethanol 1:1 used, instead of chloroform, because of solubility problems. Nonetheless, by using this mixture in the thin film hydration method, the characteristics of the liposomes containing $(TEAC_{12})_4$, $(Im_{12})_4$, and $(Im_3Im_{16})_2$ were similar to those observed using chloroform.

The composition (mole ratio) of the liposomes prepared is summarized in Table 1.

Table 1. Composition (mole ratio) of the CAL/DOPE liposomes prepared.

α	$(TEAC_{12})_4$	$(Im_{12})_4$	$(Im_3Im_{16})_2$	$(Im_{16})_4$
0.1	-	-	1:9	1:9
0.2	1:4	1:4	1:4	1:4
0.3	1:2.33	1:2.33	1:2.33	1:2.33
0.4	1:5	1:5	1:5	-
0.45	1:1.22	1:1.22	-	-
0.5	1:1	1:1	1:1	-

Mole ratio is expressed with respect to the n_{CAL}.

2.3. Preparation of Lipoplexes

The lipoplexes were prepared by mixing appropriate volumes of the liposome solution and of the aqueous ctDNA (or p-EFGP-C1) HEPES 10 mM solutions in order to obtain the desired L/D ratio. For each α value, the mass ratio L/D is given by the expression:

$$\frac{L}{D} = \frac{m_{CAL} + m_{DOPE}}{m_{DNA}} \tag{2}$$

where m_{DOPE}, m_{CAL}, and m_{DNA} are the masses of the zwitterionic phospholipid, of the calixarene, and of the DNA, respectively, in the solution. In all the liposome solutions investigated, the mass of DNA was kept constant at 10^{-4} g (the concentration was 1.0 mg/mL or 8.1×10^{-5} mol L^{-1} given in base-pairs). The calixarene/DOPE/ctDNA lipoplexes will be named CAL/DOPE/DNA lipoplexes.

The stability of the lipoplexes was followed by changes in their size and polydispersity with time. The size remained unchanged for more than 48 h. The authors also checked the stability of the lipoplexes, of different compositions, after dilution with buffer HEPES 10 mM. No variations in their size were observed.

2.4. Zeta Potential Measurements

Zeta-potential, ζ, values were calculated measuring the electrophoretic mobility of the liposomes and of the lipoplexes from the velocity of the particles, using a laser Doppler velocimeter (LDV). A Zetasizer Nano ZS Malvern Instrument Ltd. (Malvern, Worcestershire, UK) was used. Temperature was kept at 303.0 ± 0.1 K, and DTS1060 polycarbonatecapillary cells were utilized. ctDNA concentration in the buffered solutions of liposomes was 8.1×10^{-5} M. Data are expressed as mean $\pm$ SD from at least three separate experiments, $n = 9$.

2.5. Dynamic Light Scattering, DLS, Measurements

A Zetasizer Nano ZS Malvern Instrument Ltd. (Worcestershire, UK) was used to estimate the hydrodynamic diameter, d_H (Z average), and the polydispersity index, PDI, of the lipoplexes using DLS measurements. A scattering angle of 90° was used. A fixed concentration of 8.1×10^{-5} M of ctDNA was present in all the liposome solutions investigated. Data are expressed as mean $\pm$ SD from at least three separate experiments, $n = 9$. Temperature was maintained at 303.0 ± 0.1 K.

2.6. Agarose Gel Electrophoresis

Agarose gel (1%) was prepared in a TAE buffer (40 mM Tris-acetate, 1 mM EDTA) in a total volume of 180 µL and stained with the dye Red Safe (10 µL) for the visualization of the nucleic acid bands. The ctDNA concentration was kept constant at 8.1×10^{-5} M. The method was as follows: (i) 20 µL of the buffered liposome solution was mixed with 5 µL of $5 \times$ DNA loading buffer. (ii) After homogenization, the resulting solution was added in each well. Electrophoresis was performed at 90 V for 90 min. A detector Ultima 16si (Hoefer Inc., Holliston, MA, USA) was used for visualizing the nucleic acid bands by irradiation with UV light (254 nm).

2.7. Circular Dichroism, CD, Spectra

A Biologic Mos-450 spectropolarimeter (Cambridge, UK) was used to register the CD spectra. Scans were taken from 220 to 310 nm with a standard quartz cell of 10 mm path length. Three independent experiments were done. Each spectrum was obtained from an average of 10 runs, with a 5 min equilibration before each scan, at 303.0 ± 0.1 K. The ctDNA concentration was kept constant at 8.1×10^{-5} M in the lipoplex solutions. All solutions were prepared in 10 mM HEPES buffer, pH = 7.4.

2.8. Atomic Force Microscopy, AFM

Atomic force microscopy was used to study the structures of the lipoplexes. A resonance frequency of around 240 KHz and a nominal force constant of 42 N/m were the working conditions in a Molecular Imaging PicoPlus 2500 AFM (Agilent Technologies, Santa Clara, CA, USA). Silicon cantilevers (Model Pointprobe, Nanoworld, Neufchâtel, Switzerland) were used. The images were recorded in air and in tapping mode. Data collection (256×256 pixels) was registered with scan speeds about 0.5 Hz. The ctDNA concentration in the buffered HEPES 10 mM liposome solutions, pH = 7.4, was 0.6 µM.

Images of the buffered liposome solutions were obtained using the following method: (a) In order to prepare a modified mica surface, 0.1% (*v/v*) APTES aqueous solution was dropped onto a freshly cleaved mica surface. It was washed with ultra-pure water after 20 min and air dried. (b) A 30 µL droplet of the lipoplex solution was deposited on the modified mica surface and incubated for 30 min. (c) Subsequently, the mica surface was washed with pure water and air dried for AFM imaging.

2.9. Electron Transmission Microscopy, TEM

TEM images of the lipoplexes were obtained in a Zeiss Libra 120 scanning electron microscope (Carl Zeiss AG, Oberkochen, Germany), at 80 kV. Samples were prepared by impregnation, using a 300 mesh copper grid coated collodion that, subsequently, was stained with a solution of uranyl acetate (2.0%). Images were processed with a bottom-mounted TEM CCD camera and recorded with a resolution of 2048 × 2048 pixels. ImageJ (National Institutes of Health (NIH), Bethesda, MD, USA) bundled with 64-bit Java 1.8.0_172 was used to analyze TEM images from independent experiments, for each of the lipoplex solutions investigated.

2.10. In Vitro Cytotoxicity Assays

The cytotoxicity of the CAL/DOPE liposomes with $\alpha = 0.5$ for $(TEAC_{12})_4$, $(Im_{12})_4$, and $(Im_{16}Im_3)_2$, and with $\alpha = 0.3$ for $(Im_{16})_4$, at different L/D values, was estimated in vitro using the MTT assay [43]. These are the maximum α values that could be prepared for the different calixarenes, and they were chosen to carry out the cell viability assays because they correspond to the highest content of the cationic calixarene within the liposomes. The cell lines used were RPE-1 (normal cell line), A549 (adenocarcinomic human alveolar basal epithelial cell line), HepG2 (human liver cancer cell line), LS180 (adenocarcinomic human colonic epithelial cell line), and MCF7 (breast cancer cell line). They were a gift from different research groups from the IBIS (the Institute of Biomedicine of Seville). In any case, all cell lines used were from commercial suppliers. Cell lines were plated out into 96-well plates at a density of 3000 cells per plate. The next day, the liposome solutions were added to the wells, and the plate wa s returned to the incubator for 4 days more. Later, they were pulsed with MTS (ROCHE, Basilea, Switzerland). According to the manufacturer instructions, cell viability was measured by luminometry in a Varioskan Flash (Thermo Fisher Scientific, Waltham, MA, USA). Each liposome concentration was measured in triplicate.

2.11. Transfection Assays

The cell line chosen to carry out these experiments was the U2OS, from human osteosarcoma, because these cells are suitable for transfection experiments. The non-viral vectors investigated were $(TEAC_{12})_4$/DOPE liposomes, containing the plasmid pEGFP-C1. Liposomes of $(TEAC_{12})_4$ were selected because they are the least toxic of all the CAL/DOPE liposomes investigated. On the other hand, pEGFP-C1 is a plasmid carrying an enhanced GFP coding sequence with the required regulatory elements for constitutive expression of the gene in human cells. The method used to carry out the transfection experiments was as follows: 3 µg of pEGFP-C1 was added to a solution containing 180 µL of Opti-MEM (Gibco, Thermo Scientific, Waltham, MA, USA), and the amount of liposome buffered solution (HEPES 10 mM) necessary to obtain the L/D ratio for each α value was investigated. The resulting mixture was incubated at room temperature for 20 min, and afterwards it was added to a 50% confluent 6 cm plate with 3 mL of DMEM medium (Sigma Aldrich, Darmstadt, Germany).

The cells were transfected with a mixture of transfection reagent and Opti-MEM (not pEGFP-C1 included) as negative control. As positive control, FuGENE 6 transfection reagent (E2311, from Promega Corporation, Madison, WI, USA) was used, according to the manufacturer's protocol (i.e., 3 µg of pEGFP-C1 in 200 µL Opti-MEM plus 9 µL of

FuGENE 6). Transfection efficiency was evaluated by flow cytometry with a FACSCalibur (BD Biosciences, Franklin Lakes, NJ, USA) 24 h after transfection.

2.12. UV–Visible Spectroscopy

The doxorubicin concentration was determined by UV–visible spectroscopy, using a Hitachi UV-visible 3900 (Chiyoda, Tokyo, Japan) by measuring absorbance at 490 nm. Temperature was kept using a Lauda (Stuttgart, Baden-Würtenberg, Germany) flow cryostat connected to the cell compartment.

2.13. Encapsulation Efficiency Measurements

A dialysis method was used to estimate the doxorubicin, DOX, encapsulation efficiency of the $(TEAC_{12})_4/DOPE$ liposomes. A total of 900 µL of drug-loaded liposomes was added to a Spectra/Por® 3 (MWCO 3.5 kDa) from Spectrum Laboratories, Inc. (Rancho Dominguez, CA, USA) The final concentration of antibiotic was 2×10^{-4} M. The dialysis membrane was plunged into a beaker containing 30 mL of 10 mM HEPES buffer (pH = 7.4), the same used for liposome hydration. The liposomes' stabilization was ensured by keeping the temperature at 4 °C throughout the process in order to avoid doxorubicin degradation [32]. An aliquot of 1 mL from the beaker was taken at different time intervals. These aliquots were replaced each time by an equal volume of buffer HEPES 10 mM in order to keep constant the total volume of buffer in the beaker. Dialysis was followed for at least 24 h. The encapsulation efficiency, EE, was calculated by using Equation (3):

$$EE(\%) = \frac{[DOX]_{Liposomes}}{[DOX]_T} \times 100 = \frac{[DOX]_T - [DOX]_{buffer}}{[DOX]_T} \tag{3}$$

where $[DOX]_{Liposomes}$, $[DOX]_T$, and $[DOX]_{buffer}$ are the DOX concentration encapsulated into the calixarene-based liposomes, the total DOX concentration present in the system, and the DOX concentration in the buffer solution, respectively. All the concentrations are referred to the total volume of solutions.

The loading capacity, LC, can be calculated using the following expression:

$$LC = \frac{n_{DOX.enc.}}{n_{DOX,total} + n_{CAL+DOPE}} \times 100 \tag{4}$$

where $n_{DOX.enc.}$, $n_{DOX,total}$, and $n_{CAL+DOPE}$ represent the moles of the encapsulated drug, total drug, and total lipids, respectively.

Quantification of doxorubicin concentration was done by UV–visible spectroscopy ($\lambda = 490$ nm). Each experiment was performed in triplicate.

2.14. In Vitro Drug Release

After applying the dyalisis method, the doxorubicin loaded $(TEAC_{12})_4/DOPE$ liposomes were suspended in buffer HEPES 10 mM, pH = 7.4, in a glass vial under continuous magnetic stirring, 200 rpm, at 37.4 °C (the human body temperature). A sample was removed at determined time intervals and, subsequently, replaced with an equal amount of buffer. This is a way to simulate the in vivo removal of a drug into a systemic circulation. The concentration of the antibiotic was estimated by UV–visible spectroscopy, measuring the absorbance at 490 nm. The absorbance data were corrected from the dilution effect. Three separate experiments were done. The precision was close to 7%.

2.15. Statistical Analysis

Values are expressed as the mean ± standard errors of independent experiments. Statistical analysis was performed with Student's t-test and one-way analysis of variance (ANOVA). When $p < 0.05$ (95% confidence) the differences were considered as significant.

3. Results and Discussion

3.1. Calixarene-Based Liposomes

Calixarene-based liposomes were prepared at several cationic CAL molar fractions α ($\alpha = n_{CAL}/(n_{CAL} + n_{DOPE})$). In particular, this magnitude was varied within the interval from 0.1 to 0.5. Molar ratios higher than 0.5 were not investigated because these liposome solutions showed a high polydispersity index, PDI (PDI > 0.8). Figure 1 shows the dependence of the hydrodynamic diameter of the calixarene-based liposomes on the calixarene molar ratio, α, in these nanostructures. The interval of α studied for each calixarene was limited by solubility problems. In the case of $(Im_{16})_4$, it was particularly narrow. Figure 1 shows that the size of the CAL/DOPE liposomes initially decreased upon increasing α, but a subsequent increase in the molar ratio led to an increase in the liposome sizes. This trend was observed for all CAL/DOPE liposomes with the exception of $(Im_{16})_4$, for which no clear trend was observed due to the narrow α interval studied.

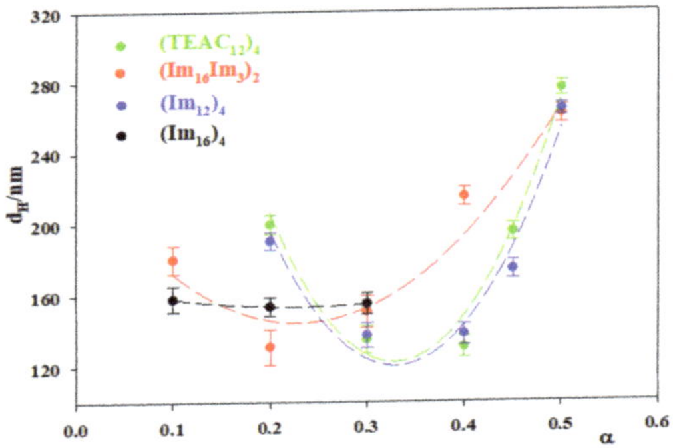

Figure 1. Dependence of the hydrodynamic diameter, d_H, on the cationic calixarene molar fraction, α. T = 303.1 ± 0.1 K.

The experimental observations could be explained by considering the main interactions controlling the liposome sizes. On one hand the hydrophobic interactions are the driving force for the formation of liposomes, and they favor a diminution in the liposome size [44]. On the other hand, the electrostatic repulsions between the positively charged head groups of the calixarenes in the bilayer make an increase in the liposome size more favorable. At low α values the hydrophobic interactions mainly control the liposome sizes, and an increment in α results in a diminution of the hydrodynamic diameter, d_H. A further increment in α means an increase in the amount of cationic CAL within the liposomes, and, as a consequence, the electrostatic repulsions will augment. At a given α value, depending on the calixarene nature, the electrostatic repulsions overcome the hydrophobic interactions, and an increase in α is followed by an increase in d_H since the cationic head groups tend to separate. The hydrophobic interactions are expected to be stronger in the case of $(TEAC_{12})_4$ and $(Im_{12})_4$ than for $(Im_3Im_{16})_2$. The latter has alternate chains, which makes the interaction between the two 16 C tails more difficult. Besides, two of the chains are short (three C atoms), this resulting in weak hydrophobic interactions when compared to those with 12 C atoms. Therefore, the minimum in the plots of d_H vs. α is expected to be observed for higher α values in the case of $(TEAC_{12})_4$ and $(Im_{12})_4$ than in the case of $(Im_3Im_{16})_2$, as in fact is observed. Following the same reasoning, $(Im_{16})_4$ should present the minimum at the highest α value. However, this is not observed.

Apart from the solubility problems, a possible explanation would be that the long hexadecyl chains could fold towards themselves, resulting in steric hindrance [45], which would make smaller liposomes more favorable.

The cytotoxicity of the CAL/DOPE liposomes was estimated by using the MTT assay. Since DOPE is considered a biocompatible lipid, the cell viability in several cell lines was determined for the highest α value, corresponding to the liposomes with the largest amount of cationic calixarenes. Figure 2 shows the results obtained, from which some conclusions can be reached. $(TEAC_{12})_4$/DOPE liposomes are the less toxic of all. From the comparison of the data corresponding to the $(TEAC_{12})_4$/DOPE liposomes to those of $(Im_{12})_4$, one can say that the substitution of quaternary ammonium groups by imidazolinium ones results in an increase in the cytotoxicity of the liposomes. A similar result was observed by Rodik et al. in a previous work [46]. On the other hand, an increase in the length of the hydrophobic chains attached to the lower rim of the calixarenes caused a diminution in the cell viability. This was particularly evident for the $(Im_{16})_4$/DOPE liposomes. This behavior can be attributed to a higher lipophilicity, which results in a higher ability to penetrate and disrupt the cell membrane. It was particularly evident for the $(Im_{16})_4$/DOPE liposomes. A similar trend has been previously observed by other authors [47]. Besides, Figure 2 shows that $(Im_3Im_{16})_2$/DOPE liposomes present a specific selectivity towards tumor cells, while $(Im_{12})_4$/DOPE liposomes act particularly on MCF7 mammary cells and, at high concentrations, on HepG2 and A549 cells.

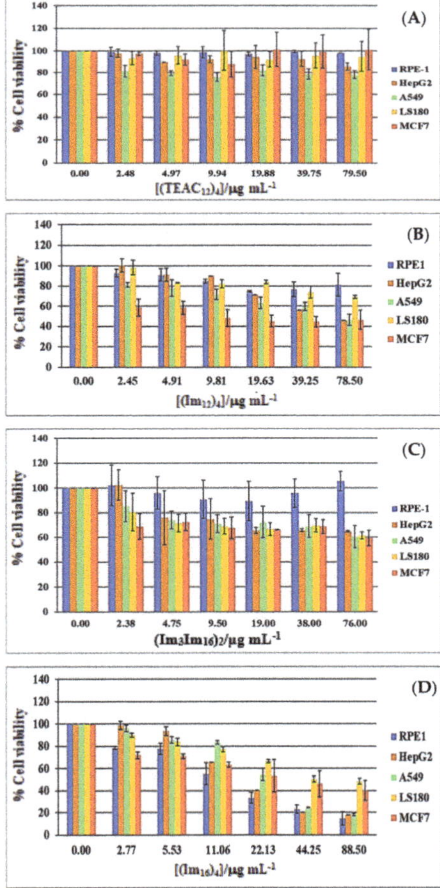

Figure 2. Cell viability values in the presence of different liposome concentrations at α = 0.5 for $(TEAC_{12})_4$, $(Im_{12})_4$, and $(Im_3Im_{16})_2$, and at α = 0.3 for $(Im_{16})_4$. The results are the average of three independent experiments: (**A**) $(TEAC_{12})_4$; (**B**) $(Im_{12})_4$; (**C**) $(Im_3Im_{16})_2$; (**D**) $(Im_{16})_4$.

3.2. CAL/DOPE/DNA Lipoplexes

In order to study the characteristics of the CAL/DOPE/DNA lipoplexes, the zeta potential, ζ, was measured for a given molar ratio, α, at different mass ratios L/D. Figure 3 shows the results obtained for the different calixarenes studied in this work. In all cases when L/D increases, the charge of the lipoplex goes from negative to positive. The solutions were stable, and no turbidity was observed in any system. This is an expected result since, for a given α, an increase in L/D means an increment in the amount of cationic lipid, CAL, in the liposomes. The charge inversion observed indicates that the polynucleotide interacts with the cationic calixarene within the lipoplex. From the data in Figure 3 it is possible to calculate $(L/D)_\Phi$, the value of the mass ratio corresponding to a zeta potential equal to zero; that is, when the lipoplexes are neutral. $(L/D)_\Phi$ values can also be calculated theoretically using Equation (5). The deduction of this equation is described in the Supplementary Information.

$$\left(\frac{L}{D}\right)_\Phi = \frac{q_{ADN}^-}{q_{CAL}^+} \times \frac{M_{CAL}}{M_{bp}} \times \frac{(n_{CAL}M_{CAL} + n_{DOPE}M_{DOPE})}{(n_{CAL}+n_{DOPE})} \times \frac{1}{\frac{n_{CAL}}{(n_{CAL}+n_{DOPE})}}$$
$$= \frac{q_{ADN}^-}{q_{CAL}^+} \times \frac{M_{CAL}}{M_{bp}} \times \frac{\alpha M_{CAL}+(1-\alpha)M_{DOPE}}{\alpha} \tag{5}$$

where q_{ADN}^- and q_{CAL}^+ are the charges of the polynucleotide and of the CAL, respectively. The charge of calf thymus DNA is considered to be -2 per base-pairs [48]. M_{CAL} and M_{DOPE} are the molecular weights of the cationic calixarene and non-ionic lipid, respectively, M_{bp} being the polynucleotide molecular weight per base-pair. n_{CAL} and n_{DOPE} are the number of moles the CAL and DOPE, respectively. The rest of the symbols have been previously defined. Table 2 summarizes the experimental and the theoretical $(L/D)_\Phi$ values. One can see that, within experimental errors, the theoretical and the experimental values agree quite well. It is important to know the charge of the different lipoplexes prepared because one of the requirements for an efficient cellular uptake is that the charge of the nanocarrier (the lipoplexes in this work) has to be positive in order to cross the negatively charged cellular membrane [49].

Table 2. Theoretical and experimental $(L/D)_\Phi$ values.

	$(L/D)_\Phi$							
α	$(TEAC_{12})_4$		$(Im_{12})_4$		$(Im_{16}Im_3)_2$		$(Im_{16})_4$	
0.10	-		-		6.6 [a]	6.6 [b]	6.9 [a]	6.9 [b]
0.20	3.8 [a]	3.6 [b]	3.7 [a]	3.6 [b]	3.6 [a]	3.7 [b]	3.9 [a]	3.9 [b]
0.30	2.8 [a]	2.6 [b]	2.7 [a]	2.6 [b]	2.6 [a]	2.6 [b]	2.9 [a]	2.9 [b]
0.40	2.3 [a]	2.2 [b]	2.2 [a]	2.0 [b]	2.1 [a]	2.1 [b]	-	-
0.45	2.1 [a]	1.8 [b]	2.0 [a]	1.8 [b]	-	-	-	-
0.50	2.0 [a]	1.8 [b]	1.9 [a]	1.7 [b]	1.8 [a]	1.8 [b]	-	-

[a] Theoretical values; [b] Experimental values.

Charge inversion of DNA in the lipoplexes can also be investigated by gel electrophoresis. Figure S1 (Supplementary Information) shows the results obtained by using this technique. In this figure a migration to the anode is observed for L/D values lower than $(L/D)_\Phi$, for the systems investigated, although a diminution in the mobility of the band is found when L/D approximated to $(L/D)_\Phi$. Once this value is reached, the mobility is hindered, this pointing out that the charge of the polynucleotide has been inverted from negative to positive. These results are in agreement to those found by zeta potential measurements (see Figure 3).

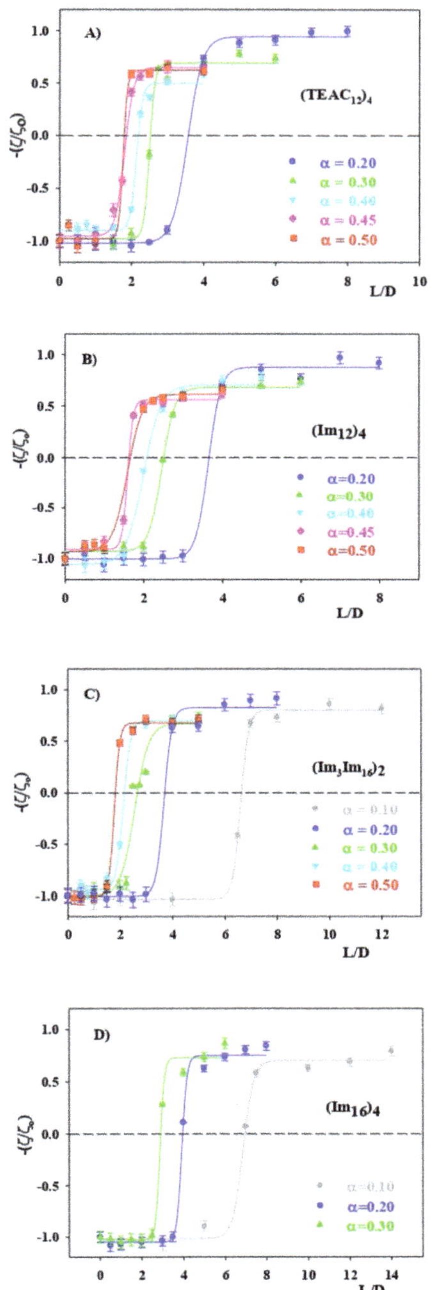

Figure 3. Dependence of the relative zeta potential, $-(\zeta/\zeta_o)$, of the CAL/DOPE/DNA lipoplexes on L/D for different molar ratios α. T = 303.0.1 ± 0.1 K. (**A**) $(TEAC_{12})_4$; (**B**) $(Im_{12})_4$; (**C**) $(Im_3Im_{16})_2$; (**D**) $(Im_{16})_4$.

Another important magnitude is the size of the lipoplexes. Recent studies indicated that there is a particular size range adequate for cellular uptake [18,50]. Although cellular uptake depends on the type of cells and on the different barriers making this process difficult [49], the appropriate nanocarrier size is usually considered to be a few hundred nanometers. Figure 4 shows the dependence of the hydrodynamic diameter, d_H, on the mass ratio L/D, for a given α value, for the calixarenes investigated. In all cases a Gaussian dependence of d_H on L/D was observed. This dependence can be explained as follows. At low as well as at high L/D values, the lipoplexes are negatively and positively charged, respectively (see Figure 3). That is, there are repulsive forces among them that kept them apart and maintained a stable size distribution. When L/D is approaching $(L/D)_\Phi$, the charge of the lipoplexes is moving closer to zero. As a consequence, the lipoplexes do not repel each other and an aggregation process occurs, a steep increment in d_H being observed. This explanation is supported by Figure S2 (Supplementary Material), which shows the dependence of the relative zeta potential, (ζ/ζ_0), and of the hydrodynamic diameter, d_H, of CAL/DOPE/DNA lipoplexes on L/D for $\alpha = 0.2$. Apart from the L/D values close to $(L/D)_\Phi$, the lipoplex sizes observed for the different molar ratios α, and for all the CAL studied, are within the hundred nanometers size range.

A way of getting information about the conformational changes of the DNA in the lipoplexes, when the mass ratio L/D varies, is by circular dichroism, CD. Figure 5 shows the CD spectra of the CAL/DOPE/DNA lipoplexes for different α and L/D values. First, it was checked that the liposomes in the absence of DNA did not contribute to the spectra. All CD spectra were run taking as reference an aqueous buffer solution HEPES 10 mM, at pH = 7.4. Figure 5 shows the CD spectrum corresponding to pure DNA in aqueous buffered solution of HEPES 10 mM. This spectrum presents a negative band, at about 247 nm, due to the right-handed helicities of the polynucleotide, and a positive band, close to 280 nm, coming from the π–π stacking interactions between the bases. This spectrum is in agreement with that expected for the right-handed B form of the double-stranded ctDNA [51]. For all the CAL/DOPE/DNA lipoplexes investigated, a diminution in the positive band intensity upon increasing L/D was observed. The $(Im_{16})_4$/DOPE/DNA lipoplexes could not be studied at higher L/D values because of solubility problems. This makes the comparison of the results obtained for this calixarene with those corresponding to the rest of the macrocycles more difficult.

The dependence of the positive band intensity on L/D could be explained considering the attractive electrostatic interactions between the DNA phosphate groups and the positively charged calixarenes within the lipoplexes. These interactions could cause the opening of the DNA double strand and conformational changes in the polynucleotide. An increment in L/D is accompanied by an increase in the amount of cationic calixarene in the lipoplexes. Therefore, it would be expected that the diminution in the positive band intensity was larger, for a given α value, the higher L/D is, as is observed.

The displacement of the inflection point (observed at 260 nm for pure DNA) towards higher wavelengths as well as the increase in the negative band intensity are usually related to the DNA denaturation and to DNA conformational changes [52,53]. Bombelli et al. found a similar dependence of the DNA spectrum in gemini surfactants/DOPE/DNA lipoplexes on L/D [54]. These authors proposed that the gemini surfactant/DNA interactions in the liposomes causes a conformational DNA change from a B form to a more condensed Ψ phase, where the polynucleotide molecules are partially inserted within an inverted hexagonal lipid rearrangement, which gives the DNA a certain spatial organization and a fixed directionality.

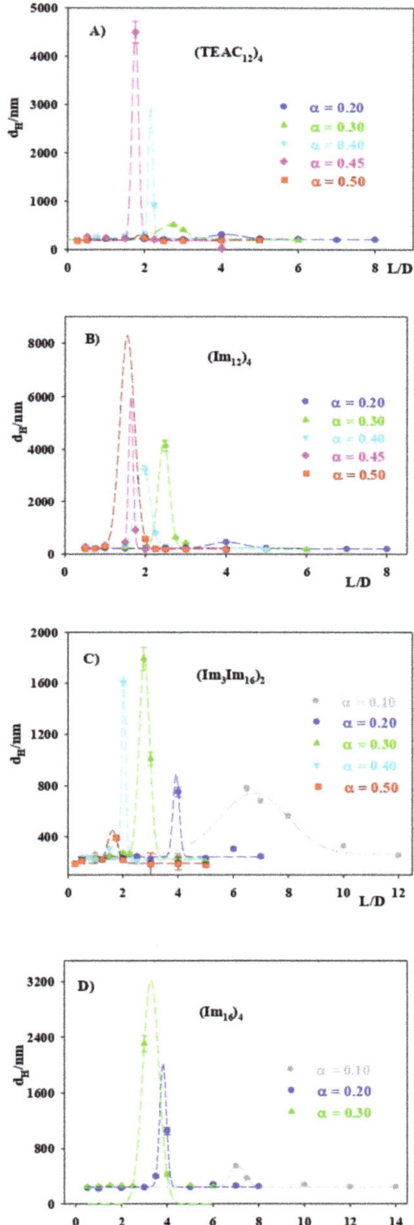

Figure 4. Dependence of the hydrodynamic diameter, d_H, of the CAL/DOPE/DNA lipoplexes on L/D for different molar ratios α. T = 303 $\pm$ 0.1 K. (**A**) (TEAC$_{12}$)$_4$; (**B**) (Im$_{12}$)$_4$; (**C**) (Im$_3$Im$_{16}$)$_2$; (**D**) (Im$_{16}$)$_4$.

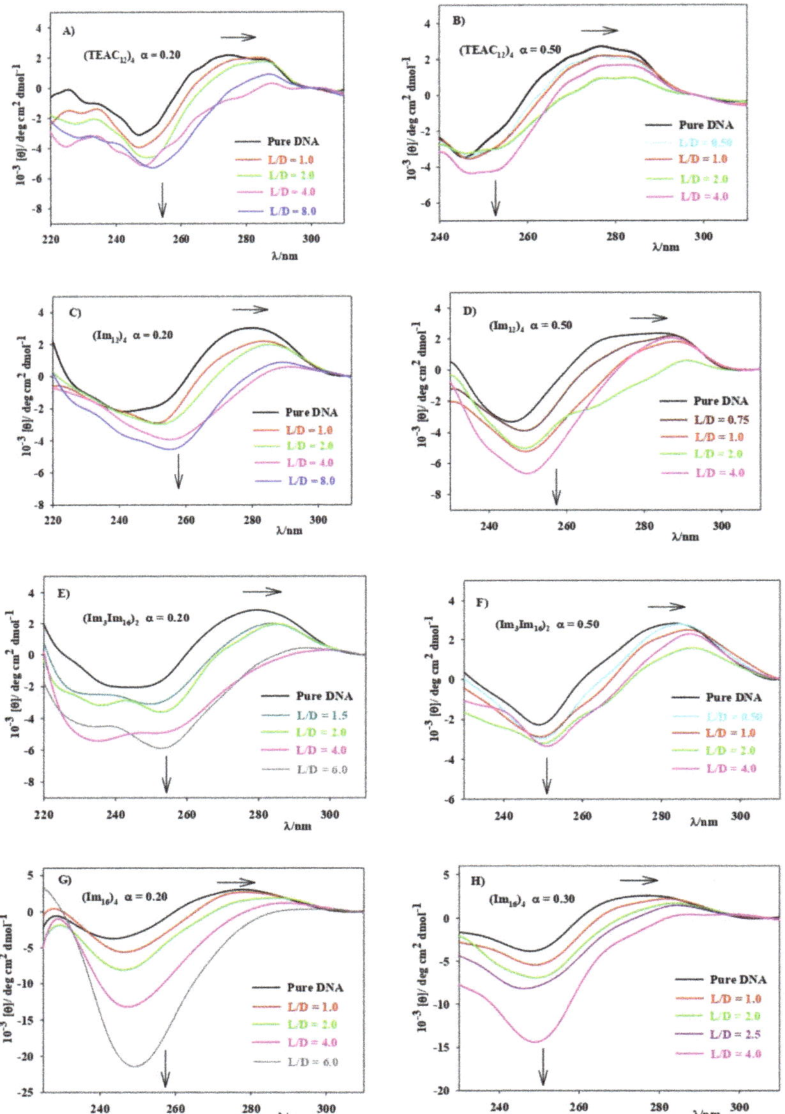

Figure 5. Dependence of the circular dichroism spectra of the CAL/DOPE/DNA lipoplexes on L/D for different molar ratios α. T = 303 ± 0.1 K.; (**A**) α= 0.20 for $(TEAC_{12})_4$; (**B**) α = 0.50 for $(TEAC_{12})_4$; (**C**) α = 0.20 for $(Im_{12})_4$; (**D**) α = 0.50 for $(Im_{12})_4$; (**E**) α = 0.20 for $(Im_3Im_{16})_2$; (**F**) α= 0.50 for $(Im_3Im_{16})_2$; (**G**) α = 0.20 for $(Im_{16})_4$; and (**H**) α = 0.30 for $(Im_{16})_4$.

The displacement of the inflection point wavelength diminishes when α increases. This could be due to the effects of the presence of DOPE on the DNA conformation, as was pointed out by Marty et al. [55] in different liposome formulations. It is also observed in Figure 4 that the wavelength displacement is lower for $(Im_{12})_4$ than for $(TEAC_{12})_4$ (14 nm vs. 6 nm, respectively). Both calixarenes have the same hydrophobic tail length. This observation was explained considering that the imidazolinium groups were intercalated, at least partially, between the DNA base pairs, this stabilizing the B form of the DNA [56].

At this point it is worth noting that there are not enough experimental results for $(Im_{16})_4$ that permit the comparison with the other CAL.

The circular dichroism results seem to indicate that the lipoplex formation results in DNA conformational changes. In order to support this hypothesis, atomic force microscopy measurements, AFM, were carried out. Figure 6 shows the AFM images corresponding to pure DNA (Figure 6A), together with those of CAL/DOPE/DNA lipoplexes of different compositions, for two of the calixarenes investigated (Figure 6B–E). In the absence of liposomes, the pure DNA presents an elongated form (see Figure 6A). For $(Im_3Im_{16})_2$/DOPE/DNA and $(TEAC_{12})_4$/DOPE/DNA lipoplexes at $\alpha = 0.2$ for L/D = 1 (Figure 6B,D), which is lower than $(L/D)_\Phi$, some globular structures are observed, and the length of the DNA seems to be somewhat shorter, although no substantial conformational variations are observed. Yan et al. [57] explained the formation of the globular structures, linked across DNA strands, on the basis of an increased bending of the DNA double helix, this leading to the formation and stabilization of intramolecular loops. The separation of the double DNA strand into single strands, due to electrostatic attractions between the DNA and the cationic calixarenes, favors this process. For L/D = 7, (L/D) > $(L/D)_\Phi$, the number of globular structures present in the images increases. However, a full condensation of the polynucleotide is not reached (Figure 6C,E) since some elongated fragments of DNA are still observed. These experimental observations are in agreement with the CD spectra.

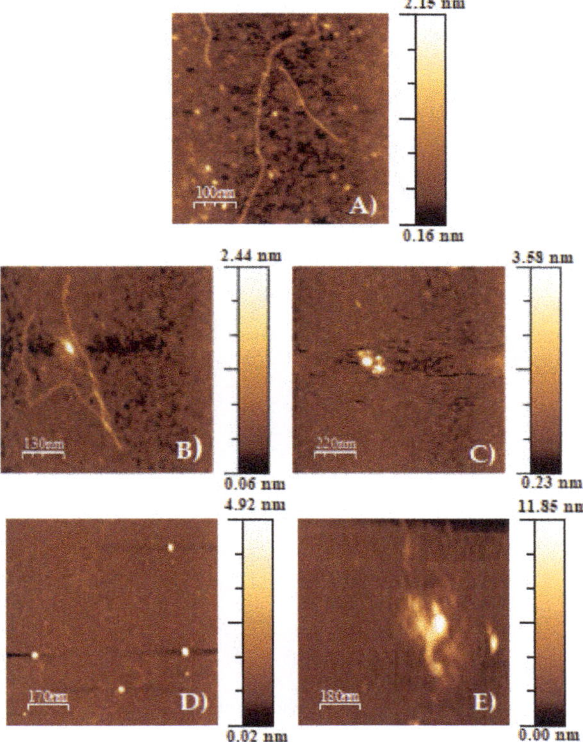

Figure 6. AFM topographic images of CAL/DOPE/DNA lipoplexes in buffered solutions, 10 mM HEPES (pH = 7.4), adsorbed on APTES modified mica surface. (**A**) Pure DNA; (**B**) $(Im_3Im_{16})_2$ $\alpha = 0.2$ and L/D = 1; (**C**) $(Im_3Im_{16})_2$ $\alpha = 0.2$ and L/D = 7; (**D**) $(TEAC_{12})_4$ $\alpha = 0.2$ and L/D = 1; (**E**) $(TEAC_{12})_4$ $\alpha = 0.2$ and L/D = 7.

With the goal of visualizing the morphology of the CAL/DOPE liposomes and CAL/DOPE/DNA lipoplexes, transmission electron microscopy measurements were carried out. Figure 7 shows the TEM images obtained for some of the systems investigated. One can see that a spherical morphology is observed for both the liposomes and the lipoplexes studied. The molar ratio $\alpha = 0.3$ was chosen because Figure 1 shows that, for the two calixarenes studied, the minimum size is found close to this molar ratio value. It is interesting to indicate that $(TEAC_{12})_4$/DOPE and $(Im_{12})_4$/DOPE liposomes are the less cytotoxic among the calixarene-based liposomes investigated. For the mass ratio L/D = 5, the charge inversion of the polynucleotide is complete for $(TEAC_{12})_4$/DOPE/DNA and $(Im_{12})_4$/DOPE/DNA lipoplexes, at $\alpha = 0.3$ (see Figure 3). Figure 7 shows that the two lipolexes not only are spherical, but their size is in the order of a few hundred nanometers. This is important in relation with the use of lipoplexes for gene delivery since it has been shown that a spherical morphology and a small size are two characteristics of the genetic material nanocarriers that favor transfection efficiency [49]. Besides, the liposome and lipoplex sizes measured using DLS (Figure 4) and TEM are in agreement, as one can see in Table 3.

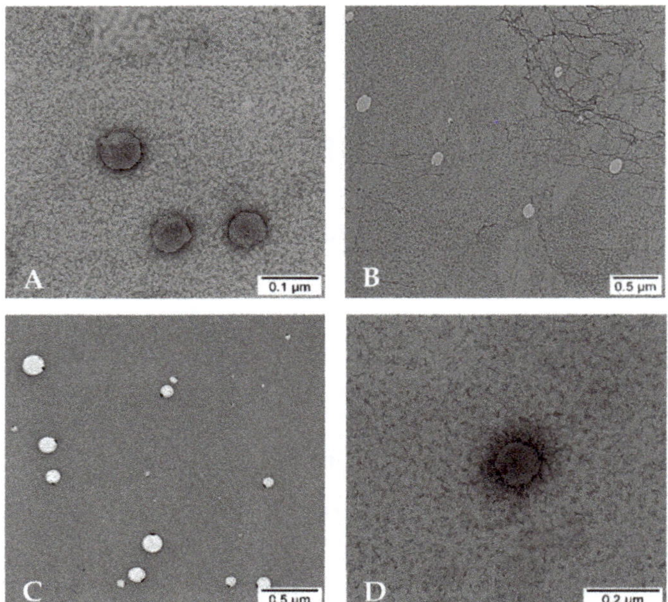

Figure 7. TEM images of the following systems: (**A**) $(TEAC_{12})_4$/DOPE liposomes with $\alpha = 0.3$; (**B**) $(Im_{12})_4$/DOPE liposomes with $\alpha = 0.3$; (**C**) $(TEAC_{12})_4$/DOPE/DNA lipoplexes with $\alpha = 0.3$ and L/D = 5; (**D**) $(Im_{12})_4$/DOPE/DNA lipoplexes with $\alpha = 0.3$ and L/D = 5.

Table 3. Sizes of various liposomes and lipoplexes, with a molar ratio $\alpha = 0.3$, measured by dynamic light scattering, DLS, and electronic transmission microscopy.

	Hydrodynamic Diameter (DLS)/nm		Diameter (TEM)/nm	
	$(TEAC_{12})_4$	$(Im_{12})_4$	$(TEAC_{12})_4$	$(Im_{12})_4$
Liposomes	147 ± 4	157 ± 4	130 ± 32	140 ± 40
Lipoplexes (L/D = 5)	183 ± 9	276 ± 8	183 ± 39	223 ± 47

3.3. Transfection Efficiency of CAL/DOPE/pDNA Lipoplexes

The transfection experiments were carried out for the lipoplexes containing the least cytotoxic calixarene: $(TEAC_{12})_4$. Before carrying out these measurements, the cell viability of the liposomes for the human bone osteosarcoma epithelial cells U2OS using the MTT assay was carried out. This is the cell line used in the transfection experiments because it is considered an easy-to-transfer cell line. Figure 8 shows the results obtained. In this figure, the cytotoxicity of the $(TEAC_{12})_4$/DOPE liposomes in the presence of additional DOPE (+1/4 of the DOPE amount present in the liposomes) is also presented. These systems were investigated because in the transfection experiments the addition of DOPE could make the delivery of genetic material more efficient [58]. Therefore, given that the transfection efficiency, TE, of the lipoplexes was studied in the presence of this phospholipid, the cell viability experiments in the presence of different cell lines for $(TEAC_{12})_4$/DOPE liposomes + DOPE was also carried out.

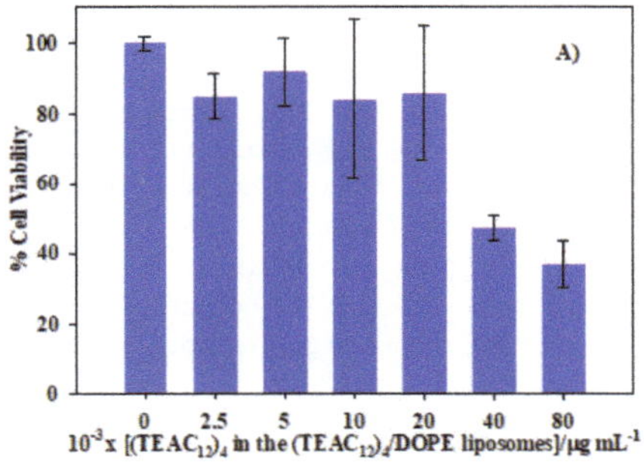

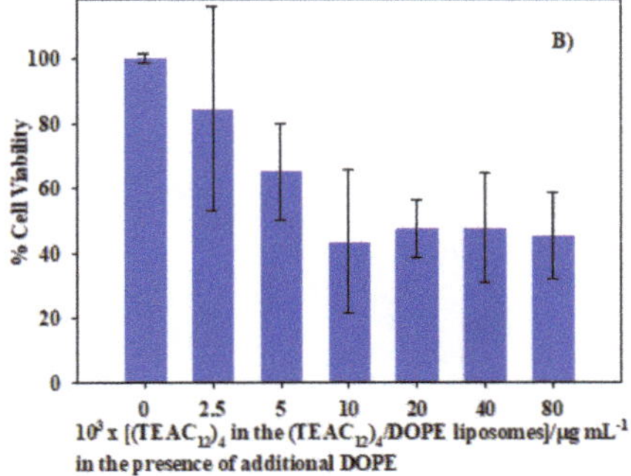

Figure 8. Dependence of the cell viability (%) of $(TEAC_{12})_4$/DOPE liposomes in U2OS cancer cell line, at 48 h, on the concentration of $(TEAC_{12})_4$ within the liposomes, for a constant cationic lipid molar ratio α = 0.3. The experiments were done in triplicate. (**A**) In the absence of additional DOPE; (**B**) in the presence of additional DOPE (+1/4 of the DOPE amount present in the liposomes), which is not forming part of the $(TEAC_{12})_4$/DOPE liposomes.

One can see that, for α = 0.3, the presence of additional DOPE, which is not present in the liposomes, substantially diminishes the cell viability of $(TEAC_{12})_4$/DOPE liposomes. In the absence of additional DOPE, the cell viability of the $(TEAC_{12})_4$/DOPE liposomes is lower than 60% for concentrations of the cationic calixarene $[(TEAC_{12})_4] \geq 20$ µg mL^{-1}. However, the addition of DOPE to the system caused the cell viability of the $(TEAC_{12})_4$/DOPE liposomes to be lower than 60% for cationic calixarene concentrations $[(TEAC_{12})_4] > 5$ µg mL^{-1}. That is, keeping the cationic calixarene molar ratio α equal to

0.3, the cell viability substantially decreases when the [(TEAC$_{12}$)$_4$] within the liposomes increases if additional DOPE is present.

The transfection process of the plasmid pEGFP-C1 was carried out on the U2OS cells. The TE of the (TEAC$_{12}$)$_4$/DOPE/pEGFP-C1 lipoplexes, with α = 0.3, within a L/D range between 9 and 90, in the presence as well as in the absence of additional DOPE, was estimated. Expression of GFP is frequently used to follow transfection, and it does not require any additional manipulation of the sample since GFP is an intrinsically fluorescent protein. Therefore, its fluorescence can be readily measured directly. The lowest L/D value studied was 9, with the idea of assuring that the lipoplexes have a positive charge, a requirement to cross the cell membrane. On the other hand, L/D values higher than 90 were not investigated because this would mean a high amount of cationic calixarene present in the liposomes, this increasing cytotoxicity (see Figure 2). In regard to the introduction of additional DOPE in order to improve the TE, amounts of DOPE up to $^1/_4$ of that present in the lipoplexes were investigated. Higher additional phospholipid amounts were not studied to avoid a further increase in cytotoxicity (see Figure 8).

Figure 9 shows that for the mass ratio L/D = 9 no transfection was observed in the absence as well as in the presence of additional DOPE. Negligible TE values were also found for L/D values lower than 90. For this reason, they are not shown in Figure 9. However, for L/D = 90 a low TE was observed, close to 3%. This TE is much lower than that of the FuGENE 6 reagent. When DOPE was added ($^1/_4$ of the amount of phospholipid present in the lipoplexes), the TE for L/D = 90 increased up to 16%. For additional DOPE amounts lower than $^1/_4$ no changes in the TE for L/D=9 were observed. The increment in the TE caused by the addition of the helper lipid DOPE could be due to an increase in the stabilisation of the interactions cationic lipid/p-EGFP-C1 [59,60]. Besides, the addition of DOPE could also make the transfer of the genetic material in the context of endosomal escape more favorable, due to its fusogenic character [61].

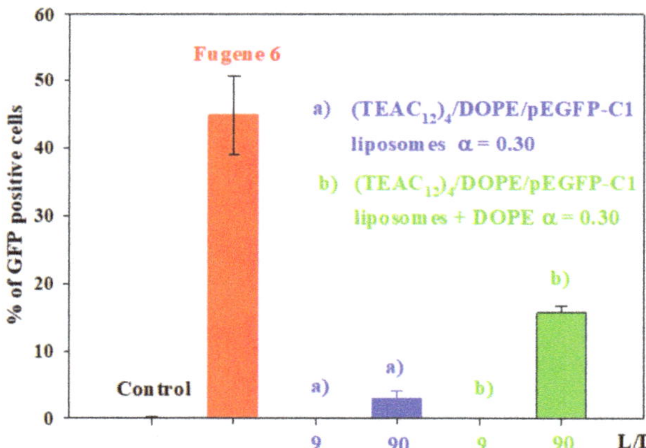

Figure 9. Percentage of GFP-positive cells after transfection with 3 µg of p-EGFP-C1, 24 h post transfection; 10,000 cells were analyzed per condition by flow cytometry (FACS). (a) (TEAC$_{12}$)$_4$/DOPE/pEGFP-C1 liposomes at α= 0.30; and (b) (TEAC$_{12}$)$_4$/DOPE/pEGFP-C1 liposomes + DOPE at α = 0.30.

Figure 10 shows the FACS analyses of cells transfected with 3 µg of p-EGFP-C1 with the indicated reagents. Representative images of GFP-positive cells after transfection are shown in Figure 11.

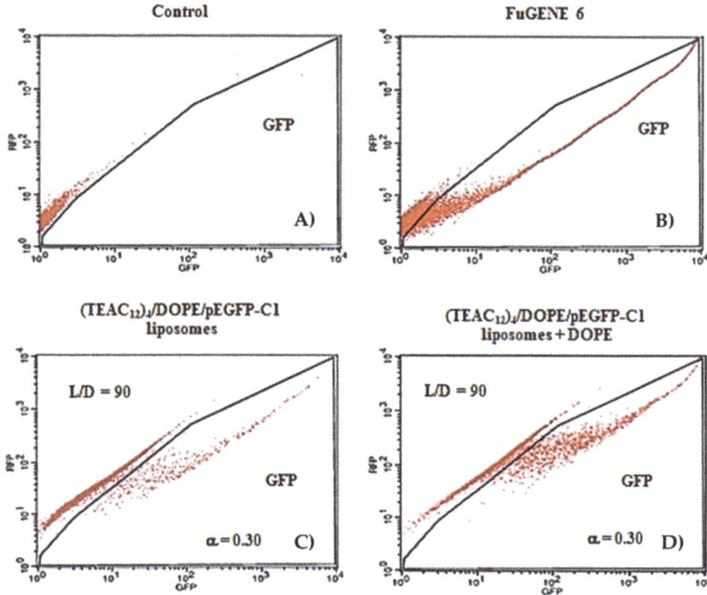

Figure 10. FACS analysis of cells transfected with 3 µg of p-EGFP-C1 for the different reagents. The charts show green fluorescent emission (X axis) vs. red fluorescent emission (Y axis) of 10,000 live cells analyzed per condition 24 h post-transfection. The GFP gate defines the area where cells with a clear increase in their green fluorescent emission are observed without a parallel increase in their red fluorescent emission. (**A**) Control; (**B**) FuGENE 6; (**C**) (TEAC$_{12}$)$_4$/DOPE/pEGFP-C1 liposomes at α = 0.30 for L/D = 90; and (**D**) (TEAC$_{12}$)$_4$/DOPE/pEGFP-C1 liposomes + DOPE at α = 0.30 for L/D = 90.

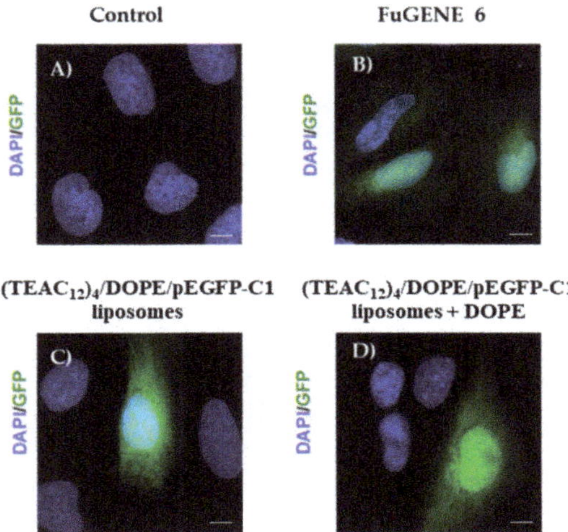

Figure 11. Representative images of GFP positive cells after transfection with 3 µg of pEPFG-C1. (**A**) Control; (**B**) FuGENE 6; (**C**) (TEAC$_{12}$)$_4$/DOPE/pEGFP-C1 lipoplexes at α = 0.3 and L/D = 90; (**D**) (TEAC$_{12}$)$_4$/DOPE/pEGFP-C1 lipoplexes at α = 0.3 and L/D = 90 in the presence of additional DOPE. Scale bar: 10 µm.

3.4. Encapsulation of Doxorubicin

Doxorubicin (see Scheme 2), DOX, is an antineoplastic drug displaying a strong antitumoral activity against a wide spectrum of human cancers. It is used in the treatment of various lung, breast, or ovarian cancers. It is also used in chemotherapy for leukemia and lymphomas [62–64]. DOX intercalates between the base-pairs of the DNA, this resulting in the inhibition of the synthesis and transcription of the genetic material. The result is the blocking of the enzyme topoisomerase II, which hindered the division and growing of cells. Besides, the interactions DOX/DNA cause variations in the chromatin structure, which triggers apoptosis in cells [65]. In spite of the beneficial DOX activity, its clinical use is limited by its side effects. Gastrointestinal toxicity, stomatitis, myelosuppression, or cardiotoxicity are some of the most frequent side effects caused by the treatment with DOX [66–68].

The study of the encapsulation of drugs within nanocarriers is of great interest because it could permit the transportation of the drug towards its therapeutic target but, simultaneously, diminishing the drug side effects. One of the most frequently used nanocarriers are liposomes, particularly in the case of doxorubicin [69–71]. In fact, there is a commercial liposome preparation for DOX administration in chemotherapy called Doxil® [72]. Other types of nanovehicles have also been used to administer doxorubicin [73–76]. In this work, the $(TEAC_{12})_4$/DOPE liposomes were used to study the encapsulation and the release of doxorubicin. These liposomes were chosen because of the low cytotoxicity they present, which is one of the requirements of nanocarriers for biomedical applications.

Before studying the encapsulation of doxorubicin within the calixarene-based liposomes, the stability of these nanostructures was investigated. Stability was followed by DLS measurements, through the dependence of the hydrodynamic diameter, d_H, and the polydispersity, PDI, on time, at 310 K (simulating the human body temperature). Figure 12 shows the results. One can see that the liposomes were stable during approximately 6 days. After that time the size as well as the PDI increased, this indicating that the system was not stable for longer times. The results observed in Figure 12 could be explained by the fragmentation of the lipid membrane of the liposomes due to the hydrolytic decomposition of the phospholipid molecules. As a consequence, their structure will vary, and an increment in the surface of the liposome membranes can occur, this causing an increase in both the hydrodynamic diameter and the polydispersity.

The encapsulation of doxorubicin within the $(TEAC_{12})_4$/DOPE liposomes was done during the hydration process of the lipid bilayer (thin lipid film method). A DOX buffered solution (HEPES 10 mM, pH = 7.4) was added to the dry lipid bilayer. Afterwards, the vortex-sonication cycles were carried out, followed by the extrusion process. The final DOX concentration was 2×10^{-4} mol L^{-1} in all cases. In order to estimate the amount of DOX encapsulated, the drug-loaded liposomes were dialyzed (see Section 2.12). The doxorubicin concentration was determined by UV–visible spectroscopy measuring absorbance at 490 nm. The temperature was kept at 277 K in order to avoid doxorubicin degradation. The results obtained are summarized in Table 4.

Table 4. Encapsulation efficiency, EE%, and loading capacity, LC, of doxorubicin in the $(TEAC_{12})_4$/DOPE liposomes. T = 277 K. Three independent experiments were carried out for each system studied.

α	EE%	LC%
0.2	72 ± 5	1.85 ± 0.18
0.4	83 ± 1	4.21 ± 0.09

Table 4 shows that for the two molar fractions α investigated, the encapsulation efficiency was high. EE% increased when α augmented. This experimental observation could be explained by considering the interactions between $(TEAC_{12})_4$ and doxorubicin, which were investigated in a previous study [30]. An increase in the amount of the antineoplastic drug in the liposome, an increase in α, will favor these interactions, this leading to an increment in the encapsulation efficiency. The loading capacity also followed the same trend as EE%, and the explanation is similar to that given above.

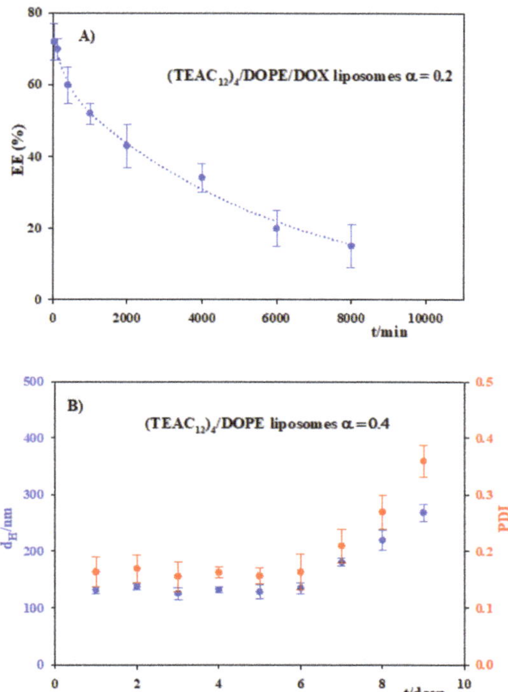

Figure 12. Variation of the hydrodynamic diameter, d$_H$, and of the polydispersity, PDI, with time for (TEAC$_{12}$)$_4$/DOPE liposomes at 310 K. (**A**) α = 0.2; (**B**) α = 0.4.

The size and the polydispersity, PDI, of the (TEAC$_{12}$)$_4$/DOPE liposomes with and without loaded DOX were compared. The measurements were done by DLS, and the data are listed in Table 5. One can see that the hydrodynamic diameter as well as the PDI of the liposomes were similar in the absence and in the presence of doxorubicin.

Table 5. Hydrodynamic diameter, d$_H$, and polydispersity, PDI, of (TEAC$_{12}$)$_4$/DOPE liposomes and doxorubicin loaded (TEAC$_{12}$)$_4$/DOPE liposomes. T = 310 K. The results are the average of three independent experiments.

α	d$_H$/nm [a]	PDI [a]	d$_H$/nm [b]	PDI [b]
0.2	200 ± 7	0.23 ± 0.02	186 ± 6	0.18 ± 0.02
0.4	131 ± 6	0.163 ± 0.012	134 ± 6	0.22 ± 0.03

[a] (TEAC$_{12}$)$_4$/DOPE liposomes. [b] DOX loaded (TEAC$_{12}$)$_4$/DOPE liposomes.

When a drug is loaded in a nanocarrier, it is important to study the release time of the drug. The method used to investigate the release was described in the Experimental section, and it was carried out also at 310 K, in order to mimic the human body temperature. Figure 13 shows the variations of EE% against time for doxorubicin loaded (TEAC$_{12}$)$_4$/DOPE liposomes with molar fractions 0.2 and 0.4. The concentration of doxorubicin was estimated measuring the absorbance at 490 nm. From the variations of EE% with time, it is possible to deduce that the release follows a pseudo-first-order kinetics.

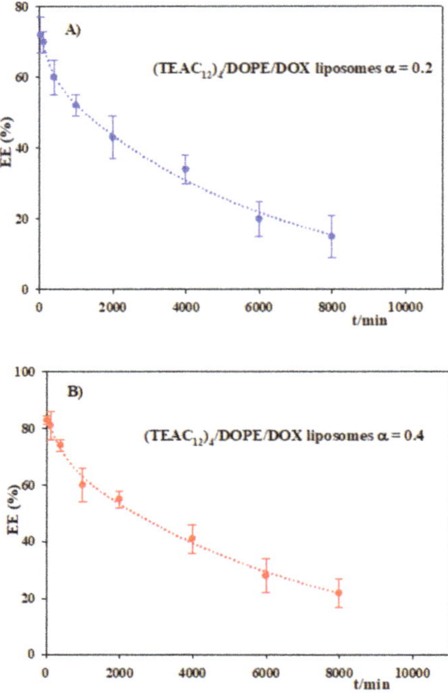

Figure 13. Release of doxorubicin from DOX loaded (TEAC$_{12}$)$_4$/DOPE liposomes at 310 K. (**A**) α = 0.2; (**B**) α = 0.4. The values are the average of three independent experiments.

The following kinetic rate constants were estimated: 1.9×10^{-4} min^{-1} and 1.6×10^{-5} min^{-1} for α = 0.2 and α = 0.4, respectively. Figure 13 shows that not all the doxorubicin was released from the liposomes. This could be explained by considering that, once the liposomes are fragmented, CAL molecules free from the lipid bilayer will be present in the solution. (TEAC$_{12}$)$_4$ can form different aggregates at low CAL concentrations, and DOX could be bound to them [30]. This hypothesis is in agreement with the results shown in Figure 13, since less doxorubicin was released in the case of α = 0.4 than of α = 0.2. An increase in the molar ratio α means an increment in the amount of calixarene present in the liposomes. Therefore, once the liposomes are broken, there will be a larger number of free CAL molecules to associate with the doxorubicin and, consequently, the amount of antineoplastic drug released from the liposomes will be lower.

The results obtained indicate that (TEAC$_{12}$)$_4$/DOPE liposomes can be used as noncytotoxic nanocarriers for the antineoplastic drug doxorubicin. Besides, even when liposomes are fragmented, the DOPE as well as the (TEAC$_{12}$)$_4$ molecules are non-cytotoxic [30]. The half-lives of the release were 2.5 and 3 days for α = 0.2 and α = 0.4, respectively, this pointing out that the side effects of the doxorubicin could be diminished due to a controlled release of the drug encapsulated as compared to the use of the naked drug.

4. Conclusions

In this work the formation of calixarene-based liposomes, CAL/DOPE, was investigated. Calixarenes with hydrophobic chains of different length attached to their lower rim were considered. The nature of the hydrophilic head present in their upper rim was also changed. The phospholipid DOPE was used, together with the cationic calixarene, for forming the lipid bilayer of the liposomes. The liposomes were characterized using several techniques. TEM images showed their spherical morphology. Cell viability experiments permitted the estimation of the cytotoxicity of the CAL/DOPE liposomes of different compositions, the results showing that the $(TEAC_{12})_4$ liposomes are the least cytotoxic.

The formation and characterization of CAL/DOPE/DNA lipoplexes of different compositions was investigated and the nanostructures characterized. They have a spherical geometry and a size on the order of a few hundred nanometers. Subsequently, transfection experiments were carried out only for the least cytotoxic calixarene, $(TEAC_{12})_4$. Results showed that some of these $(TEAC_{12})_4$/DOPE/p-EFGP-C1 lipoplexes can transfect, although the transfection efficiency, TE, is low. However, the presence of an additional amount of DOPE substantially increases the TE. This could be explained considering that the presence of DOPE can stabilize the interactions between the cationic lipid and the plasmidic DNA. Besides, the addition of DOPE could also make the transfer of the genetic material in the context of endosomal escape more favorable, due to its fusogenic character.

The antineoplastic agent doxorubicin was encapsulated in the $(TEAC_{12})_4$/DOPE liposomes with high encapsulation efficiencies. The liposomes were stable for close to 6 days, at 310 K (the human body temperature). The drug release was studied, and the results showed that the liposomes can be utilized for a controlled release of the drug. Their use could suppose a diminution in its side effects.

The results obtained also show that the calixarene-based liposomes seem to be better nanocarriers, for both nucleic acids and doxorubicin, than their aggregates (micelles and vesicles), formed by the naked calixarenes.

Future investigations will be oriented to the design and preparation of new non-cytotoxic amphiphilic calixarenes with the goal of using the CAL/DOPE/DNA lipoplexes as nanocarriers for the delivery of genetic materials, with a high transfection efficiency. The CAL/DOPE liposomes can also be checked as nanovehicles for different drugs.

Supplementary Materials: The following are available online at https://www.mdpi.com/article/10.3390/pharmaceutics13081250/s1, Deduction of Equation (5); Figure S1: Electrophoretic mobility shift assay on an agarose gel (1%) for CAL/DOPE/DNA lipoplexes; Figure S2: Dependence of the relative zeta potential, (ζ/ζ_0), and of the hydrodynamic diameter, d_H, of CAL/DOPE/DNA lipoplexes on L/D for $\alpha = 0.2$. T = 303.0.1 $\pm$ 0.1 K.

Author Contributions: Conceptualization, M.L.-L., P.L.-C. and M.L.M.; methodology, F.J.O., J.A.L., C.B.G.-C., I.V.R., M.L.-L., F.R.B., P.H., P.L.-C. and M.L.M.; software, J.A.L., F.J.O., E.B. and P.H.; validation, M.L.-L., P.L.-C., P.H. and M.L.M.; formal analysis, F.J.O., J.A.L., E.B., M.L.-L., P.L.-C. and M.L.M.; investigation, F.J.O., J.A.L., C.B.G.-C., I.V.R., R.V.R., V.I.K., F.R.B., E.B., M.L.-L., P.L.-C., P.H. and M.L.M.; resources, P.L.-C., P.H. and M.L.M.; data curation, F.J.O., J.A.L. and M.L.M.; writing—original draft preparation, F.J.O., M.L.-L., P.L.-C., P.H. and M.L.M.; writing—review and editing, F.J.O., M.L.-L., P.L.-C., P.H., R.V.R., V.I.K. and M.L.M.; visualization, F.J.O., M.L.-L., P.L.-C., P.H. and M.L.M.; supervision, F.J.O., M.L.-L., P.L.-C. and M.L.M.; project administration, M.L.-L., P.L.-C. and M.L.M.; funding acquisition, P.L.-C., P.H., V.I.K. and M.L.M. All authors have read and agreed to the published version of the manuscript.

Funding: This work was financed by the Consejería de Conocimiento, Innovación y Universidades de la Junta de Andalucía (FQM-206, FQM-274, and PY20-01234), the VI Plan Propio Universidad de Sevilla (PP2019/00000748), RTI2018-100692-B-100; P18-RT-1271; PI18-0005-2018; VI-PP AY.SUPLEM-2019; RYC-2015-18670, The R+D+I grant PID2019-104195G from the Spanish Ministry of Science and Innovation-Agencia Estatal de Investigación/10.13039/501100011033 (P.H.) and the European Union (Feder Funds). The authors thank the University of Seville for the grant VPPI-US. J.A.L. also thanks the Fundación ONCE funded by the Fondo Social Europeo.

Institutional Review Board Statement: Not applicable.

Informed Consent Statement: Not applicable.

Data Availability Statement: Not applicable.

Conflicts of Interest: The authors declare no conflict of interest.

References

1. Akbarzadeh, A.; Rezaei-Sadabady, R.; Davaran, S.; Joo, S.W.; Zarghami, N.; Hanifehpour, Y.; Samiei, M.; Kouhi, M.; Nejati-Koshki, K. Liposome: Classification, preparation, and applications. *Nanoscale Res. Lett.* **2013**, *8*, 102. [CrossRef] [PubMed]
2. Patil, Y.P.; Jadhav, S. Novel methods for liposome preparation. *Chem. Phys. Lipids* **2014**, *177*, 8–18. [CrossRef] [PubMed]
3. Bozzuto, G.; Molinari, A. Liposomes as nanomedical devices. *Int. J. Nanomed.* **2015**, *10*, 975–999. [CrossRef]
4. Guo, S.; Huang, L. Nanoparticles containing insoluble drug for cancer therapy. *Biotechnol. Adv.* **2014**, *32*, 778–788. [CrossRef]
5. Kapoor, B.; Gupta, R.; Gulati, M.; Singh, S.K.; Khursheed, R.; Gupta, M. The why, where, who, how, and what of the vesicular delivery systems. *Adv. Colloid Interface Sci.* **2019**, *271*, 101985. [CrossRef]
6. Balazs, D.A.; Godbey, W.T. Liposomes for use in gene delivery. *J. Drug Deliv.* **2010**, *2011*, 326497. [CrossRef] [PubMed]
7. Maione-Silva, L.; Gava de Castro, E.; Leite Nascimento, T.; Ramos Cintra, E.; Cleres Moreira, L.; SantanaCintra, B.A.; Campos Valadares, M.; Martins Lima, E. Ascorbic acid encapsulated into negatively charged liposomes exhibits increased skin permeation, retention and enhances collagen synthesis by fibroblasts. *Sci. Rep.* **2019**, *9*, 1–14. [CrossRef] [PubMed]
8. Bangham, A.D.; Standish, M.M.; Watkins, J.C. Diffusion of univalent ions across the lamellae of swollen phospholipids. *J. Mol. Biol.* **1965**, *13*, 238–252. [CrossRef]
9. Cuomo, F.; Ceglie, S.; Miguel, M.; Lindman, B. Oral delivery of all-trans retinoic acid mediated by liposome carriers. *Colloids Surf. B* **2021**, *201*, 111655. [CrossRef]
10. Logendiran, M.; Radhakrishnan, H.R.; Nagarajan, K.; Subbiah, L.; Palanisamy, S. Review on liposome based cancer theragnostics and therapy current advancements. *Int. J. Pharm. Bio Sci.* **2020**, *11*, 127–137. [CrossRef]
11. Azadi, Y.; Ahmadpour, E.; Ahmadi, A. Targeting strategies in therapeutic applications of toxoplasmosis: Recent advances in liposomal vaccine delivery systems. *Curr. Drug Targets* **2020**, *21*, 541–558. [CrossRef]
12. Ahmed, K.S.; Hussein, S.A.; Ali, A.H.; Korma, S.A.; Qiu, L.; Chen, J. Liposome: Composition, characterization, preparation, and recent innovation in clinical applications. *J. Drug Targets* **2019**, *27*, 742–761. [CrossRef]
13. Bayat, F.; Hosseinpour-Moghadam, R.; Mehryab, F.; Fatahi, Y.; Shakeri, N.; Dinarvand, R.; Ten Hagen, T.L.M.; Haeri, A. Potential application of liposomal nanodevices for non-cancer diseases: An update on design, characterization and biopharmaceutical evaluation. *Adv. Colloid Interface Sci.* **2020**, *277*, 102121. [CrossRef]
14. Daeihamed, M.; Dadashzadeh, S.; Haeri, A.; Akhlaghi, M.F. Potential of liposomes for enhancement of oral drug absorption. *Curr. Drug Deliv.* **2017**, *14*, 289–303. [CrossRef]
15. Nam, J.; Beales, P.A.; Vanderlick, T.K. Giant phospholipid/block copolymer hybrid vesicles: Mixing behavior and domain formation. *Langmuir* **2011**, *27*, 1–6. [CrossRef]
16. Matsumura, A.; Tsuchiye, K.; Torisoe, K.; Sakai, K.; Abe, M. Photochemical control of molecular assembly formation in a catanionic surfactant system. *Langmuir* **2011**, *27*, 1610–1617. [CrossRef]
17. Devi, U.; Brown, J.R.D.; Almond, A.; Webb, S.J. Pd(II)-mediated assembly of porphyrin channels in bilayer membranes. *Langmuir* **2011**, *7*, 1448–1456. [CrossRef]
18. Schuhle, D.T.; van Rijn, P.; Laurent, S.; Elst, L.V.; Muller, R.N.; Stuart, M.C.A.; Schatze, J.; Peters, J.A. Liposomes with conjugates of a calix[4]arene and a Gd-DOTA derivative on the outside surface; an efficient potential contrast agent for MRI. *Chem. Commun.* **2010**, *46*, 4399–4401. [CrossRef] [PubMed]
19. Tian, H.-W.; Pan, Y.-C.; Guo, D.S. Assemblyenhanced molecular recognition of calix[6]arene. *Supramol. Chem.* **2018**, *30*, 562–567. [CrossRef]
20. Sanabria Español, E.; Maldonado Villamil, M. Calixarenes: Generalities and their role in improving the solubility, biocompatibility, stability, bioavailability, detection, and transport of biomolecules. *Biomolecules* **2019**, *9*, 90. [CrossRef] [PubMed]
21. Gutsche, C.D. *Calixarenes: An Introduction*, 2nd ed.; RSC Publishing: Cambridge, UK, 2008. [CrossRef]
22. Sansone, F.; Balduino, L.; Casnati, A.; Ungaro, R. Calixarenes: From biomimetic receptors to multitalented ligands for biomolecular recognition. *New J. Chem.* **2010**, *34*, 2715–2728. [CrossRef]
23. Arduini, A.; Demuru, D.; Pochini, A.; Secchi, A. Recognition of quaternary ammonium cations by calix[4]arene derivatives supported on gold nanoparticles. *Chem. Commun.* **2005**, 645–647. [CrossRef]
24. Arduini, A.; Ciesa, F.; Fragassi, M.; Pochini, A.; Secchi, A. Selective synthesis of two constitutionally isomeric oriented calix[6]arene-based rotaxanes. *Angew. Chem. Int. Ed.* **2005**, *44*, 278–281. [CrossRef]
25. Rudkevich, D.M. Progress in supramolecular chemistry of gases. *Eur. J. Org. Chem.* **2007**, *2007*, 3255–3270. [CrossRef]
26. Rudkevich, D.M. Nanoscale molecular containers. *Bull. Chem. Soc. Jpn.* **2002**, *75*, 393–413. [CrossRef]
27. Vicens, J.; Harrowfield, J. (Eds.) *Calixarenes in the Nanoworld*; Springer: Dordrecht, The Netherlands, 2007. [CrossRef]
28. Sansone, F.; Dudic, M.; Dpnofrio, G.; Rivetti, C.; Baldini, L.; Casnati, A.; Cellai, S.; Ungaro, R. DNA condensation and cell transfection properties of guanidinium calixarene: Dependence on macrocycle lipophilicity, size, and confrmation. *J. Am. Chem. Soc.* **2006**, *128*, 14528–14536. [CrossRef] [PubMed]
29. Rodik, R.V.; Anthony, A.S.; Kalchenki, V.I.; Mely, Y.; Klymchenko, A.S. Cationic amphiphilic calixarenes to compact DNA into small nanoparticles for gene delivery. *New J. Chem.* **2015**, *39*, 1654–1664. [CrossRef]
30. Ostos, F.J.; Lebrón, J.A.; López-Cornejo, P.; López-López, M.; García-Calderón, M.; García-Calderón, C.B.; Rosado, I.V.; Kalchenko, V.I.; Rodik, R.V.; Moyá, M.L. Self-aggregation in aqueous solution of amphiphilic cationic calix[4] arenes. Potential use as vectors and nanocarriers. *J. Mol. Liq.* **2020**, *304*, 112724. [CrossRef]

31. Narkhede, N.; Uttam, B.; Kandi, R.; Rao, C.P. Silica-calix hybrid composite of allyl calix[4]arene linked to MCM-41 nanoparticles for sustained release of doxorubicin into cancer cells. *ACS Omega* **2018**, *3*, 229–239. [CrossRef]

32. Ostos, F.J.; Lebrón, J.A.; Moyá, M.L.; López-López, M.; Sánchez, A.; Clavero, A.; García-Calderón, C.G.; Rosado, I.V.; López-Cornejo, P. P-Sulfocalix[6]arene as nanocarrier for controlled delivery of doxorubicin. *Chem. Asian J.* **2017**, *12*, 679–689. [CrossRef]

33. Lebrón, J.A.; Ostos, F.J.; López-López, M.; Moyá, M.L.; Kardell, O.; Sánchez, A.; Carrasco, C.J.; García-Calderón, M.; García-Calderón, M.B.; Rosado, I.V.; et al. Preparation and characterization of metallomicelles of Ru(II): Cytotoxic activity and use as vectors. *Colloid Surf. B* **2019**, *175*, 116–125. [CrossRef] [PubMed]

34. Lebrón, J.A.; López-López, M.; Moyá, M.L.; Ostos, F.J.; Moreno, L.; García, D.; Moreno-Gordillo, P.; Rosado, I.V.; López-Cornejo, P. Metallo-liposomes derived from the [Ru(bpy)3]2+ complex as nanocarriers of therapeutic agents. *Chemosensors* **2021**, *9*, 90. [CrossRef]

35. Barrán-Berdón, A.; Yélamos, B.; García-Río, L.; Domenech, O.; Aicart, E.; Junquera, E. Polycationc macrocyclic scaffolds as potential non-viral vectors of DNA: A multidisciplinary study. *ACS Appl. Mater. Interfaces* **2015**, *7*, 14404–14414. [CrossRef] [PubMed]

36. Mochizuki, S.; Nishina, K.; Fujii, S.; Sakurai, K. The transfection efficiency of calix[4]arene-based lipids: The role of the alkyl chaim length. *Biomater. Sci.* **2015**, *3*, 317–322. [CrossRef]

37. Drakalska, E.; Momekova, D.; Manolova, Y.; Budurova, D.; Momekov, G.; Genova, M.; Antonov, L.; Lambov, N.; Rangelov, S. Hybrid liposomal PEGylated calix[4]arene systems as drug delivery platforms for curcumin. *Int. J. Pharm.* **2014**, *472*, 165–174. [CrossRef]

38. Liu, Y.-J.; Chao, H.; Tan, L.-F.; Yuan, L.-X.; Ji, L.-N. Ruthenium(II) mixed-ligand complexes containing 2-(6-methyl-3-chromonyl)-imidazol[4,5-f][1,10]-phenanthroline: Synthesis, DNA-binding and photoclavage studies. *Inorg. Chim. Acta* **2006**, *359*, 3807–3814. [CrossRef]

39. Secco, F.; Venturini, M.; Biver, T.; Sanchez, T.; Prado-Gotor, R.; Grueso, E. Solvent effects on the kinetics of the interaction of 1-pyrenecarboxaldehyde with calf thymus DNA. *J. Phys. Chem. B* **2010**, *114*, 4686–4691. [CrossRef]

40. Asayama, S.; Nohara, A.; Negishi, Y.; Hawakami, H. Alkylimidazolium end-modified poly(ethylene glycol) to form the mono-ion complex with plasmid DNA for in vivo gene delivery. *Biomacromolecules* **2014**, *15*, 997–1001. [CrossRef]

41. Zhang, H. Thin-Film hydration followed by extrusion method for liposome preparation. *Methods Mol. Biol.* **2017**, *1522*, 17–22. [CrossRef]

42. Moyá, M.L.; López-López, M.; Lebrón, J.A.; Ostos, F.J.; Pérez, D.; Camacho, V.; Beck, V.; Merino-Bohórquez, V.; Camean, M.; Madinabeitia, N.; et al. Preparation and characterization of new liposomes. Bactericidal activity of cefepime encapsulated into cationic liposomes. *Pharmaceutics* **2019**, *11*, 69. [CrossRef]

43. Meerloo, J.; Kaspers, G.J.L.; Cloos, J. Cell sensitivity assays: The MTT assay. *Methods Mol. Biol.* **2011**, *731*, 237–245. [CrossRef] [PubMed]

44. Israelachvili, J.; Mitchell, D.; Niham, B.W. Theory of self-assembly of hydrocarbon amphiphiles into micelles and bilayer. *J. Chem. Soc. Faraday Trans. 2* **1976**, *72*, 1525–1568. [CrossRef]

45. Nagarajan, R.; Wang, C.-C. Theory of surfactant aggregation in water-ethylene glycol mixed solvents. *Langmuir* **2000**, *16*, 5242–5251. [CrossRef]

46. Rodik, R.V.; Klymchenko, A.S.; Jain, N.; Miroshnichenko, S.I.; Richet, L.; Kalchenko, V.I.; Mely, I. Virus-sized DNA nanoparticles for gene delivery based on micelles of cationic calixarenes. *Chem. Eur. J.* **2011**, *17*, 5526–5538. [CrossRef]

47. Giuliani, M.; Morbioli, I.; Sansone, F.; Casnati, A. Moulding calixarenes for biomacromolecule targeting. *Chem. Commun.* **2015**, *51*, 14140–14159. [CrossRef]

48. Múñoz-Úbeda, M.; Misra, S.K.; Barrán-Berdón, A.L.; Aicart-Ramos, C.; Sierra, M.B.; Biswas, J.; Kondaiah, P.; Junquera, E.; Bhattacharya, S.; Aicart, E. Why is less cationic lipid required to prepare lipoplexes from plasmid DNA than linear DNA in gene therapy? *J. Am. Chem. Soc.* **2011**, *133*, 18014–18017. [CrossRef]

49. Ahmed, T.; Kamel, A.O.; Wettig, S.D. Interactions between DNA and Gemini surfactants: Impact on gene therapy. Part I. *Nanomedicine* **2016**, *11*, 289–306. [CrossRef]

50. Kirby, A.J.; Camilleri, P.; Engberts, J.F.B.N.; Feiters, M.N.; Nolte, R.J.M.; Söderman, O.; Bergsma, M.; Bell, P.C.; Fielden, M.L.; García Rodríguez, C.I.; et al. Gemini surfactants: New synthetic vectors for gene transfection. *Angew. Chem. Int. Ed.* **2003**, *42*, 1448–1457. [CrossRef]

51. Neidle, S. *Nucleic Acid Structure and Recognition*; Oxford University Press: New York, NY, USA, 2002; pp. 89–138, ISBN 109780198506355.

52. Pietralik, Z.; Kumita, J.R.; Dobson, M.C.; Kozak, M. The influence of novel gemini surfactants containing cycloalkylside-chains on the structural phases of DNA in solution. *Colloid Surf. B* **2015**, *131*, 83–92. [CrossRef] [PubMed]

53. Bombelli, S.; Borocci, M.; Diociaiuti, F.; Faggioli, L.; Galantini, P.; Luciani, P.; Mancini, G.; Sacco, M.G. Role of the spacer of cationic gemini amphiphiles in the condensation of DNA. *Langmuir* **2005**, *21*, 10271–10274. [CrossRef]

54. Bombelli, C.; Faggioli, F.; Luciani, P.; Mancini, G.; Sacco, M.G. Efficient transfection of DNA by liposomes formulated with cationic gemini amphiphiles. *J. Med. Chem.* **2005**, *48*, 5378–5382. [CrossRef]

55. Marty, R.; N'soukpoe-Kossi, C.N.; Charbonneau, D.; Weinert, C.M.; Kreplack, L.; Tajmir-Riahi, H.-A. Conformational analysis of DNA complexation with cationic lipids. *Nucleic Acids Res.* **2009**, *37*, 849–857. [CrossRef] [PubMed]

56. Chaires, J.B. A thermodynamic signature for drug-DNA binding mode. *Arch. Biochem. Biophys.* **2006**, *453*, 26–31. [CrossRef]
57. Yan, L.; Iwasaki, H. Thermal denaturation of plasmid DNA observed by atomic force microscopy. *Jpn. J. Appl. Phys.* **2002**, *41*, 7556–7559. [CrossRef]
58. Fisicaro, E.; Compari, C.; Bacciottini, F.; Contardi, L.; Pongiluppi, E.; Barbero, N.; Viscardi, G.; Quagliotto, P.; Donofrio, G.; Krafft, M.P. Non-viral gene delivery by highly florinated gemini bispyridinium surfactants-based DNA nanoparticles. *J. Colloid Interface Sci.* **2017**, *102*, 182–191. [CrossRef] [PubMed]
59. Liu, F.; Yang, J.; Huang, L.; Liu, D. Effect of non-ionic surfactants on the formation of DNA/emulsion complexes and emulsion-mediated gene transfer. *Pharm. Res.* **1996**, *13*, 1642–1646. [CrossRef] [PubMed]
60. Hara, T.; Liu, F.; Liu, D.; Huang, L. Emulsion formulations as a vector for gene delivery in vitro and vivo. *Adv. Drug Deliv. Rev.* **1997**, *24*, 265–271. [CrossRef]
61. Chen, Y.; Lu, Y.; Tao, C.; Huang, J.; Zhang, H.; Yu, Y.; Zou, H.; Gao, J.; Zhong, Y. Complexes containing cationic and anionic pH-sensitive liposomes: Comparative study of factors influencing plasmid DNA gene delivery to tumors. *Int. J. Nanomed.* **2013**, *8*, 1573–1593. [CrossRef]
62. Carvalho, C.; Santos, R.; Cardoso, S.; Correia, S.; Oliveira, P.; Santos, M.; Moreira, P. Doxorubicin: The good, the bad and the ugly effect. *Curr. Med. Chem.* **2009**, *16*, 3267–3285. [CrossRef]
63. Harada, M.; Bobe, I.; Saito, I.; Shibata, N.; Tanaka, R.; Hayashi, T.; Kato, Y. Improved anti-tumor activity of stabilized anthracycline polymeric micelle formulation, NC-6300. *Cancer Sci.* **2011**, *102*, 192–199. [CrossRef]
64. Weber, P.; Wagner, M.; Schneckenburger, M. Cholesterol dependent uptake and interaction of doxorubicin MCF-7 breast cancer cells. *Int. J. Mol. Sci.* **2013**, *14*, 8358. [CrossRef]
65. Walsby, E.J.; Coles, S.J.; Knapper, S.; Burnett, A.K. The topoisomerase II inhibitor bórdelo in causes cell cycle arrest and apoptosis in myeloid leukemia cells and act in synergy with cytarabine. *Haematologica* **2011**, *96*, 393–399. [CrossRef]
66. Thirumaran, R.; Prendergast, G.C.; Gilman, P.B. Cytotoxic chemotherapy in clinical treatment of cancer. In *Cancer Immunotherapy*; Prendergast, G.C., Jaffee, E.M., Eds.; Academic Press: Chicago, IL, USA, 2007; Chapter 7, pp. 101–116.
67. Lefrak, E.A.; Pitha, J.; Rosenheim, S.; Gottlieb, J.A. A clinical pathological analysis of adryamicin cardiotoxycity. *Cancer* **1973**, *32*, 302–314. [CrossRef]
68. Sinha, B.K.; Mimnaugh, E.G. Free radicals and anticancer drug resistance: Oxygen free radicals in the mechanisms of drug cytotoxicity and resistance by certain tumors. *Free Radic. Biol. Med.* **1990**, *8*, 567–581. [CrossRef]
69. Storm, G.; van Bloois, L.; Steerenberg, P.A.; van Etten, E.; de Groot, G.; Crommelin, D.J.A. Liposome encapsulation of doxorubicin: Pharmaceutical and therapeutic aspects. *J. Control. Release* **1989**, *9*, 215–229. [CrossRef]
70. Gokhale, P.C.; Radhakrishnan, B.; Husain, S.R.; Abernethy, D.R.; Sacher, R.; Dritschilo, A.; Rahman, A. An improved method of encapsulation of doxorubicin in liposomes: Pharmacological, toxicological and therapeutic evaluation. *Br. J. Cancer* **1996**, *74*, 43–48. [CrossRef] [PubMed]
71. Cowens, L.W.; Creaven, P.J.; Greco, W.R.; Brenner, D.E.; Tung, Y.; Petrelli, N. Initial clinical (Phase I) trial of TLC-99 (Doxorubicin encapsulated liposomes). *Cancer Res.* **1993**, *53*, 2796–2802. [PubMed]
72. Barenholz, Y.C. Doxil—The first FDA-approved nanodrug: Lessons learned. *J. Control. Release* **2012**, *160*, 117–134. [CrossRef]
73. Patil, R.R.; Guhagarkar, S.A.; Devarajan, P.V. Engineered nanocarriers of doxorubicin: A current update. *Crit. Rev. Ther. Drug Carr. Syst.* **2008**, *25*, 1–61. [CrossRef]
74. Norouzi, M.; Yathindranath, V.; Thliveris, J.A.; Kopec, B.M.; Siahaan, T.J.; Miller, D.W. Doxorubicin-loaded iron oxide nanoparticles for glioblastoma therapy: A combinational approach for enhanced delivery of nanoparticles. *Sci. Rep.* **2020**, *10*, 11292. [CrossRef]
75. Chao, Y.; Liang, Y.; Fang, G.; He, H.; Yao, Q.; Xu, H.; Chen, Y.; Tang, X. Biodegradable polymersomes as nanocarriers for doxorubicin hydrochloride: Enhanced cytotoxicity in MCF-7/ADR cells and prolonged blood circulation. *Pharm. Res.* **2017**, *34*, 610–618. [CrossRef] [PubMed]
76. Sagnella, S.M.; Duong, H.; NacMillan, A.; Boyer, C.; Whan, R.; Carroll, J.A.; Davis, T.P.; Kavallaris, M. Dextean-based doxorubicin nanocarriers with improved tumor penetration. *Biomacromolecules* **2014**, *15*, 262–275. [CrossRef] [PubMed]

 *pharmaceutics*

Article

Poly(L-Lactic Acid)-co-poly(Butylene Adipate) New Block Copolymers for the Preparation of Drug-Loaded Long Acting Injectable Microparticles

Vasiliki Karava [1], Aggeliki Siamidi [1], Marilena Vlachou [1,*], Evi Christodoulou [2], Nikolaos D. Bikiaris [2], Alexandra Zamboulis [2], Margaritis Kostoglou [3], Eleni Gounari [4,5] and Panagiotis Barmpalexis [6]

[1] Department of Pharmacy, Section of Pharmaceutical Technology, Zografou Campus, National and Kapodistrian University of Athens, 15784 Athens, Greece; vaso.karava111@gmail.com (V.K.); asiamidi@pharm.uoa.gr (A.S.)

[2] Department of Chemistry, Laboratory of Polymer Chemistry and Technology, Aristotle University of Thessaloniki, 54124 Thessaloniki, Greece; evicius@gmail.com (E.C.); nbikiaris@gmail.com (N.D.B.); azampouli@chem.auth.gr (A.Z.)

[3] Laboratory of Chemical and Environmental Technology, Aristotle University of Thessaloniki, 54124 Thessaloniki, Greece; kostoglu@chem.auth.gr

[4] Biohellenika Biotechnology Company, Leoforos Georgikis Scholis 65, 57001 Thessaloniki, Greece; egounari@biohellenika.gr

[5] Department of Biochemistry, School of Medicine, Faculty of Health Sciences, Aristotle University of Thessaloniki, 54124 Thessaloniki, Greece

[6] Department of Pharmaceutical Technology, School of Pharmacy, Aristotle University of Thessaloniki, 54124 Thessaloniki, Greece; pbarmp@pharm.auth.gr

* Correspondence: vlachou@pharm.uoa.gr; Tel.: +30-210-727-4674

Citation: Karava, V.; Siamidi, A.; Vlachou, M.; Christodoulou, E.; Bikiaris, N.D.; Zamboulis, A.; Kostoglou, M.; Gounari, E.; Barmpalexis, P. Poly(L-Lactic Acid)-co-poly(Butylene Adipate) New Block Copolymers for the Preparation of Drug-Loaded Long Acting Injectable Microparticles. *Pharmaceutics* **2021**, *13*, 930. https://doi.org/10.3390/pharmaceutics13070930

Academic Editors: Francisco José Ostos, José Antonio Lebrón and Pilar López-Cornejo

Received: 19 May 2021
Accepted: 21 June 2021
Published: 23 June 2021

Publisher's Note: MDPI stays neutral with regard to jurisdictional claims in published maps and institutional affiliations.

Abstract: The present study evaluates the use of newly synthesized poly(L-lactic acid)-co-poly(butylene adipate) (PLA/PBAd) block copolymers as microcarriers for the preparation of aripiprazole (ARI)-loaded long acting injectable (LAI) formulations. The effect of various PLA to PBAd ratios (95/5, 90/10, 75/25 and 50/50 w/w) on the enzymatic hydrolysis of the copolymers showed increasing erosion rates by increasing the PBAd content, while cytotoxicity studies revealed non-toxicity for all prepared biomaterials. SEM images showed the formation of well-shaped, spherical MPs with a smooth exterior surface and no particle's agglomeration, while DSC and pXRD data revealed that the presence of PBAd in the copolymers favors the amorphization of ARI. FTIR spectroscopy showed the formation of new ester bonds between the PLA and PBAd parts, while analysis of the MP formulations showed no molecular drug–polyester matrix interactions. In vitro dissolution studies suggested a highly tunable biphasic extended release, for up to 30 days, indicating the potential of the synthesized copolymers to act as promising LAI formulations, which will maintain a continuous therapeutic level for an extended time period. Lastly, several empirical and mechanistic models were also tested, with respect to their ability to fit the experimental release data.

Keywords: long acting injectables; poly(L-lactic acid); poly(butylene adipate); block copolymers; aripiprazole; microparticles; sustained release

1. Introduction

In the past few decades synthetic polymers that degrade under physiological conditions (i.e., biodegradable polymers) have become increasingly common in medical and pharmaceutical applications [1–3]. Especially, in the case of particulate drug formulations (such as nano- or microparticles) an increasing number of polymeric materials, and especially polyesters, have been introduced and implemented for drug delivery [4–6]. Amongst them, the preparation of long-acting injectable (LAI) formulations is probably the most intensively studied application for such polyester-based systems. In general, LAIs' are being utilized to reduce drug administration frequency, resulting to higher patient compliance.

Compared to other formulations, LAIs can provide a prolonged and constant therapeutic effect, enhance the biological half-life of drugs, improve bioavailability and protect the active pharmaceutical ingredients (APIs) against harsh environmental conditions [7–9]. In addition, compared to other materials, such as lipids, polyesters can be customized more easily in order to adapt to any specific type of drug [4]. However, despite the numerous scientific reports, the fact that only about 20 different LAI products are available in the market, suggests that the design of such drug formulations is a rather difficult task [10].

In this context, perhaps the most widely explored polyester used is poly(L-lactide) (PLA) and its copolymer with glycolic acid (i.e., poly (lactic-co-glycolic acid), PLGA) [11,12]. In general, PLA or PLGA LAI formulations have been investigated as suitable matrix/carriers to deliver a variety of APIs, including small molecules, peptides and proteins for periods ranging from one week to several months [13–19]. In the last two decades several PLA or PLGA-based products have been brought into the market [11,20,21]. The characteristics of the polymer, such as molecular weight (MW), copolymer composition, terminal groups functionality and glass-transition temperature (T_g) are the key factors affecting its biodegradability and, hence, release kinetics. However, despite their inherent flexibility, PLA and PLGA-based LAIs face a number of challenges, including initial burst release, enhanced lag-time, incomplete drug dissolution and poor drug stability during both production and storage [10,11]. In an attempt to overcome these limitations, a wide range of suitable biodegradable polymers, including poly-ε-caprolactone, polyorthoesters, polydioxanones, polyphosphazenes, polyanhydrides, poly(acyanoacrylates), polyiminocarbonates, polyoxalates and polyurethanes, have been proposed as alternatives. Among them, poly(alkylene adipate) derivatives were only recently introduced showing promising results [22,23].

In general, poly(alkylene adipate)s, derived from dicarboxylic acids and different aliphatic diols, such as poly(ethylene adipate) (PEAd)), poly(propylene adipate) (PPAd) and poly(butylene adipate) (PBAd), seem to be a promising PLA or PLGA substitute in terms of ecological and economic (balance of cost–benefit) factors [24–27]. In a recently published attempt, the use of poly(alkylene adipate)s, as sole matrix/carriers for the preparation of drug LAI microparticle formulations showed promising results, in terms of efficacy, although incomplete drug dissolution was recorded in addition to rather short sustained action (a plateau was reached in dissolution at 3 days) [22]. These results indicate that a certain amount of tunning is needed in order for this type of polyesters to be suitable for LAI formulations. Similar results were also obtained from another study, where the combination of poly(butylene adipate) (PBAd) with PLA in the form of a physical blend was utilized in order to prepare novel electrospun nanofibrous matrices for the sustained delivery of the immunomodulatory drug, teriflunomide [23]. In this study, a controlled release pattern of the drug was achieved and varied analogous to the proportion of the PBAd and the drug content.

In view of these findings, we recently published a study on the synthesis of a new block copolymer with enhanced physicochemical and mechanical performance, based on butylene adipate segments [28]. Specifically, block copolymers of PBAd combined with PLA (Figure 1a) were synthesized via a two stage polycondensation and analyzed in regard to thermal and mechanical properties. The results showed that the continuity of the two polymers throughout the copolymer volume and the semicrystalline morphology were both easily tuned by either the preparation method conditions and the ratio of PBAd to PLA to the drug. Based on these features it can be assumed that the new prepared block copolymer may be a promising candidate for the preparation of drug-loaded LAI microparticles (MPs).

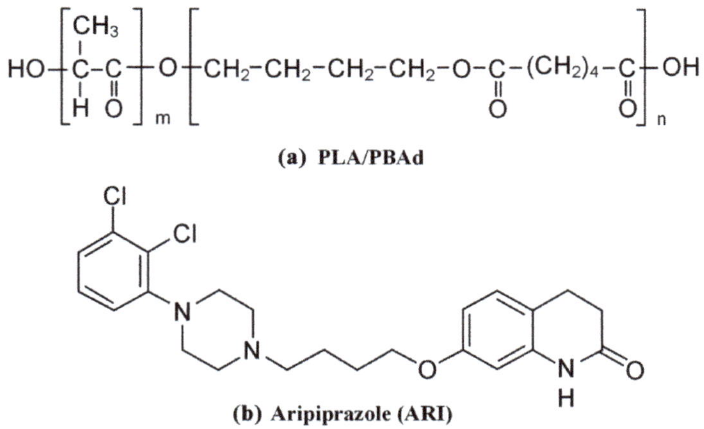

(a) PLA/PBAd

(b) Aripiprazole (ARI)

Figure 1. Chemical structures of (**a**) PLA/PBAd copolymers and (**b**) aripiprazole (ARI).

In the present study the use of the recently synthesized PLA/PBAd block copolymer was evaluated as a suitable LAI carrier. Aripiprazole (ARI, Figure 1b), a second-generation antipsychotic, is used as a model drug, which according to the FDA, is one of the several antipsychotic drugs marketed as a LAI formulation (please see Abilify Maintena®), which, however, presents a long initial lag-time (this is why oral ARI is simultaneously given for 14 consecutive days after the initial LAI injection) and is administrated rather frequently (i.e., once per month) [29–31]. Hence, within the set of the present study, after the initial evaluation of the biodegradation and the cytotoxicity profile of the neat PLA/PBAd block copolymers, ARI-loaded PLA/PBAd MPs were prepared via the emulsification/solvent evaporation method and thoroughly evaluated in terms of physicochemical and pharmacotechnical properties.

2. Materials and Methods

2.1. Materials

ARI (7-(4-[4-(2,3-Dichlorophenyl) piperazin-1-yl]butoxy)-3,4-dihydroquinolin-2(1*H*)-one) form III crystals were kindly donated by Pharmathen S.A. (Athens, Greece). Adipic acid (ACS reagent, ≥99.0%), 1,4-butanediol (99%), tetrabutyl titanate (TBT) (97%) and the Tin(II) 2-ethylhexanoate (TEH) (96%) catalysts were obtained from the Sigma-Aldrich (Saint Louis, MO, USA). L-Lactide (98%) and (*S*,*S*)-3,6-dimethyl-1,4-dioxane-2,5-dione were purchased from Alfa Aesar Chemicals (Kandel, Germany). *Rhizopus delemar* and *Pseudomonas cepacia* lipases were purchased from Fluka BioChemika, Steinheim, Germany. All other solvents and reagents used were of analytical or pharmaceutical grade and were used as received.

2.2. Synthesis of PBAd and PLA/PBAd Block Copolymers

PBAd and PLA/PBAd copolymers were prepared by the method we previously published [28]. Briefly, PBAd was prepared via a two-stage esterification and polycondensation. During the first stage (esterification), accurately weighed amounts of adipic acid and 1,4-butanediol, in a 1/1.1 molar ratio, were placed in a round-bottom flask and the polymerization mixture was degassed and purged with nitrogen several times, before heating to 180 °C under constant stirring and then gradually heating up to 220 °C over a period of three hours. After removal of the water formed, the nitrogen flow was stopped and 400 ppm of TBT (0.05 g mL^{-1} in toluene) was added to the mixture under high vacuum (5.0 Pa), in order to avoid excessive foaming. The temperature was then increased to 240 °C and the polycondensation reaction was carried out for another two hours.

After the preparation of the neat PBAd the PLA/PBAd copolymers were prepared via ring opening polymerization of L-lactide. Briefly, proper amounts of L-lactide and PBAd (corresponding to a final copolymer weight ratio of 95/5, 90/10, 75/25 and 50/50 *w/w* PLA to PBAd) were placed in round bottom flasks along with the THE (used as a catalyst at 400 ppm based on the L-lactide concentration). After nitrogen purging, the mixture was heated up to 200 °C and the reaction was initiated and carried out under constant mechanical stirring for one hour. The MW of the prepared copolymers was increased by heating up to 220 °C, under high vacuum (5.0 Pa) for 15 min. Then, the flasks were cooled to room temperature and the copolymers were purified by dissolving them in chloroform and precipitating in cold methanol twice, prior to using them. The precipitates were filtered and dried in a vacuum oven at 50 °C for 24 h. All samples were collected, placed in hermetically sealed vials, after purging with N_2, and stored at 5 °C before further use.

2.3. Characterization of PLA/PBAd Block Copolymers

Following their preparation, the newly synthesized block copolymers were characterized in terms of biodegradation and cytotoxicity profiles, characteristics that are extremely significant when preparing drug LAI formulations. Results, in terms of structural characterization, MW, physical state, thermal properties and molecular mobility, are given in a previous study of ours (Table 1) [28].

Table 1. Values of interest for the synthesized copolymers: molecular weight values, weight averaged (M_w) and number averaged (M_n), and polydispersity index (PDI), estimated by SEC, and crystallization, melting and glass transition temperatures (T_c, T_m and T_g accordingly), estimated by DSC.

Technique	SEC			DSC					
					PBAd			PLA	
Sample	M_w (g/mol)	M_n (g/mol)	PDI	T_c (°C)	T_g (°C)	T_m (°C)	T_c (°C)	T_g (°C)	T_m (°C)
PLA	130k	73k	1.79	-	-	-	-	55	149/155
PLA/PBAd 95/05	98k	61k	1.60	21/32	-	53	-	53	148/155
PLA/PBAd 90/10	97k	59k	1.65	15/32	-	52	-	54	148/155
PLA/PBAd 75/25	95k	57k	1.66	2/31	−64	52/55	-	54	147/154
PLA/PBAd 50/50	98k	57k	1.73	28	−60	52/55	-	54	148/154
PBAd	90k	49k	1.85	29	−62	55	-	-	-

2.3.1. Enzymic Hydrolysis

PLA/PBAd enzymatic hydrolysis was performed based on a previously employed method [32]. Briefly, the neat PBAd and PLA and the PBAd/PLA copolymers were prepared in the form of films, using an OttoWeber Type PW 30 hydraulic press (Paul-Otto Weber GmbH, Remshalden, Germany). The films were placed in petri dishes and 5 mL of phosphate buffer solution (0.2 M, pH 7.4) was added, containing 0.09 mg/mL of *Rhizopus delemar* lipase and 0.01 mg/mL of *Pseudomonas cepacia* lipase. The petri dishes were kept at 37.0 ± 1.0 °C in an oven for twenty days, while the media were replaced every 24 h. After predetermined time intervals, the films were removed from the lipase solution, washed thoroughly with distilled water and dried at 40 °C in vacuo, until constant weight. Every measurement was repeated three times. The degree of enzymatic hydrolysis was estimated from the weight loss, as compared to the initial weight of the samples.

2.3.2. Size-Exclusion Chromatography

The molecular weights of all samples after enzymatic hydrolysis were estimated by size-exclusion chromatography (SEC). The analysis was performed by means of SEC equipment consisting of a Waters 600 high pressure liquid chromatographic pump (Waters, Milford, MA, USA), Waters Ultrastyragel columns (HR-1, HR-2, HR-4 and HR-5) and a Shimadzu RID-10A refractive index detector (Shimadzu Corporation, Kyoto, Japan). Column calibration was performed using polystyrene standards (1–300 kg/mol in molecular weight). The concentration of the prepared solutions was 20 mg/1000 mL, the injection volume was 150 mL and the flow rate was 1 mL/min, operating at 60 °C.

2.3.3. Cytotoxicity Studies

Human adipose-derived mesenchymal stem cells culture (hAMSCs): For the evaluation of neat polymer and copolymers cytotoxicity, hAMSCs were provided from Biohellenika S.A. (Thessaloniki, Greece) after adipose tissue isolation from healthy volunteer donors. Experimentally, after liposuction adipose tissue washed twice in PBS (phosphate buffered saline) (1X, pH 7.4) (BIOWEST, Nuaillé, France). Overnight digestion was performed with 5 mg collagenase type I (Sigma-Aldrich, Saint Louis, MO, USA) per 10 g of adipose tissue after overnight incubation. The mixture was filtered using a 70 μm cell strainer (CORNING, Glendale, AZ, USA) and centrifuged at $850 \times g$ for 10 min. The pellet was resuspended in Dulbecco's modified Eagle's medium (DMEM)(BIOWEST, Nuaillé, France) supplemented with 10% fetal bovine serum (FBS)(BIOWEST, Nuaillé, France) and 2% penicillin/streptomycin and plated in culture flasks for 72 h until cells' adherence to the plastic surface (37 °C incubation with 5% CO2). The cell culture medium was replaced every 2–3 days until 80–90% confluence was reached. Cells were used in the experiments between passage 4 and 5. Every cells' detachment was performed with 0.05% trypsin–EDTA (BIOWEST, Nuaillé, France).

Sterilization of the materials and cell seeding: All the materials were sterilized in gradually reduced ethanol concentrations (100%, 70% and 50% in ddH$_2$O) and, after washing twice with ddH$_2$O, were left to air dry for 5 h under sterile conditions. Fibrin glue was prepared after the blood sampling of a healthy volunteer donor. A total of 10 μL of fibrin glue per film were placed in the bottom of a 24-well plate and the materials were seeded using a sterile pincher from above by applying minimal manual pressure and were left to air dry overnight under sterile conditions.

hAMSCs were detached using trypsin–EDTA 1x in PBS. A total of 3.5×10^5 cells were resuspended in the DMEM full medium and were subsequently placed above the films of each condition. A total of 3.5×10^5 cells were also plated in a plastic surface without any material and used as a control group. Upon air drying for 4 h in the incubator 1 mL of the DMEM full medium was added per well for the culture initiation. After 48 h, the cytotoxic effect of the materials was determined with an MTT (3-[4,5-dimethylthiazol-2-yl]-2,5 diphenyl tetrazolium bromide) assay.

In vitro cytotoxicity assays: The MTT cell proliferation assay, which employs the reduction of tetrazolium salts by metabolically active cells for examining cellular viability, was used for in vitro cytotoxicity assessment (Trevigen, Gaithersburg, MD, USA 4890-025-K). After 48 h of coincubation with the formulations, the medium was removed and cells were washed once with PBS before adding fresh medium including the 1/10 MTT reagent (Sigma-Aldrich, Saint Louis, MI, USA). Upon the removal of the MTT, 1 mL/well of DMSO was introduced for one additional hour of incubation. The optical density of MTT formazan deposits was quantified by a spectrophotometer at a 570 nm and 630 nm wavelength (PerkinElmer, Boston, MA, USA). All experiments were conducted in triplicate.

2.4. Preparation of ARI MPs

ARI MPs were prepared using PLA and PBAd polymers and their copolymers using an emulsification/solvent evaporation method. Briefly, 250 mg of polymer (pure PLA, pure PBAd and copolymers) were initially dissolved in 5 mL of dichloromethane and stirred with a magnetic stirrer. Then 50 mg of ARI were added to the solution and sonicated for 1 min until complete dispersion. The aqueous phase (50 mL of deionized H$_2$O and 50 mL of 1% *w/v* PVA solution) was then added to the dispersion phase, homogenized and left under stirring (1200 rpm), at room temperature, until the solvent was completely evaporated. When the microspheres were formed, they were separated from the rest of the solution by centrifugation at 4500 rpm for 10 min. Possible solvent or emulsifier residue was removed by three consecutive washes with deionized water. The microspheres were then freeze-dried in order to remove any water residue. For the preparation of non-drug loaded MPs, the same procedure as the one described above was followed without the addition of the API. All final samples were subsequently stored at 4 °C using hermetically sealed amber glass vails before further use.

2.5. Characterization of MPs

2.5.1. Differential Scanning Calorimetry (DSC)

DSC studies were conducted, using a Perkin–Elmer, Pyris Diamond DSC. In brief, accurately weighed samples (5.0 ± 0.1 mg) of the raw materials (i.e., neat PLA, neat PBAd and neat ARI) and the PLA-ARI, PBAd-ARI and PLA/PBAd-ARI MPs were hermetically sealed in aluminum pans and placed in the DSC sample holder. Then the samples were heated from 25 to 180 °C with a heating rate of 20 °C/min and the various thermal events were recorded using the Pyris Diamond software. The melting points (T_{melt}) were determined as the peak temperature and the glass-transition temperature (T_g) was determined as the inflection point temperature, while the enthalpy of fusion (ΔH_f) was determined as the integrated area of the heat flow curve in all cases. Nitrogen flow (50 mL/min) was applied in order to provide a constant thermal blanket within the DSC cell. The instrument was calibrated for temperature using high purity benzophenone, indium and tin, while the enthalpic response was calibrated using indium. All measurements were conducted in triplicate. The standard deviations of temperatures and enthalpies determined, in this work, were not higher than 1.0 °C and 3.0 J/g, respectively.

2.5.2. Wide Angle Powder X-ray Diffractometry (pXRD)

pXRD patterns of the raw materials (i.e., neat PLA, neat PBAd and neat ARI) and the PLA-ARI, PBAd-ARI and PLA/PBAd-ARI MPs were recorded using an XRD-diffractometer (Rigaku-Miniflex II, Chalgrove, Oxford, UK) with a CuKα radiation for crystalline phase identification ($\lambda = 0.15405$ nm for CuKα). All samples were scanned from 5 to 50° with a scanning rate of 1 °/min.

2.5.3. Scanning Electron Microscopy (SEM)

The morphology of the neat PLA, PLBAd and PLA/PBAd copolymers in the form of film before and after the enzymatic hydrolysis study and the ARI-loaded PLA, PBAD and PBAd MPs before and after the completion of the dissolution study was examined in a SEM system (JEOL JMS-840, JEOL USA Inc., Peabody, MA, USAmanufacturer, city, country). All samples (either in the form of thin films or MPs) were covered with carbon in order to provide good conductivity of the electron beam. All SEM images were collected with the following operating conditions: (1) accelerating voltage 20 kV, (2) probe current 45 nA and (3) counting time 60 s.

2.5.4. Fourier-Transformed Infrared Spectroscopy (FTIR)

The chemical structure and the formation of molecular interactions in PLA/PBAd copolymers and the PLA-ARI, PBAd-ARI and PLA/PBAd-ARI MPs was elucidated by FTIR spectroscopy. FTIR spectra of the samples were received with an FTIR spectrophotometer (model FTIR-2000, Perkin Elmer, Dresden, Germany) using KBr discs (thickness of 500 μm). The spectra were collected in the range from 4000 to 400 cm^{-1} at a resolution of 2 cm^{-1} (total of 64 coadded scans) and were baseline corrected and converted into the absorbance mode.

2.5.5. Yield, Encapsulation Efficiency and Drug Loading

MPs' yield, drug loading and encapsulation efficiency (EE) were determined by applying the following equations:

$$\text{Yield (\%)} = [\text{weight of MPs}]/[\text{initial weight of polymers and ARI}] \times 100 \tag{1}$$

$$\text{Drug loading (\%)} = [\text{weight of ARI in MPs}]/[\text{total weight of MPs}] \times 100 \tag{2}$$

$$EE\ (\%) = [\text{weight of ARI in MPs}]/[\text{initial weight of ARI}] \times 100 \tag{3}$$

Microspheres equivalent to 10 mg of aripiprazole were dissolved in the minimum quantity of dichloromethane and then diluted with the mobile phase: H_2O pH 3.5: acetonitrile 60:40 (*v/v*). The resulting solution was filtered through 0.45 μm filter paper and the filtrate was assayed for ARI using a Shimadzu Prominence HPLC system (Shimadzu Corporation, Kyoto, Japan), consisting of a degasser (Model DGU-20A5), a pump (Model LC-20AD), an automatic sampler (Model SIL-20AC), an ultraviolet–visible variable detector (Model SPD-20A) (λ_{max} = 254 nm) and a thermostatic oven (Model CTO-20AC). A reverse phase C18 column (250 mm × 4.6 mm I.D., 5 μm particle size) was used for chromatographic analysis. The flow rate was adjusted to 1 mL/min and the infusion volume was 20 μL. The chromatograms obtained were processed with the LC Solution software (v1.2, Shimadzu Corporation, Kyoto, Japan). All measurements were conducted in triplicate.

2.5.6. In Vitro Dissolution Test

The MPs (having 10 mg of ARI) were suspended in 2 mL of PBS and inserted in a dialysis tubing cellulose membrane bag (D9402-100FT; Sigma-Aldrich, Steinheim, Germany) with a molecular weight cut-off of 12,000–14,000, which was then sealed and placed into the dissolution basket (Distek Inc., North Brunswick Township, NJ, USA, model 2100C Dissolution Test System), equipped with an automatic sampler (Evolution 4300 Dissolution Sampler). The dissolution studies were performed under sink-conditions, using 400 mL of a phosphate-buffered saline (PBS) solution (pH 7.4), at 50 rpm/37 ± 0.5 °C. The solubility of ARI in PBS was 0.3 mg/mL (measured at 37 °C with the shaking flask method). Samples (2 mL) were withdrawn at predetermined time intervals, filtered and the concentration of ARI was determined using the above described validated HPLC method. Additionally, after the completion of the dissolution experiments the remaining MPs were withdrawn from the dialysis tubes, dried and analyzed for ARI content, using the HPLC method described in Section 2.5.5. All experiments were conducted in triplicate.

In order to evaluate the drug release mechanism, in vitro dissolution results were fitted to the following release kinetics models [33]:

$$\text{Zero order model: } D_t = D_0 + k_0 t \tag{4}$$

$$\text{First order model: } \log D_t = \log D_0 + k_1 t/2.303 \tag{5}$$

$$\text{Higuchi square root model: } D_t = D_0 + k_H t^{1/2} \tag{6}$$

$$\text{Hixon-Crowell model: } D_t^{1/3} = D_0^{1/3} - k_{HC} t \tag{7}$$

$$\text{Korsmeyer-Peppas model: } D_t/D_\infty = D_0 + k_P t^n \tag{8}$$

where, D_t is the amount of drug released at time t, D_0 is the initial amount of drug released, D_t/D_∞ is the fraction of drug released at time t, k_0 is the zero-order release constant, k_1 is the first-order release constant, k_H is the Higuchi release constant, k_{HC} is the Hixson–Crowell release rate constant, k_P is the Peppas release constant and n is the release exponent respectively.

2.6. Statistical Analysis

Statistical significance in the differences of the means was evaluated by using Student's *t*-test or Dunnett's test for the single or multiple comparisons of experimental groups, respectively. A difference with a *p*-value (p^*) < 0.05 was considered statistically significant.

3. Results and Discussion

3.1. Evaluation of neat PLA/PBAd Block Copolymers

As stated in the Introduction, the present study attempts to build upon the previously published promising results regarding the thermal and mechanical properties of the newly synthesized PLA/PBAd block copolymers and to evaluate their use as matrix/carriers for the preparation of ARI loaded LAI MPs. In this context, cytotoxicity and enzymatic hydrolysis of the neat copolymers are initially evaluated, since these two features are extremely important before proceeding with the preparation and evaluation of the LAI MPs.

3.1.1. Cytotoxicity Results

In general, polymers or copolymers that will be used to prepare such drug delivery systems should possess low cytotoxicity. Polyesters, based on PLA, poly (glycolic acid) (PGA) and polycaprolactone (PCL) and their copolymers, have been widely used as such biomaterials with a low cytotoxicity profile [34–40]. However, the cytotoxicity arising from the biodegradation of the newly prepared PLA/PBAd is unknown, and, hence, systematic evaluation is needed in order to verify their safety. The MW of the newly synthesized copolymers (measured by size-exclusion chromatography) varied from 98k–95k, while the polydispersity index (PDI) was below 2.0 in all cases (1.60–1.85) [28].

Figure 2 illustrates the cytotoxicity effect of the prepared block copolymers on hAM-SCs, where the y-axis shows the reduction of yellow 3-(4,5-dimethythiazol2-yl)-2,5-diphenyl tetrazolium bromide (MTT) by mitochondrial succinate dehydrogenase.

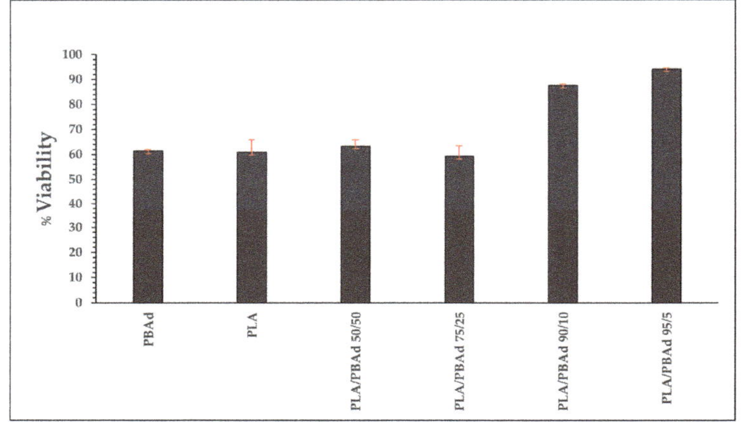

Figure 2. Cytotoxic effect on hAMSCs after incubation with neat PBAd, neat PLA and the newly prepared PLA/PBAd block copolymers.

During this study, the MTT, which enters hAMSCs, passes through the mitochondria where it is reduced to formazan. Subsequently, the cells are solubilized and the formazan content is measured spectrophotometrically. Since MTT reduction can only happen in metabolically active cells, the degree of activity is a measure of the cells' viability. Generally, in order for a material to be classified as toxic a reduction in the measured absorbance should be more than 50% as compared to the control sample. Hence, based on the obtained results, all studied materials can be considered as non-toxic since the max reduction in the measured absorbance was 40%. Specifically, in the case of PLA/PBAd block copolymers the obtained results showed a similar (for PLA/PBAd 50/50 and 75/23 w/w) or a significantly better (for PLA/PBAd 90/10 and 95/5 w/w) cytotoxicity profile as compared to the neat PLA, which is considered to be a non-cytotoxic biopolymer. Furthermore, results showed that the metabolic activity of the cancer cell line was dependent on the PLA to PBAd ratio within the copolymer, with samples higher in PLA showing a remarkably lower toxicity. Therefore, based on the MTT assay results it can be said that all prepared copolymers are non-toxic and, hence, are suitable candidates for the preparation drug LAI formulation.

3.1.2. Enzymatic Hydrolysis

In addition to non-toxicity, evaluation of the enzymatic hydrolysis profile of a polymer (or copolymer) is needed in order to clarify whether this material can be used as a LAI matrix/carrier, including, of course, MP based formulations. Generally, enzymatic hydrolysis (i.e., the path to polymer's degradation) is controlled by various factors related to the structure, the solid and thermal properties, etc. Among them, the mobility of polymer (or copolymer) segments, the crystalline morphology (including spherulite size), the ratio and the balances of hydrophilic/hydrophobic segments, the molecular weight, the T_{melt} and T_g are all factors that significantly affect the hydrolysis rate and extent [41–45].

In the present study, the enzymatic hydrolysis of the raw materials (i.e., the neat PBAd and PLA) and the newly synthesized PLA/PBAd block copolymers were evaluated in solutions containing a mixture of *R. delemar* and *Pseudomonas cepacia* lipases, at 37 °C and pH 7.4. Figure 3 shows the calculated hydrolysis in terms of weight loss vs. time profiles.

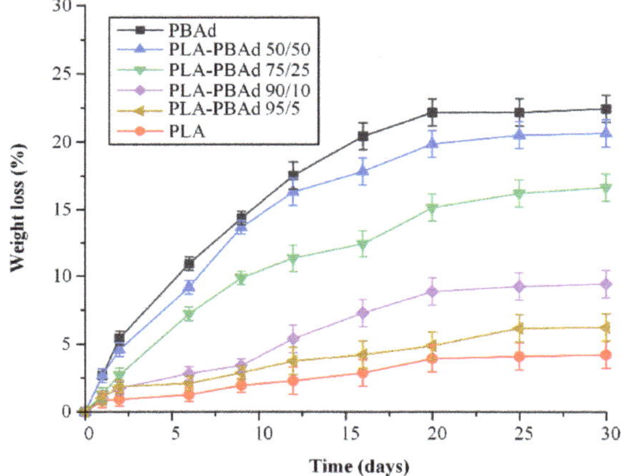

Figure 3. The enzymatic hydrolysis profile measured as % weight loss vs. time plots for the neat PBAd, the neat PLA and the various PLA/PBAd block copolymers.

In the case of the neat PLA, the results showed an extremely slow enzymatic hydrolysis rate reaching 3% within the first six days of testing. This slow degradation for PLA may be attributed to the polymer's high hydrophobic nature and to its high degree of crystallinity and its rather high T_{melt} and T_g (i.e., 150 °C and 55 °C, respectively [28]). In contrast to PLA, PBAd showed a substantially higher degree of enzymatic hydrolysis. This higher hydrolytic rate is in agreement with previous results [23] and can be attributed to the polymer's low T_g (approximately −55 °C) and T_{melt} onset (40 °C), which allow the polymer's segments to move around more freely, thus, enabling water to penetrate and hydrolyze the PBAd ester bonds more easily. In the case of PLA/PBAd samples, results showed an increase in the copolymer's hydrolysis, which was proportional to the PBAd content. Specifically, as the content of PBAd increased, the copolymer's degradation (measured in terms of weight loss) also increased. Hence, based on the obtained results, it should be noted that the prepared block copolymers also show highly tunable enzymatic hydrolysis characteristics. This is extremely important since, depending on the pharmacological properties of the API, the specific disease features and patients' individual characteristics, the proposed new biomaterials may be received as a universal solution for tailored drug or patient treatment.

However, despite the above presented significant findings, regarding the hydrolysis rate and extent of the prepared copolymers, in depth analysis of the degradation process is also needed in order to gain a true insight into the enzymatic biodegradation phenomena. In this context, the morphology of the prepared samples, before and after enzymatic hydrolysis, was evaluated via SEM. The results, presented in Figure 4, showed that the neat PLA remained almost unaffected after six days of testing, while neat PBAd showed an extensive mass degradation, which was dispersed uniformly along the whole surface of the sample. Similarly, the PLA/PBAd copolymers showed increased mass loss, as the content of PBAd increased, while, according to all collected images, it was obvious that the degradation mechanism of the copolymers, during their enzymatic hydrolysis at 37 °C, was related initially to surface erosion. This was also confirmed by SEC measurements after the first six days of study, which showed that molecular weight values remained practically unchanged compared to the initial samples, while weight loss was taking place (Table 2). However, even in this case, hydrolysis is a dynamic procedure. It has been found that the hydrolytic chain cleavage proceeds preferentially in the amorphous regions of polyesters, leading initially to the increase in polymer crystallinity [46]. Due to the interconnections of amorphous fractions, hydrolysis becomes also a bulk erosion process after a period of time.

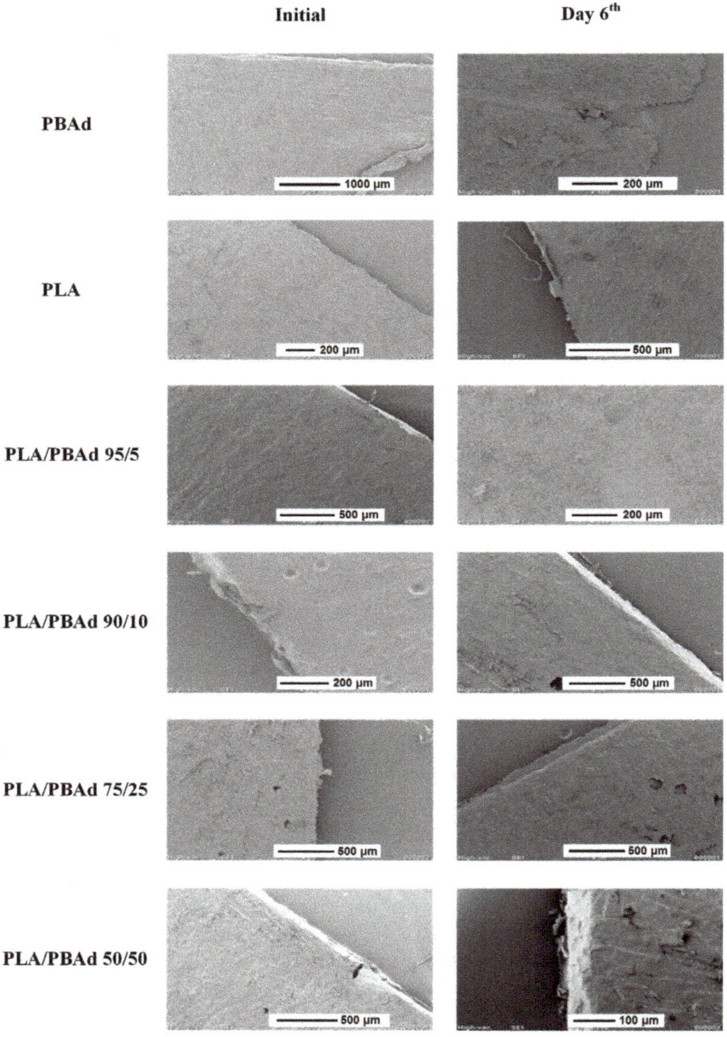

Figure 4. SEM micrographs of the neat PBAd and PLA and the various PLA/PBAd block copolymers during enzymatic hydrolysis at zero time (initial) and after six days.

Table 2. SEC estimated molecular weight values, M_w and M_n, and % weight loss, after six days of enzymatic hydrolysis.

Sample	Mw (g/mol)	Mn (g/mol)	% Weight Loss
PLA	129.7k	72.8k	1.267
PLA/PBAd 95/05	97.8k	60.8k	2.109
PLA/PBAd 90/10	97.1k	58.6k	2.837
PLA/PBAd 75/25	93.9k	56.9k	7.230
PLA/PBAd 50/50	97.9k	57.1k	9.178
PBAd	88.9k	47.9k	10.912

3.2. Evaluation of ARI-loaded MPs

Based on the previously obtained results, the newly synthesized PLA/PBAd block copolymers show a good cytotoxicity profile and highly tunable enzymatic hydrolysis characteristics, features that makes them good candidates as LAI matrix/carriers. Hence, in the following section the preparation of such drug-loaded formulations (in the form of MPs) will be thoroughly evaluated.

3.2.1. MPs Morphology Evaluation Via SEM

As stated previously, in the present study the newly prepared PLA/PBAd block copolymers (at several PLA to PBAd ratios) were tested as suitable biopolymers for the preparation of ARI LAI MPs. In this set framework, the effect of the PLA and PBAd content on the size and the morphological characteristics of the drug-loaded MPs was initially investigated via SEM. Results in Figure 5 showed the formation of spherical MPs with a smooth exterior surface, while in all cases no particle agglomeration was observed.

Specifically, regarding the MPs prepared with the initial polymeric raw materials (i.e., PLA and PBAd), results showed the formation of significantly larger particles in the case of PLA with more spherical shape and uniform size distribution, while some defects were also observed on the surface of the said MPs. These differences can be attributed to the more hydrophobic nature of PLA (as compared to PBAd), which leads to a better homogenization and consequently more controlled solvent removal processes. Additionally, a significant role also plays the notable differences in the thermal properties of the two tested biopolymers, with PBAd's lower T_{melt} and T_g values enabling the 'softening' of the just formed MPs during the solvent removal phase, leading in this way to the formation of smaller drug-loaded spherical MPs (as compared to PLA). In the case of PLA/PBAd, results showed that as the PBAd content increased within the block copolymer the particle size of the obtained MPs decreased. Specifically, the average particle size (measured as d_{50}) of the prepared ARI-loaded MPs, measured from at least ten SEM images, was estimated as 58.2 ± 15 µm, 43.3 ± 10 µm, 30.15 ± 10 µm and 18.8 ± 5 µm, for the MPs prepared with PLA/PBAd 95/5, 90/10, 75/25 and 50/50, respectively. Considering the previously published results on the thermal properties of the prepared neat block copolymers [28], where it was found that the melting properties of the two monocomponents (i.e., PLA and PBAd) are retained in the newly prepared biomaterial, the obtained results indicate that the addition of PBAd in the block copolymer chain (and its more hydrophilic nature and lower melting temperature) is responsible for the reduction of the resultant ARI-loaded MP's size.

3.2.2. MPs Yield, Drug Loading and EE

Table 3 summarizes the yield, drug loading and EE values for the prepared ARI-loaded MPs.

Figure 5. SEM images for the ARI-loaded MPs prepared using PLA, PBAd and the newly synthesized PLA/PBAd block copolymers as the matrix/carriers.

Table 3. Yield, drug loading and EE of the prepared ARI MPs.

Sample	Average Particle Size (d_{50}) (μm)	Yield (%)	Drug Loading (%)	EE (%)
PLA	56.3 ± 15	77.73 ± 2.84	16.35 ± 1.75	42.73 ± 2.08
PLA/PBAd 95/5	58.2 ± 15	89.32 ± 2.03	14.56 ± 1.86	44.84 ± 2.87
PLA/PBAd 90/10	43.3 ± 10	92.51 ± 2.48	13.19 ± 2.83	38.17 ± 3.54
PLA/PBAd 75/25	30.2 ± 10	98.60 ± 1.38	12.48 ± 2.16	39.78 ± 2.86
PLA/PBAd 50/50	18.8 ± 5	97.40 ± 1.24	11.30 ± 3.14	32.67 ± 3.07
PBAd	21.3 ± 5	60.35 ± 3.68	17.48 ± 2.47	36.52 ± 4.23

Based on the obtained results, the yield in all MPs ranged from 60.35 to 98.6% with the PLA-PBAd copolymers showing much more improved yields, as compared to the neat PLA and PBAd MPs. Additionally, a closer look at the obtained results revealed that the higher yield values were recorded in the case of MPs containing higher amounts of PBAd (i.e., PLA/PBAd 75/25 and 50/50), indicating that the presence of PBAd in the backchain of the prepared copolymer results in a more efficient (at least in terms of yield productivity) MP's preparation process. However, in contrary to the previous findings, results regarding MP's drug loading and EE showed the opposite effect. Specifically, as the content of PBAd increased in the PLA/PBAd copolymer, the resultant ARI-loaded MPs showed lower drug loadings and EE values. This indicates that, contrary to MP's productivity, the presence of PBAd results in droplets that were harder to solidify, and hence there was much more time for the API molecules to diffuse away from the droplet into the aqueous phase medium, resulting in the preparation of MPs with a lower drug content.

3.2.3. MPs' Thermal Properties and Physical State Evaluation Via DSC

The thermal properties and the physical state of the drug-loaded MPs as compared to the neat raw materials were evaluated with the aid of DSC (Figure 6a). In the case of the neat ARI, results showed an initial endothermic peak at 140.6 °C, corresponding to the melting of the ARI form III crystals, followed by a small recrystallization exotherm (at 144 °C) and a second endothermic peak at 151. 9 °C corresponding to the ARI form I crystals melting. These results indicate that the initially used ARI was in the form of polymorph III crystals, while during its melting a phase transition from polymorph III to polymorph I was recorded. This behavior is in agreement with previous studies evaluating the phase transition phenomena occurring during ARI's DSC heating [47]. Regarding the neat initial polymers, the results in the case of PBAd showed a broad DSC endotherm with a peak at 63.3 °C, corresponding to its melting, while PLA showed a T_g transition point at 69.9 °C and a melting endotherm at 153.4 °C, both indicative of its semicrystalline nature.

Looking at the DSC thermograms of the drug-loaded PBAd- and PLA-MPs, similar thermal events were detected, as is in the case of pure (neat) polymers. Specifically, in the case of ARI-PBAd MPs, a broad endothermic peak was recorded at 58.9 °C corresponding to the melting of the crystalline part of the polymer, while for ARI-PLA MPs a T_g (with an endothermic overshoot due to the molecules' relaxation) was recorded at 67.5 °C followed by an endothermic melting peak at 143.9 °C, corresponding to the melting of the PLA. It is important to note that in all thermograms a small drop in the obtained thermal events was recorded (as compared to the neat polymeric raw materials), which is attributed to the presence of the API and the remaining solvents (used for the preparation of the MPs) that act as plasticizers to the whole system.

Additionally, it should be pointed out that in the case of ARI-loaded PBAd MPs, no thermal events were recorded in respect to the API, indicating that probably the drug was amorphously dispersed within the polymeric matrix, although in situ solubilization of the ARI crystals during the DCS heating scan cannot be excluded. On the contrary, results from the DSC thermograms of the drug-loaded PLA MPs, showed a small melting endotherm at 137 °C, which is probably attributed to some of the remaining ARI form III crystals. Similar results were also obtained for the drug-loaded MPs prepared with the newly synthesized PLA/PBAd block copolymers, where the DSC endotherm corresponding to the API melting was decreasing as the PBAd content increased, while at the higher PBAd content used (i.e., PLA/PBAd 75/25 and 50/50) no such API melting peaks were recorded. Hence, based on the DSC results it seems that the presence of PBAd in the PLA/PBAd block copolymers favors the amorphization of the API leading to its complete amorphization in ratios higher that 75/25 *w/w* PLA to PBAd.

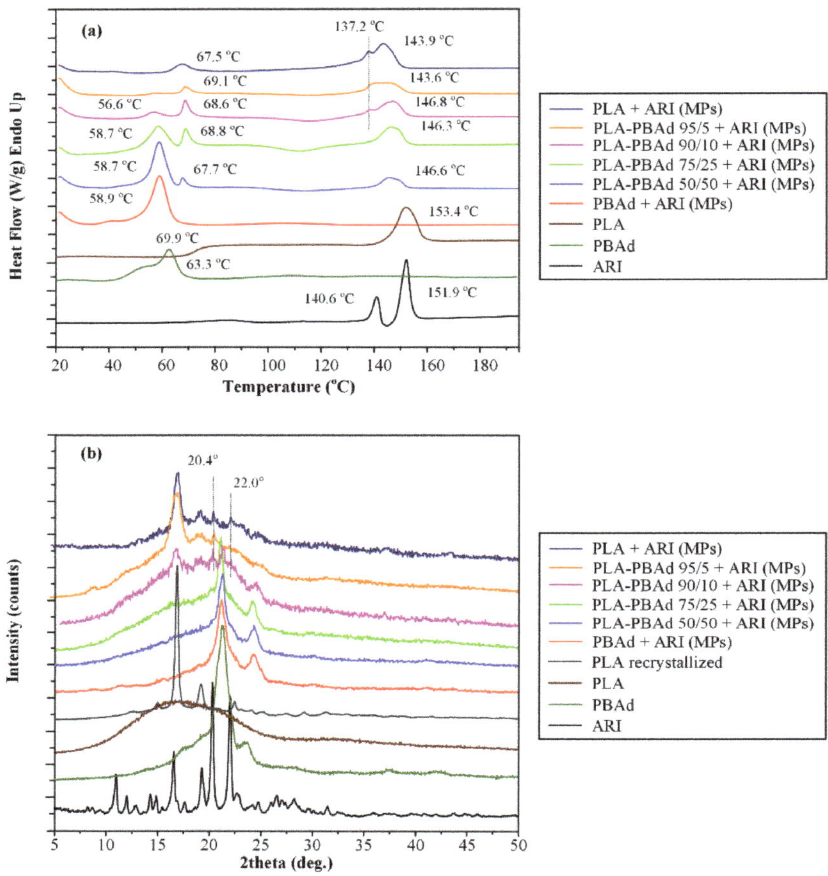

Figure 6. DSC thermograms (**a**) and XRD diffractograms (**b**) of the neat raw materials (i.e., ARI, PLA and PBAd) and the prepared drug-loaded MPs.

3.2.4. Physical State Verification Via pXRD

The physical state of the API after the preparation of the drug-loaded MPs was also evaluated via pXRD in order to verify the results suggested by DSC. Figure 6b shows the pXRD diffractograms of the raw materials, the recrystallized neat PLA (after solubilizing in dichloromethane, i.e., the solvent used for the preparation of the MPs) and the respective MPs. In the case of the neat ARI, results showed several sharp pXRD diffractogram peaks at 2θ of 10.9, 16.6, 19.3, 20.3 and 22.0°, which were all characteristic of the ARI form III crystals [48]. In the case of neat PBAd, two characteristic pXRD peaks were recorded at 2θ of 21.1 and 24.2°, which were both located over a broad amorphous halo, indicating that the neat copolymer was semicrystalline in nature. Regarding the neat PLA, two different pXRD patterns were recorded before and after its recrystallization. Specifically, the polymer as received showed a characteristic amorphous halo, indictive of its highly amorphous nature, while upon its recrystallization new crystals were recorded at 2θ positions of 14.8, 16.9, 19.1 and 22.5°, respectively, all of which were also seen in the case of PLA's melt recrystallization [28]. In regard to the drug-loaded MPs using only PBAd, the recorded pXRD diffractograms showed only the characteristic peaks of the neat copolymer, indicating that the API was amorphously dispersed within the MPs' matrix/carrier. In contrast, in the case of PLA drug-loaded MPs, in addition to the polymer's characteristic pXRD pattern, two diffractogram peaks corresponding to the ARI's form III crystals (i.e., 2θ of 20.4 and 22.0°, respectively) were also recorded, indicating that the API was recrystallized during the formation of the said MPs. Similarly, in the case of MPs prepared with the newly synthesized PLA/PBAd block copolymers having high PBAd content (i.e., 75/25 and 50/50 PLA to PBAd ratio), no ARI characteristic pXRD peaks were recorded, indicating that the API was amorphously dispersed within the said matrix-carriers. On the contrary, in the rest of the samples, i.e., those using PLA/PBAd copolymers with high PLA content, two characteristic ARI form III peaks (although of low intensity) were recorded, indicating that a small portion of the API was recrystallized in these cases. Hence, based on the obtained results, the pXRD analysis verifies the previously presented DSC findings, since the increase in PLA's content within the newly synthesized PLA/PBAd block copolymers, leads, indeed, to ARI's recrystallization.

3.2.5. Evaluation of Molecular Interactions

In a further step, FTIR spectroscopy was used in order to identify the formation of molecular interactions during the preparation of the drug-loaded MPs. Initially, before proceeding with the FTIR analysis of the MPs, the spectra of the newly prepared PLA/PBAd block copolymers were evaluated (Figure 7a) in an attempt to identify the molecular interactions evolving between the two polymeric components (i.e., PLA and PBAd) during the copolymerization process. Specifically, in the case of PLA the asymmetric and symmetric vibrations of the methylene groups were recorded at 2995 cm^{-1} and 2945 cm^{-1}, respectively, while the vibrations of the carbonyl C=O and the C-O-C ester groups were recorded at 1757–1710 cm^{-1} and 1188 cm^{-1}. In the case of PBAd, the characteristic absorption peaks of the ester -COO- and the C-O-C appeared at 1735 cm^{-1} and 1100–1300 cm^{-1}, respectively, while the peaks located at 1450–1465 cm^{-1} were attributed to the C-H bending vibrations of the methylene and methyl groups. In all spectra the low intensity peaks recorded at 3300–3550 cm^{-1} can be attributed to the presence of -OH end groups. Regarding the newly synthesized copolymers, results showed increased similarities among the recorded spectra. Specifically, in all cases a strong absorption peak at 1730 cm^{-1} was recorded, due to the formation of a new ester bond between the PLA and the PBAd (responsible for the formation of the new block copolymer). Additionally, there were also several peaks in the range of 750–1100 cm^{-1} and 1100–1400 cm^{-1}, corresponding to the C-C and C-O vibrations, respectively. Finally, the presence of the methylene groups within the newly synthesized copolymers was also confirmed by the specific FTIR absorption peaks recorded in the region of 2700–3000 cm^{-1}.

Moving along with the evaluation of molecular interactions, evolving within the prepared drug-loaded MPs, Figure 7b shows the recorded FTIR spectra of all systems along with the spectrum of the neat API. In regard to ARI, results showed the presence of several characteristic FTIR absorption peaks at 3195 cm^{-1} (corresponding to the NH vibrations), 2949 and 2840 cm^{-1} (attributed to the CH), 1679 cm^{-1}, due to C=O, and 1628 cm^{-1}, due to C=C, vibrations), while the peaks at 1160 and 1123 cm^{-1} are attributed to the stretching vibration of the single C-O and C-C bonds, respectively. Before proceeding with the analysis of the MPs' FTIR spectra, it is important to note that in general, polyesters, such as those evaluated in the present study, consist mainly of ester bonds and terminal carboxylic and hydroxyl groups, which can interact, via hydrogen bonding (HB), with the ester groups or the amino groups of ARI and its two chlorine atoms located in the dichlorophenyl part of the molecule. Hence, in order to determine if such interactions exist in the prepared systems, we will focus our analysis on the characteristic peaks recorded in the region of the hydroxyl and carbonyl groups of the FTIR spectrum. Looking at the obtained MPs' spectra, the hydroxyls of the polyesters in the MPs were recorded at 3480 cm^{-1} and no obvious shifts were apparent amongst the examined systems. Additionally, in the region of carbonyls' absorption bands, all polyesters show a similar wide peak at 1730 cm^{-1}, while next to it (at 1679 cm^{-1}) the carbonyl vibration of the pure drug is recorded indicating that the API was successfully encapsulated in the prepared MPs. Finally, since, there are no differences (or shifts/displacements) in the FTIR absorption peaks, it seems that no molecular interactions are taking place between the API and the copolymer.

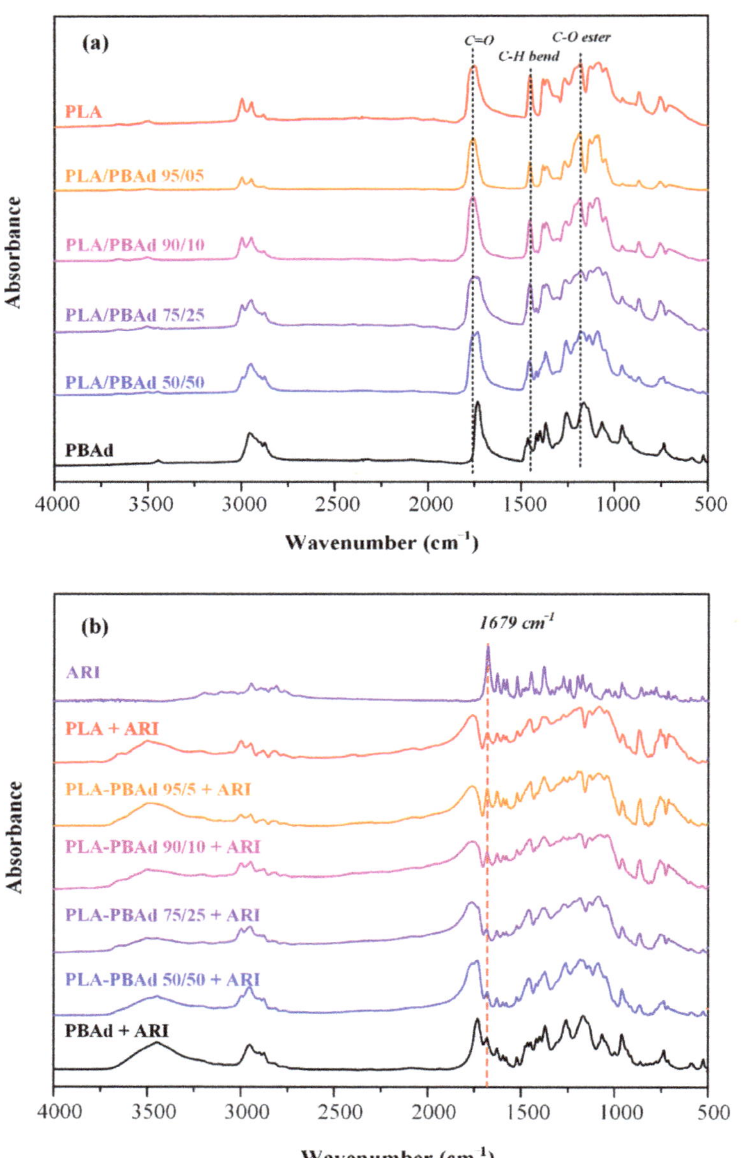

Figure 7. FTIR spectra of: (**a**) the neat PLA, PBAd and the PBA/PBAd copolymers and (**b**) the API and the prepared API-loaded MPs.

3.2.6. In Vitro Dissolution Profile

In the final step of the present work, the effect of the newly synthesized block copolymers on the in vitro dissolution characteristics of ARI were evaluated. Figure 8a depicts the dissolution profiles of the prepared drug-loaded MPs. The maximum ARI released from the PLA-PBAd MPs ranged from 37.70% (with PLA-PBAd 95/5) to 60.38% (with PLA-PBAd 50/50) indicating a wide distribution, in terms of drug release extent. This can be partially attributed to the amorphization of the drug within the MPs, induced by the presence of the PBAd, and its fine dispersion within the polymeric matrix (confirmed previously by the XRD and DSC results). Additionally, drug assay analysis of the remaining MPs after the completion of the dissolution trials (Table 4) revealed the presence of un-dissolved API still "trapped" within the polymeric structure. This may explain the incomplete delivery of the API observed in all MPs formulations. Lastly, regarding the ARI that was neither recovered from the microparticles nor released, we can postulate that this was probably lost during the withdrawal of the MPs from the dialysis tubes and the drying process conducted before the ARI content analysis. Additionally, we may assume that a small portion of the API may be "lost" due to the drug's degradation during dissolution, although in order to support/verify this hypothesis ARI solution stability at 37 °C for 30 days has to be performed.

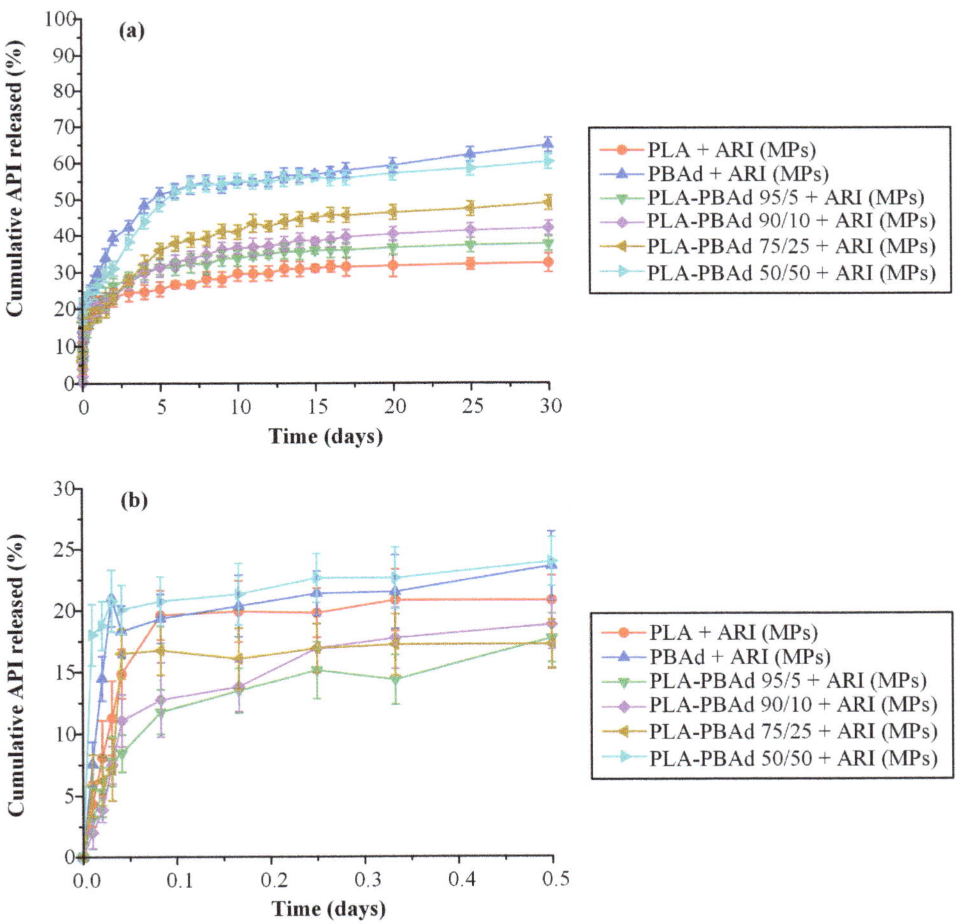

Figure 8. In vitro drug % release vs. time of the encapsulated ARI in PLA/PBAd MPs, during the 30 (**a**) and the 0.5 (**b**) days of the experiment.

Table 4. % Aripiprazole drug remained in the MPs at the end of the in vitro release study.

Sample	Remained ARI (%)
PLA	66.28 ± 0.23
PLA/PBAd 95/5	60.08 ± 2.12
PLA/PBAd 90/10	58.42 ± 1.54
PLA/PBAd 75/25	50.63 ± 2.03
PLA/PBAd 50/50	38.94 ± 1.78
PBAd	34.78 ± 2.27

The MPs prepared, using PLA, showed the lowest drug release rate, on contrary to the MPs prepared with PBAd where the maximum release rate was achieved. The rest of the formulations using the newly synthesized PLA-PBAd block copolymers showed increasing API release rate (and extend) as the PBAd content increased. Therefore, it seems that the addition of PBAd to the polymeric matrix significantly improves the hydrolysis rate of PLA and, consequently, the dissolution rate of the encapsulated API. Based on these findings it must be said that the use of the newly synthesized PLA/PBAd block copolymers as matrix/carriers for the preparation of ARI-loaded MPs, results in highly tunable extended release profiles for the API, which may be controlled for up to 30 days. Therefore, under in vivo conditions this could possibly lead to new formulations able to maintain continuous therapeutic levels for an extended time period (>30 days), with no lag-time, and hence, emerge as an alternative long-acting treatment option for the management of chronic diseases.

Looking again back to the obtained dissolution results, it is obvious that ARI's dissolution from the prepared MPs followed a biphasic release profile in all cases. Specifically, after an initial burst phase (Figure 8b) attributed to the active substance present on the surface of the MPs, a fast release phase was observed for up to approximately five (5) days, followed by a slower release phase for the remaining twenty-five (25) days. Keeping in mind that drug release from such MPs is mainly controlled by the interplay between API's diffusion from the polymeric matrices and polyester's erosion/degradation behavior, it can be assumed that in both phases (i.e., the fast and the slow) these two different mechanisms have a different impact. Therefore, in an attempt to identify the differences prevailing in each release phase, the obtained dissolution data were fitted in the various kinetic models described in Section 2.5.6. The goodness of fit (expressed by the correlation coefficient, R^2) and the k-constants for each model are summarized in Table 5.

Table 5. Dissolution data model fitting results for the employed drug release kinetic models.

Release Fitting Model	PLA		PBAd		PLA/PBAd 95/5		PLA/PBAd 90/10		PLA/PBAd 75/25		PLA/PBAd 5050	
	R^2	k-Constant	R^2	k-Constant	R^2	k-Constant	R^2	k-Constant	R^2	k-Constant	R^2	k-Constant
Fast-release phase												
Zero order	0.60	2.77 d^{-1}	0.77	8.03 d^{-1}	0.68	4.61 d^{-1}	0.67	4.32 d^{-1}	0.82	5.54 d^{-1}	0.80	7.43 d^{-1}
First order	<0.01	0.09 d^{-1}	0.58	0.20 d^{-1}	0.16	0.11 d^{-1}	0.12	0.10 d^{-1}	0.54	0.11 d^{-1}	0.57	0.17d^{-1}
Higuchi	0.31	14.84 d^{-1}	0.93	25.45 d$^{-1/2}$	0.83	16.45 d$^{-1/2}$	0.81	16.03 d$^{-1/2}$	0.94	16.67 d$^{-1/2}$	0.93	22.95 d$^{-1/2}$
Hixson–Crowell	<0.01	0.05 d^{-1}	0.69	0.06 d^{-1}	0.08	0.03 d^{-1}	0.04	0.03 d^{-1}	0.48	0.03 d^{-1}	0.49	0.05 d^{-1}
Korsmeyer–Peppas	0.99	21.90 d^{-n}	0.99	29.93 d^{-n}	0.99	20.66 d^{-n}	0.98	20.35 d^{-n}	0.98	19.06 d^{-n}	0.98	26.70 d^{-n}
Slow-release phase												
Zero order	0.77	4.20 d^{-1}	0.98	3.91 d^{-1}	0.84	3.98 d^{-1}	0.89	4.02 d^{-1}	0.87	4.08 d^{-1}	0.93	3.42 d^{-1}
First order	0.69	0.07 d^{-1}	0.71	0.03 d^{-1}	0.75	0.06 d^{-1}	0.77	0.06 d^{-1}	0.74	0.06 d^{-1}	0.80	0.04 d^{-1}
Higuchi	0.60	16.64 d$^{-1/2}$	0.53	10.58 d$^{-1/2}$	0.66	16.08 d$^{-1/2}$	0.66	15.21 d$^{-1/2}$	0.62	14.60 d$^{-1/2}$	0.66	12.79 d$^{-1/2}$
Hixson–Crowell	0.74	0.02 d^{-1}	0.76	0.01 d^{-1}	0.81	0.02 d^{-1}	0.83	0.02 d^{-1}	0.79	0.01 d^{-1}	0.85	0.01 d^{-1}
Korsmeyer–Peppas	0.76	5.21 d^{-n}	0.98	0.61 d^{-n}	0.83	4.99 d^{-n}	0.88	3.89 d^{-n}	0.85	3.31 d^{-n}	0.93	2.62 d^{-n}

Looking at the obtained results, in the case of the initial release phase (i.e., up to five days) the higher R^2 values for all samples were obtained for the Korsmeyer–Peppas equation, indicating that the said model is more suitable to describe the obtained dissolution data. In general, the Korsmeyer–Peppas model is able to describe the several mechanisms that simultaneously control the dissolution behavior in such systems, by the use of the exponent n. Specifically, n values below 0.5 suggest that the drug diffuses through the matrix and is released with a quasi-Fickian diffusion mechanism, while values between 0.5 and 1 indicate an anomalous, non-Fickian, drug diffusion and values above 1 suggest a non-Fickian, Case II, release kinetics mechanism [33]. Based on the obtained Korsmeyer–Peppas fitting results, the exponent n in the initial fast release phase was below 0.5 in all cases (i.e., $n_{(PLA)} = 0.088$, $n_{(PBAd)} = 0.337$, $n_{(PLA/PBAd\ 95/5)} = 0.268$, $n_{(PLA/PBAd\ 90/10)} = 0.256$, $n_{(PLA/PBAd\ 75/25)} = 0.366$ and $n_{(PLA/PBAd\ 50/50)} = 0.348$) indicating that the drug released from all prepared MPs in the first five days was diffusion controlled. Interestingly, in the case of the slow-release phase (i.e., starting from the 6th day and lasting up to 30 days) the fitting results in Table 5 showed that the release of the API from all prepared MPs followed a zero-order release mechanism. This is also verified by the n exponent of the Korsmeyer–Peppas model fitting, where, in all cases, was approximately one (i.e., $n_{(PLA)} = 0.917$, $n_{(PBAd)} = 1.057$, $n_{(PLA/PBAd\ 95/5)} = 0.920$, $n_{(PLA/PBAd\ 90/10)} = 0.989$, $n_{(PLA/PBAd\ 75/25)} = 1.032$ and $n_{(PLA/PBAd\ 50/50)} = 1.066$). Hence, it seems that in the latter stage of API's dissolution the initially diffusion-controlled phase is compensated by the simultaneous matrix swelling (due to the polyester's wetting) and a small portion of matrix erosion (due to the polyester's hydrolytic degradation) leading in this way to a 'balanced' zero-order release profile, which is essential in achieving stable in vivo pharmacokinetic behavior.

Morphology Evaluation after Dissolution Studies

In a further step, in order to examine the process of polyester degradation/erosion during dissolution and to correlate this with the enzymatic hydrolysis results presented in Section 3.1.2, SEM images were taken after the completion of the test (Figure 9). As evidenced, in the case of neat PLA and the two polyesters containing only a small amount of PBAd, namely PLA/PBAd 95/5 and 90/10, the surface and shape of the prepared microspheres remained practically unchanged (Figure 9). On the other hand, neat PBAd and the polymeric matrices containing high PBAd load (i.e., PLA/PBAd 75/25 and PLA/PBAd 50/50) demonstrated some clear evidence of surface erosion, presumably due to polyester hydrolysis or drug dissolution. From these images, we can thus conclude that the amount of PBAd bares a crucial role to the extent of polyester degradation and consequently the drug release rates, which is in accordance to previously discussed results from neat polymer enzymatic hydrolysis studies.

A Mechanistic Release Model

Finally, since the so-called "standard" dissolution release models used in the literature and herein (see Equations (4)–(8)) present several limitations related to the assumptions made for their implementation (for details please see Reference [33]), new, more sophisticated models were also tested for modeling the obtained results.

In general, there are two types of models to describe a physicochemical process such as the drug's dissolution. The first kind is the so-called empirical models. The physical content of these models is limited. Some of them are just equations used to describe appropriately a large amount of experimental data and some of them have a kind of qualitative information of the physical mechanism that is responsible for the process evolution. The second type of models are the so-called mechanistic models. These models include information of the underlying mechanism and in addition they can consider several mechanisms acting simultaneously. After considering exhaustively the whole toolbox of existing empirical models to describe the present data, an attempt to construct a mechanistic model of the present release process was made.

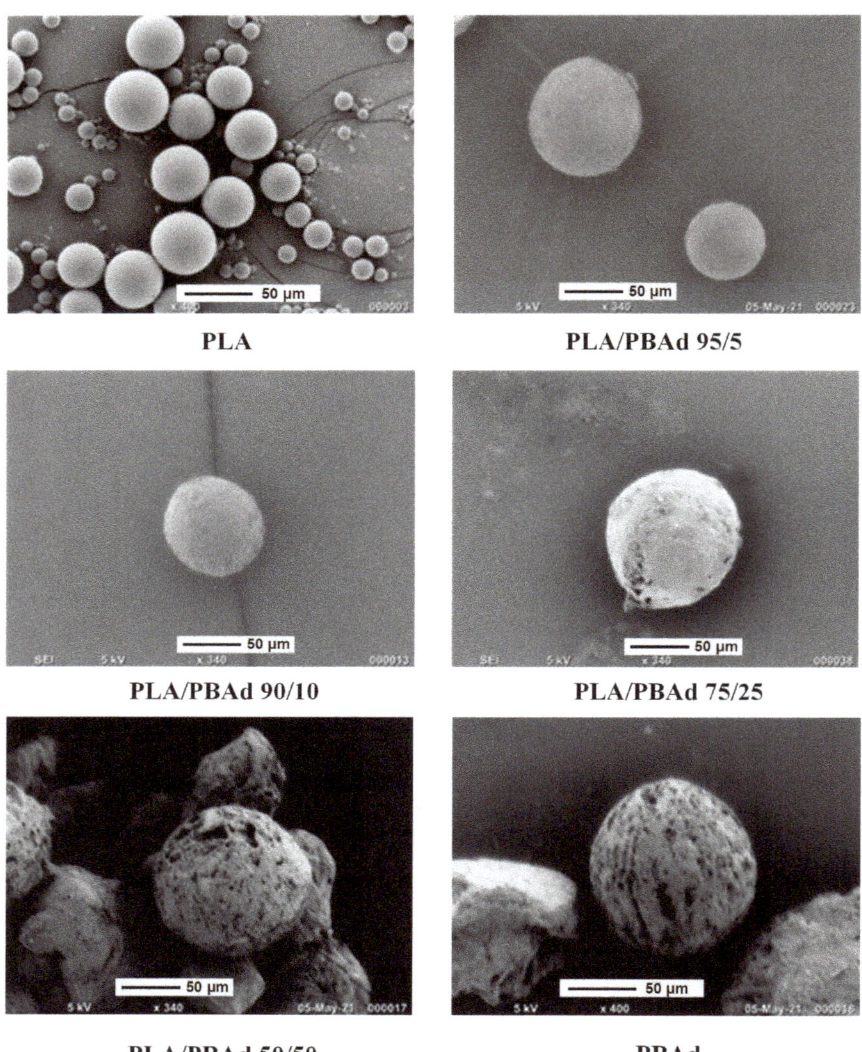

Figure 9. SEM images of the polyester erosion process after 30 days of dissolution.

Looking closely at the form of the release data, the performance of the "standard" release models and the data for hydrolysis evolution of the polymer matrices, the following scenario appears: There is an initial fast release phase that can be partially attributed to the presence of API probably in the form of a thin film layer located on, or near, the surface of the MPs. It is not clear if the mechanism of this layer release is diffusion or matrix erosion since both are equally probable. The second release phase (which is slower) is mostly controlled by Fickian diffusion, although a small erosion contribution is also there (verified by the SEM images presented in Figure 9). Assuming that a fraction of the drug in the polymer is free to move and its motion occurs through the diffusion mechanism and that the shape of the particles is approximately spherical (verified by SEM that is imaged in Figure 5), the transient partial differential equation of diffusion is probably the best model to describe the dissolution behavior of the API [49]. However, in this case a very simple exponential form, called the linear driving force approximation, can be also used to model the obtained results [50]. This same approximation was also used in the present study for modeling the dissolution kinetics of the API located on the surface layer, despite its unknown release mechanism. Finally, there is a fraction of drug immobilized in the polymer matrix. This fraction can be released only through matrix erosion (due to polyester hydrolysis). In the absence of any other information a zero order release dynamics model will be assessed for this fraction. It should be pointed out that for the limited extent of hydrolysis observed here, this approximation is quite realistic. By summarizing the above arguments, the released drug fraction evolution can be approximated by the following (uniformly valid in time) expression:

$$C_r = \varphi_1(1 - \exp(-k_1 t)) + \varphi_2(1 - \exp(-k_2 t)) + k_3 t \tag{9}$$

where C_r is the cumulative API released (%), φ_1 and k_1 are the percentage of drug in the excess layer and the corresponding kinetic constant respectively, φ_2 and k_2 are the percentage of mobile drug and the corresponding kinetic constant respectively and k_3 is the kinetic constant of the erosion process. It is noted that the linear superposition of diffusion- and erosion-induced release is allowed only because the erosion extent is small.

Figure 10 shows the comparison between the predicted and the experimentally derived points after fitting to Equation (9).

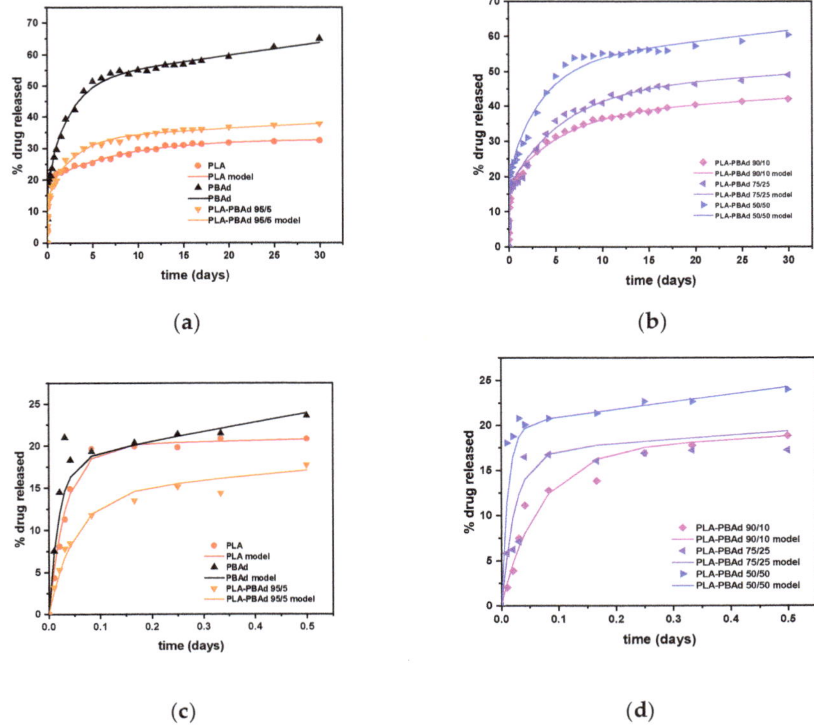

Figure 10. Comparison of experimental release data (symbols) to the mechanistic model-based Equation (9) (continuous lines). The presentation is made in two-time scales just for clarity: (**a**,**b**) 0–30 days and (**c**,**d**) 0–0.5 days.

The R^2 factor was larger than 0.99 in all cases except for composites 75/25 and 50/50 for which it was 0.98 and 0.99, respectively (probably due to a more complicated release scenario than the one described by Equation (9)). Nevertheless, and despite this small pitfall, the mechanistic model proposed herein is still more efficient compared to the "standard" empirical model tested previously, since it is able to model the dissolution profile of the API within the whole-time domain of the test (i.e., both release phases simultaneously). The values of the fitting parameter according to Equation (9) are presented in Table 6.

Table 6. Parameters derived by fitting Equation (9) to the experimental drug release data.

Material	φ_1	φ_2	k_1 (d^{-1})	k_2 (d^{-1})	k_3 (d^{-1})	$D \times 10^{17}$ (m^2/s)
PLA	20	13	30	0.12	0	7.4
PLA-PBAd 95/5	14.5	18.25	20	0.36	0.15	23.5
PLA-PBAd 90/10	17.5	20	15	0.18	0.15	6.5
PLA-PBAd 75/25	16.25	28	40	0.17	0.15	3
PLA-PBAd 50/50	20.5	31.7	80	0.28	0.3	1.9
PBAd	19	32	50	0.4	0.4	3.5

According to the obtained results, the percentage of drug in the excess layer was 20% and the corresponding parameters φ_1 and k_2 did not show any systematic correlation to the copolymer matrix composition. This is expected to be the case for a rather random procedure of accumulation of drug in the surface layer. The fraction of the mobile drug appears to increase consistently from 13% (for PLA) to 32% (for PBAd), while the corresponding kinetic constant k_2 appeared also to increase in the same order (with the exception of the 95/5 composite). Finally, the erosion constant k_3 increased as the content of PBAd increased, which is in agreement with the hydrolysis rates evaluation presented previously.

The diffusion coefficient, D, of the mobile drug presented also in Table 6 was calculated based on the following equation [50]:

$$D = k_2 r^2 / 15 \tag{10}$$

where r is the radius of the particles (presented in Table 6). Based on the obtained results, the range of values corresponding to D consisted of the drug diffusion within the polymer matrix and is in agreement with the results describing the second (and slower) release phase (depicted by k_2 constant). However, no such relation was proven in the case of the k_1 constant, since the characteristic length of diffusion in the first fast release phase is unknown. So, it can be said that the release of the drug's initial fraction (i.e., φ_1) may be either from the fast erosion of the very thin API layer located on the surface of the MPs or due to the fast initial API diffusion from this surface layer.

4. Conclusions

In the present study PLA/PBAd-based ARI-loaded LAI MPs were successfully prepared for the first time. Results regarding the highly tunable enzymatic hydrolysis profile and the low cytotoxicity of the new copolymers, amplified the previously made suggestions that these new copolymers can be considered as a quite promising candidate for the preparation of drug sustained release formulations. Evaluation in terms of morphological characteristics (via SEM), productivity (in terms of MPs' yield) and drug loading also showed extremely promising results. Physicochemical analysis of the prepared formulations revealed the amorphous API dispersion with increasing PBAd content, while no specific molecular interactions between the drug and the polyesters were recorded, based on FTIR spectroscopy. Lastly, in terms of the in vitro dissolution profile, results suggested that the newly synthesized PLA/PBAd block copolymers can successfully control the release rate and extent of the API's release from the prepared MPs, indicating that, probably, under in vivo conditions their use may lead to new formulations that will be able to maintain a continuous therapeutic level for an extended time period (>30 days), with reduced lag-time, as compared to the currently marketed ARI LAI product.

Author Contributions: Methodology, investigation, V.K.; A.S.; N.D.B.; investigation, formal analysis, writing, E.C.; methodology, formal analysis, A.Z.; release modeling, writing, M.K.; cytotoxicity studies, writing, E.G.; writing—original draft preparation, review and editing, P.B.; conceptualization, supervision, writing—review and editing, M.V. All authors have read and agreed to the published version of the manuscript.

Funding: The APC was funded by Pharmathen SA (Hellas).

Conflicts of Interest: The authors declare no conflict of interest.

References

1. Karp, J.M.; Langer, R. Development and therapeutic applications of advanced biomaterials. *Curr. Opin. Biotechnol.* **2007**, *18*, 454–459. [CrossRef] [PubMed]
2. Kenry; Liu, B. Recent advances in biodegradable conducting polymers and their biomedical applications. *Biomacromolecules* **2018**, *19*, 1783–1803. [CrossRef] [PubMed]
3. Song, R.; Murphy, M.; Li, C.; Ting, K.; Soo, C.; Zheng, Z. Current development of biodegradable polymeric materials for biomedical applications. *Drug Des. Dev. Ther.* **2018**, *12*, 3117–3145. [CrossRef] [PubMed]
4. Molavi, F.; Barzegar-Jalali, M.; Hamishehkar, H. Polyester based polymeric nano and microparticles for pharmaceutical purposes: A review on formulation approaches. *J. Control Release* **2020**, *320*, 265–282. [CrossRef] [PubMed]
5. Korzhikov, V.; Averianov, I.; Litvinchuk, E.; Tennikova, T.B. Polyester-based microparticles of different hydrophobicity: The patterns of lipophilic drug entrapment and release. *J. Microencapsul.* **2016**, *33*, 199–208. [CrossRef]
6. Sokolovskaya, E.; Rahmani, S.; Misra, A.C.; Bräse, S.; Lahann, J. Dual-stimuli-responsive microparticles. *ACS Appl. Mater. Interfaces* **2015**, *7*, 9744–9751. [CrossRef]
7. Chaudhary, K.; Patel, M.M.; Mehta, P.J. Long-acting injectables: Current perspectives and future promise. *Crit. Rev. Ther. Drug Carr. Syst.* **2019**, *36*, 137–181. [CrossRef]
8. Meyer, J.M. Converting oral to long-acting injectable antipsychotics: A guide for the perplexed. *CNS Spectr.* **2017**, *22*, 14–28. [CrossRef]
9. Pacchiarotti, I.; Tiihonen, J.; Kotzalidis, G.D.; Verdolini, N.; Murru, A.; Goikolea, J.M.; Valentí, M.; Aedo, A.; Vieta, E. Long-acting injectable antipsychotics (LAIs) for maintenance treatment of bipolar and schizoaffective disorders: A systematic review. *European Neuropsychopharmacol.* **2019**, *29*, 457–470. [CrossRef]
10. Park, K.; Skidmore, S.; Hadar, J.; Garner, J.; Park, H.; Otte, A.; Soh, B.K.; Yoon, G.; Yu, D.; Yun, Y.; et al. Injectable, long-acting PLGA formulations: Analyzing PLGA and understanding microparticle formation. *J. Control. Release* **2019**, *304*, 125–134. [CrossRef]
11. Wan, F.; Yang, M. Design of PLGA-based depot delivery systems for biopharmaceuticals prepared by spray drying. *Int. J. Pharm.* **2016**, *498*, 82–95. [CrossRef]
12. Lee, B.K.; Yun, Y.; Park, K. PLA micro- and nano-particles. *Adv. Drug Deliv. Rev.* **2016**, *107*, 176–191. [CrossRef]
13. Cun, D.; Foged, C.; Yang, M.; Frøkjær, S.; Nielsen, H.M. Preparation and characterization of poly(dl-lactide-co-glycolide) nanoparticles for siRNA delivery. *Int. J. Pharm.* **2010**, *390*, 70–75. [CrossRef]
14. Desai, K.-G.H.; Schwendeman, S.P. Active self-healing encapsulation of vaccine antigens in PLGA microspheres. *J. Control. Release* **2013**, *165*, 62–74. [CrossRef]
15. Geng, Y.; Yuan, W.; Wu, F.; Chen, J.; He, M.; Jin, T. Formulating erythropoietin-loaded sustained-release PLGA microspheres without protein aggregation. *J. Control. Release* **2008**, *130*, 259–265. [CrossRef] [PubMed]
16. Jensen, D.M.K.; Cun, D.; Maltesen, M.J.; Frokjaer, S.; Nielsen, H.M.; Foged, C. Spray drying of siRNA-containing PLGA nanoparticles intended for inhalation. *J. Control. Release* **2010**, *142*, 138–145. [CrossRef]
17. Park, W.; Na, K. Polyelectrolyte complex of chondroitin sulfate and peptide with lower pI value in poly(lactide-co-glycolide) microsphere for stability and controlled release. *Colloids Surf. B Biointerfaces* **2009**, *72*, 193–200. [CrossRef] [PubMed]
18. Zhang, Y.-m.; Yang, F.; Yang, Y.-q.; Song, F.-l.; Xu, A.-l. Recombinant interferon-alpha2b poly (lactic-co-glycolic acid) microspheres: Pharmacokinetics-pharmacodynamics study in rhesus monkeys following intramuscular administration. *Acta Pharmacol. Sin.* **2008**, *29*, 1370–1375. [CrossRef]
19. Ungaro, F.; d'Emmanuele di Villa Bianca, R.; Giovino, C.; Miro, A.; Sorrentino, R.; Quaglia, F.; La Rotonda, M.I. Insulin-loaded PLGA/cyclodextrin large porous particles with improved aerosolization properties: In vivo deposition and hypoglycaemic activity after delivery to rat lungs. *J. Control. Release* **2009**, *135*, 25–34. [CrossRef]
20. Kumar, R.; Palmieri, M.J. Points to consider when establishing drug product specifications for parenteral microspheres. *AAPS J.* **2010**, *12*, 27–32. [CrossRef] [PubMed]
21. Ye, M.; Kim, S.; Park, K. Issues in long-term protein delivery using biodegradable microparticles. *J. Control. Release* **2010**, *146*, 241–260. [CrossRef]

22. Nanaki, S.; Barmpalexis, P.; Papakonstantinou, Z.; Christodoulou, E.; Kostoglou, M.; Bikiaris, D.N. Preparation of new risperidone depot microspheres based on novel biocompatible poly (alkylene adipate) polyesters as long-acting injectable formulations. *J. Pharm. Sci.* **2018**, *107*, 2891–2901. [CrossRef] [PubMed]
23. Siafaka, P.I.; Barmbalexis, P.; Bikiaris, D.N. Novel electrospun nanofibrous matrices prepared from poly (lactic acid)/poly (butylene adipate) blends for controlled release formulations of an anti-rheumatoid agent. *Eur. J. Pharm. Sci.* **2016**, *88*, 12–25. [CrossRef] [PubMed]
24. Atkins, T.W. Fabrication of microcapsules using poly (ethylene adipate) and a blend of poly (ethylene adipate) with poly(hydroxybutyrate-hydroxyvalerate): Incorporation and release of bovine serum albumin. *Biomaterials* **1997**, *18*, 173–180. [CrossRef]
25. Brunner, C.T.; Baran, E.T.; Pinho, E.D.; Reis, R.L.; Neves, N.M. Performance of biodegradable microcapsules of poly (butylene succinate), poly (butylene succinate-co-adipate) and poly (butylene terephthalate-co-adipate) as drug encapsulation systems. *Colloids Surf. B Biointerfaces* **2011**, *84*, 498–507. [CrossRef]
26. Zorba, T.; Chrissafis, K.; Paraskevopoulos, K.M.; Bikiaris, D.N. Synthesis, characterization and thermal degradation mechanism of three poly (alkylene adipate)s: Comparative study. *Polym. Degrad. Stab.* **2007**, *92*, 222–230. [CrossRef]
27. Atkins, T.W. Biodegradation of poly (ethylene adipate) microcapsules in physiological media. *Biomaterials* **1998**, *19*, 61–67. [CrossRef]
28. Karava, V.; Siamidi, A.; Vlachou, M.; Christodoulou, E.; Zamboulis, A.; Bikiaris, D.N.; Kyritsis, A.; Klonos, P.A. Block copolymers based on poly (butylene adipate) and poly (l-lactic acid) for biomedical applications: Synthesis, structure and thermodynamical studies. *Soft Matter.* **2021**, *17*, 2439–2453. [CrossRef]
29. Burris, K.D.; Molski, T.F.; Xu, C.; Ryan, E.; Tottori, K.; Kikuchi, T.; Yocca, F.D.; Molinoff, P.B. Aripiprazole, a novel antipsychotic, is a high-affinity partial agonist at human dopamine D2 receptors. *J. Pharmacol. Exp. Ther.* **2002**, *302*, 381–389. [CrossRef] [PubMed]
30. Kane, J.M.; Sanchez, R.; Perry, P.P.; Jin, N.; Johnson, B.R.; Forbes, R.A.; McQuade, R.D.; Carson, W.H.; Fleischhacker, W.W. Aripiprazole intramuscular depot as maintenance treatment in patients with schizophrenia: A 52-week, multicenter, randomized, double-blind, placebo-controlled study. *J. Clin. Psychiatry* **2012**, *73*, 617–624. [CrossRef] [PubMed]
31. Motiwala, F.B.; Siscoe, K.S.; El-Mallakh, R.S. Review of depot aripiprazole for schizophrenia. *Patient Prefer. Adherence* **2013**, *7*, 1181–1187. [CrossRef] [PubMed]
32. Nanaki, S.; Barmpalexis, P.; Iatrou, A.; Christodoulou, E.; Kostoglou, M.; Bikiaris, D.N. Risperidone controlled release microspheres based on poly (lactic acid)-poly (propylene adipate) novel polymer blends appropriate for long acting injectable formulations. *Pharmaceutics* **2018**, *10*, 130. [CrossRef] [PubMed]
33. Costa, P.; Sousa Lobo, J.M. Modeling and comparison of dissolution profiles. *Eur. J. Pharm. Sci.* **2001**, *13*, 123–133. [CrossRef]
34. Jeong, H.; Rho, J.; Shin, J.-Y.; Lee, D.Y.; Hwang, T.; Kim, K.J. Mechanical properties and cytotoxicity of PLA/PCL films. *Biomed. Eng. Lett.* **2018**, *8*, 267–272. [CrossRef] [PubMed]
35. Balla, E.; Daniilidis, V.; Karlioti, G.; Kalamas, T.; Stefanidou, M.; Bikiaris, N.D.; Vlachopoulos, A.; Koumentakou, I.; Bikiaris, D.N. Poly (lactic Acid): A Versatile Biobased Polymer for the Future with Multifunctional Properties—from Monomer Synthesis, Polymerization Techniques and Molecular Weight Increase to PLA Applications. *Polymers* **2021**, *13*, 1822. [CrossRef]
36. Cordewener, F.W.; van Geffen, M.F.; Joziasse, C.A.P.; Schmitz, J.P.; Bos, R.R.M.; Rozema, F.R.; Pennings, A.J. Cytotoxicity of poly(96l/4d-lactide): The influence of degradation and sterilization. *Biomaterials* **2000**, *21*, 2433–2442. [CrossRef]
37. Nanaki, S.G.; Pantopoulos, K.; Bikiaris, D.N. Synthesis of biocompatible poly(ε-caprolactone)- block-poly (propylene adipate) copolymers appropriate for drug nanoencapsulation in the form of core-shell nanoparticles. *Int. J. Nanomed.* **2011**, *6*, 2981–2995. [CrossRef]
38. Karavelidis, V.; Giliopoulos, D.; Karavas, E.; Bikiaris, D. Nanoencapsulation of a water soluble drug in biocompatible polyesters. Effect of polyesters melting point and glass transition temperature on drug release behavior. *Eur. J. Pharm. Sci.* **2010**, *41*, 636–643. [CrossRef] [PubMed]
39. Díaz, E.; Puerto, I.; Sandonis, I.; Ribeiro, S.; Lanceros-Mendez, S. Hydrolytic degradation and cytotoxicity of poly (lactic-co-glycolic acid)/multiwalled carbon nanotubes for bone regeneration. *J. Appl. Polym. Sci.* **2020**, *137*, 48439. [CrossRef]
40. Elmowafy, E.M.; Tiboni, M.; Soliman, M.E. Biocompatibility, biodegradation and biomedical applications of poly (lactic acid)/poly (lactic-co-glycolic acid) micro and nanoparticles. *J. Pharm. Investig.* **2019**, *49*, 347–380. [CrossRef]
41. Abou-Zeid, D.-M.; Müller, R.-J.; Deckwer, W.-D. Biodegradation of Aliphatic Homopolyesters and Aliphatic–Aromatic Copolyesters by Anaerobic Microorganisms. *Biomacromolecules* **2004**, *5*, 1687–1697. [CrossRef]
42. Bikiaris, D.N.; Nianias, N.P.; Karagiannidou, E.G.; Docoslis, A. Effect of different nanoparticles on the properties and enzymatic hydrolysis mechanism of aliphatic polyesters. *Polym. Degrad. Stab.* **2012**, *97*, 2077–2089. [CrossRef]
43. Marten, E.; Müller, R.-J.; Deckwer, W.-D. Studies on the enzymatic hydrolysis of polyesters I. Low molecular mass model esters and aliphatic polyesters. *Polym. Degrad. Stab.* **2003**, *80*, 485–501. [CrossRef]
44. Qiu, Z.; Ikehara, T.; Nishi, T. Poly(hydroxybutyrate)/poly (butylene succinate) blends: Miscibility and nonisothermal crystallization. *Polymer* **2003**, *44*, 2503–2508. [CrossRef]
45. Tsuji, H.; Tezuka, Y. Alkaline and enzymatic degradation of L-lactide copolymers, 1. Amorphous-made films of L-lactide copolymers with D-lactide, glycolide, and epsilon-caprolactone. *Macromol. Biosci.* **2005**, *5*, 135–148. [CrossRef]
46. Jiménez, A.; Peltzer, M.; Ruseckaite, R. *Poly (Lactic Acid) Science and Technology: Processing, Properties, Additives and Applications*; Royal Society of Chemistry: London, UK, 2014.

47. Łaszcz, M.; Witkowska, A. Studies of phase transitions in the aripiprazole solid dosage form. *J. Pharm. Biomed. Anal.* **2016**, *117*, 298–303. [CrossRef] [PubMed]
48. Brittain, H.G. Chapter 1—aripiprazole: Polymorphs and solvatomorphs. In *Profiles of Drug Substances, Excipients and Related Methodology*; Brittain, H.G., Ed.; Academic Press: Cambridge, MA, USA, 2012; Volume 37, pp. 1–29.
49. Crank, J.; Crank, E.P.J. *The Mathematics of Diffusion*; Clarendon Press: Oxford, UK, 1979.
50. Tien, C. *Adsorption Calculations and Modeling*; Butterworth-Heinemann: Oxford, UK, 1994.

Article

NIR Light-Triggered Chemo-Phototherapy by ICG Functionalized MWNTs for Synergistic Tumor-Targeted Delivery

Lu Tang [1,2], Aining Zhang [1,2], Yijun Mei [1,2], Qiaqia Xiao [1,2], Xiangting Xu [1,2] and Wei Wang [1,2,*]

[1] State Key Laboratory of Natural Medicines, Department of Pharmaceutics, School of Pharmacy, China Pharmaceutical University, Nanjing 210009, China; lutang@stu.cpu.edu.cn (L.T.); zaining@stu.cpu.edu.cn (A.Z.); yjmei@stu.cpu.edu.cn (Y.M.); xiaoqiaqia@stu.cpu.edu.cn (Q.X.); xiangtingx_cpu@163.com (X.X.)

[2] NMPA Key Laboratory for Research and Evaluation of Pharmaceutical Preparations and Excipients, China Pharmaceutical University, Nanjing 210009, China

* Correspondence: wangcpu209@cpu.edu.cn

Abstract: The combinational application of photothermal therapy (PTT), chemotherapy, and nanotechnology is a booming therapeutic strategy for cancer treatment. Multi-walled carbon nanotube (MWNT) is often utilized as drug carrier in biomedical fields with excellent photothermal properties, and indocyanine green (ICG) is a near-infrared (NIR) dye approved by FDA. In addition, ICG is also a photothermal agent that can strongly absorb light energy for tumor ablation. Herein, we explored a synergistic strategy by connecting MWNT and a kind of ICG derivate ICG-NH$_2$ through hyaluronic acid (HA) that possesses CD44 receptor targeting ability, which largely enhanced the PTT effect of both MWNT and ICG-NH$_2$. To realize the synergistic therapeutic effect of chemotherapy and phototherapy, doxorubicin (DOX) was attached on the wall of MWNT via π–π interaction to obtain the final MWNT-HA-ICG/DOX nanocomplexes. Both in vitro and in vivo experiments verified the great therapeutic efficacy of MWNT-HA-ICG/DOX nanocomplexes, which was characterized by improved photothermal performance, strengthened cytotoxicity, and elevated tumor growth inhibition based on MCF-7 tumor models. Therefore, this synergistic strategy we report here might offer a new idea with promising application prospect for cancer treatment.

Keywords: multi-walled carbon nanotube; photothermal therapy; indocyanine green; synergistic strategy; cancer treatment; targeted drug delivery

Citation: Tang, L.; Zhang, A.; Mei, Y.; Xiao, Q.; Xu, X.; Wang, W. NIR Light-Triggered Chemo-Phototherapy by ICG Functionalized MWNTs for Synergistic Tumor-Targeted Delivery. *Pharmaceutics* **2021**, *13*, 2145. https://doi.org/10.3390/pharmaceutics13122145

Academic Editors: Francisco José Ostos, Pilar López-Cornejo and José Antonio Lebrón

Received: 22 November 2021
Accepted: 10 December 2021
Published: 13 December 2021

Publisher's Note: MDPI stays neutral with regard to jurisdictional claims in published maps and institutional affiliations.

1. Introduction

Cancer remains one of the deadly diseases that seriously threatens human health. Despite the encouraging progress of medical advancement, effective therapeutic methods against cancer are still insufficient. Conventional treatment modalities such as chemotherapy, radiotherapy, and surgery often undergo many drawbacks such as unavoidable side effects, severe pain, potential development of drug resistance, and inadequate effectiveness due to the instability and rapid clearance of drugs, which all cause unsatisfactory outcomes of anticancer therapy [1–3]. To overcome the limitations mentioned above, more and more attempts based on targeted drug delivery have been developed to improve cancer treatment efficacy. In this context, nanomaterial-mediated platforms have been widely explored in anticancer drug delivery, which serve as effective carriers of both therapeutic agents and diagnostic agents due to their distinctive properties and unique advantages, such as increased drug stability, reduced systemic toxicity, improved pharmacokinetics, elevated bioavailability, precise drug transportation capability, and controlled drug release ability [4–7].

From all the nanomaterials, carbon nanotubes (CNTs) have captured many researchers' attention due to their multiple application possibilities in cancer theranostics [8]. CNTs possess tiny tubular shapes composed of carbon atoms that are ordered to form a honeycomb nanostructure with many unique physicochemical characteristics [9]. Generally, CNTs can be sorted into either single-wall carbon nanotubes (SWNTs) or multi-wall carbon nanotubes (MWNTs) according to the sheet number of carbon atoms, both playing a significant role in anticancer therapy [10]. Due to their multifunctionality, CNTs are widely investigated in cancer treatment through various therapeutic modalities. For instance, the strong absorption of CNTs in near-infrared (NIR) regions make them ideal candidates in phototherapy. In addition, CNTs can transform the laser energy to acoustic signals and display excellent resonant Raman scattering and photoluminescence in NIR region, which are all advantageous to their application in cancer imaging [11]. Moreover, many studies have reported that CNTs can be taken up by various cell types due to their needle-like architecture, which enhances their deep tumor penetration to act as ideal drug delivery platforms in anticancer therapy [12]. Additionally, the ultra-high surface area of CNTs for drug loading also benefits their utilization in anticancer drug delivery. However, there are still some obstacles that limit their broad application in biomedical fields. For example, the unique nanostructure of CNTs promotes their hydrophobicity and causes cytotoxicity [13]. Therefore, functionalizing CNTs to increase their hydrophilicity as well as attenuate their inherent cytotoxicity is a key point to improve their biocompatibility for safe application in cancer therapy [14].

Phototherapy is an emerging therapeutic modality that takes advantage of laser energy to eliminate the target tumor due to its high selectivity [15,16]. Photothermal therapy (PTT) and photodynamic therapy (PDT) are two typical phototherapeutic approaches, and NIR light is always used as the light source for phototherapy due to its deep tissue penetration capability [17]. The conversion of absorbed photon energy into thermal energy to cause hyperthermia for tumor ablation is known as PTT, while the absorbance of specific light energy to produce cytotoxic reactive oxygen species (ROS) is termed as PDT, both of which play critical roles in cancer phototherapy [18]. Moreover, in recent decades, light-triggered therapies have been developed as safe treatment modalities to ablate numerous tumors with great effectiveness [18,19]. PTT, one of the representative phototherapy methods, has gained considerable attention in cancer treatment due to its various merits such as minimal trauma, easy implementation, and fewer side effects, which can effectively inhibit solid tumor growth through localized thermal destruction [20]. Indocyanine green (ICG) is a NIR dye approved by FDA for application in phototherapy due to its strong light absorbance in the NIR window [21,22]. However, some intrinsic limitations such as poor solubility, instability, concentration-dependent aggregation, and rapid clearance largely impede the effectiveness of ICG in phototherapy [23]. With the aim of overcoming these aforementioned drawbacks, construction of an appropriate drug carrier to load ICG is of great necessity [24].

In this work, a synergistic chemo-phototherapy was achieved by adopting MWNT to carry photothermal agent ICG-NH$_2$ and anticancer drug doxorubicin (DOX) (Scheme 1). Briefly, ICG-NH$_2$ was conjugated with hyaluronic acid (HA) by an amide bond to form the HA-ICG conjugate, which improved the water solubility of ICG-NH$_2$. Then, MWNT-HA-ICG was synthesized via an ester bond between the carboxyl groups on acidified MWNT and hydroxyl groups on HA, which significantly enhanced the photothermal performance compared to MWNT or ICG-NH$_2$ alone. In addition, the connection through HA could not only elevate the targetability of MWNT-HA-ICG due to its affinity with CD44 receptors that are over-expressed on the membrane of many tumor cells, but also endow the whole drug delivery system with good dispersity and biocompatibility [25–27]. Furthermore, to realize synergistic chemo-photothermal therapy, DOX was attached on the surface of MWNT by a non-covalent π–π bond to obtain the final MWNT-HA-ICG/DOX nanocomplexes (MWNT-HA-ICG/DOX). The targeted delivery of DOX through MWNT-HA-ICG greatly enhanced the therapeutic efficacy of DOX compared to free administration, while causing reduced side effects due to its non-specificity [28,29]. The novelty of this work is that we combined the drug carrier MWNT with good optical property and photothermal agent ICG-NH$_2$

in one platform to achieve synergistic photothermal therapeutic effect. Meanwhile, HA was used a targeting ligand in this nanosystem, which not only increased the targetability of the whole delivery platform, but also elevated the overall biocompability. To further improve the therapeutic efficacy, the classical anticancer drug DOX was employed in the constructed delivery system to take chemotherapy effect. Altogether, this integrated nanoplatform using combined chemo-phototherapy provides a promising strategy for precise cancer therapy.

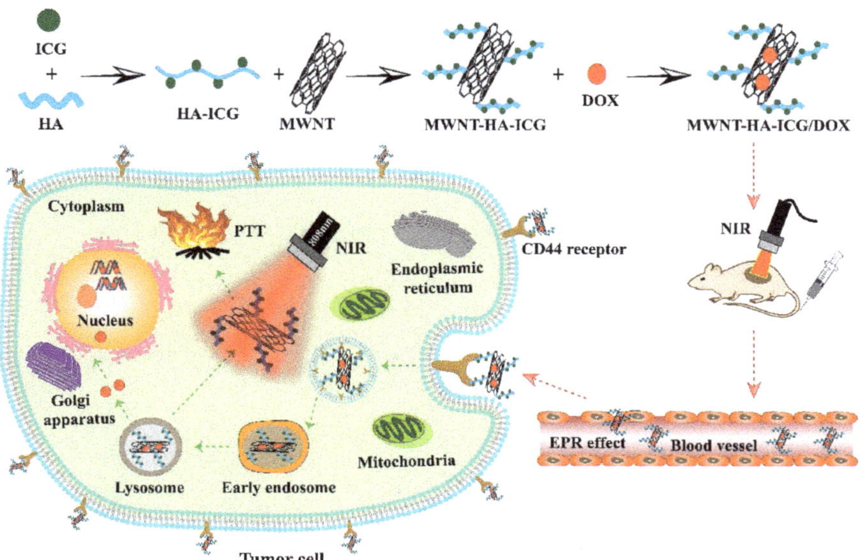

Scheme 1. Schematic illustration of the construction of MWNT-HA-ICG/DOX nanocomplexes for synergistic cancer chemo-phototherapy.

2. Materials and Methods

2.1. Materials, Cell Lines and Animals

MWNTs (diameter: 10–20 nm, length: 5–15 μm) were purchased from Shenzhen Nanotechnologies Port Co., Ltd. (Shenzhen, China). DOX·HCl was purchased from Nanjing Chemlin Chemical Industry Co., Ltd. (Nanjing, China). ICG-NH$_2$ was synthesized by Prof. Dun Wang in Shenyang Pharmaceutical University. Sodium hyaluronic acid (MW = 5 kDa), N-(3-dimethylaminopropyl)-N'-ethylcarbodiimide hydrochloride (EDC·HCl), and N-hydroxysuccinimide (NHS) were obtained from Aladdin Reagent Database Inc. (Shanghai, China). Concentrated sulfuric acid (98%), concentrated nitric acid (65%), 4-dimethylaminopyridine (DMAP), and N,N'-Carbonyldiimidazole (CDI) were purchased from Aladdin Reagent Database Inc. (Shanghai, China). 4′, 6-diamidino-2-phenylindole (DAPI) were obtained from Beyotime Institute of Biotechnology (Shanghai, China). All other reagents were of analytical grade and were commercially available.

Human breast cancer cell line MCF-7 were purchased from the Cell Bank of Shanghai Institute of Biochemistry and Cell Biology, Chinese Academy of Sciences (Shanghai, China) and cultured in DMEM medium (Hyclone) supplemented with 10% FBS (Hyclone), 100 U/mL penicillin, and 100 μg/mL streptomycin in a humidified atmosphere of 5% CO$_2$ at 37 °C.

Female BALB/c nude mice (4–6 weeks, 18–20 g) were purchased from Qinglongshan Animal Farm (Nanjing, China) and were given free access to food and water. All mice were cared for in compliance with the National Institute of Health (NIH) Guidelines for the Care and Use of Laboratory Animals and was approved by the Ethics Committee of China Pharmaceutical University (Ethics Code: 2021-12-002). The tumor xenograft models were established by injecting MCF-7 cell suspensions (1×10^6 cells) subcutaneously into the right flank region of nude mice. A caliper was used to measure tumor sizes, and the tumor volume was calculated as follows: (tumor length) $\times$ (tumor width)^2/2.

2.2. Preparation of MWNT-HA-ICG/DOX Nanocomplexes

2.2.1. Carboxylation of MWNTs

Pristine MWNTs (300 mg) were suspended in mixed acid (H_2SO_4/HNO_3, v/v = 3:1, 140 mL) by ultrasonication for 18 h. Then, the suspension was diluted with deionized water and filtered through a 0.22 μm micro-porous membrane, followed by repeated washing through deionized water until the pH of filtrate was neutral. The solid product on the membrane was then redispersed in deionized water and lyophilized to obtain the final MWNT-COOH, which was stored at room temperature for further use.

2.2.2. Synthesis of HA-ICG

Firstly, HA (60 mg) and NHS (43.79 mg) were dissolved in the mixture of water and DMF (1:3), after which EDC solution was added into the HA solution and stirred for 1 h to activate the carboxyl groups of HA. Then, ICG-NH$_2$ (44.4 mg) was dissolved in the mixture of water and DMF (1:3) and added to the HA solution dropwise, followed by stirring for 24 h at room temperature. The resultant solution was dialyzed (MWCO: 3500) using 50% DMF for 24 h to remove excess ICG-NH$_2$ and then dialyzed using deionized water for another 48 h to remove DMF. Finally, the solution was lyophilized to obtain the HA-ICG and stored at $-20\ ^\circ$C for further use.

2.2.3. Synthesis of MWNT-HA-ICG

MWNT-COOH (10 mg) and CDI (12 mg) were dissolved in 6 mL formamide and stirred for 1 h to activate the carboxyl groups of MWNT-COOH. Afterwards, the activated MWNT-COOH was dropped into HA-ICG solution in formamide with DMAP (15 mg) and stirred for 24 h under the protection of nitrogen. Next, the mixed solution was dialyzed (MWCO: 8000–14000) using 50% DMF for 24 h to remove the excess CDI and DMAP and then dialyzed using deionized water for another 48 h to remove the organic solvent. The final MWNT-HA-ICG was lyophilized and stored at $-20\ ^\circ$C for further use.

2.2.4. Synthesis of MWNT-HA-ICG/DOX

First, 10 mg MWNT-HA-ICG and 10 mg DOX·HCl were dissolved in deionized water, respectively. Then, DOX solution was added into MWNT-HA-ICG solution and stirred for 24 h at room temperature. Afterwards, the solution was centrifuged (12,000 rpm, 10 min), and the precipitate was washed with PBS (pH 7.4) to remove free DOX by repeated centrifugation. Finally, the resultant precipitate was MWNT-HA-ICG/DOX and was lyophilized and stored at $-20\ ^\circ$C for further use.

2.3. Characterization of MWNT-Based Formulations

Solid samples of raw MWNT and MWNT-COOH were prepared to obtain their Raman spectra using confocal micro-Raman spectroscopy (LabRam HR800, Paris, France). The successful synthesis of HA-ICG and the grafting percentage of ICG on HA were confirmed by ^{1}H-NMR spectroscopy (Avance™ 600, Bruker, Germany, 300 MHz), UV, and fluorescence spectroscopy. For the measurement of ^{1}H-NMR spectra, HA and HA-ICG were dissolved in D$_2$O, and ICG-NH$_2$ was dissolved in DMSO-d_6. Then, the measurement was carried out using 300 MHz under 20 $^\circ$C. UV spectra were collected by dissolving ICG-NH$_2$, DOX, MWNT-COOH, MWNT-HA-ICG, and MWNT-HA-ICG/DOX into deionized water, following by the scanning of their wavelength from 200–900 nm. The fluorescence spectra of samples were obtained by dissolving ICG-NH$_2$, DOX, and MWNT-HA-ICG/DOX into deionized water, which was then irradiated by 760 nm and 490 nm excitation light to obtain their fluorescence spectra. In addition, thermogravimetric analysis (TGA) was performed to analyze raw MWNT, MWNT-COOH, and MWNT-HA-ICG (heat from 30 °C to 700 °C, nitrogen, heating rate at 10 °C/min). The particle size and zeta potential of various MWNT-based formulations were measured by dynamic light scattering (DLS) using a Malvern Zetasizer (Nano ZS-90, Malvern Instruments Ltd., Malvern, UK). The morphology of raw MWNT, MWNT-COOH, and MWNT-HA-ICG/DOX was verified by transmission electron microscope system (TEM, Hitachi, Japan, 80 kV).

2.4. In Vitro Release of DOX from MWNT-HA-ICG/DOX

In vitro drug release profiles of DOX from MWNT-HA-ICG/DOX were performed at 37 °C in three different pH phosphate-buffered solutions (PBS). Briefly, 1 mg MWNT-HA-ICG/DOX was dispersed in 2 mL release medium with pH of 7.4, 6.5, or 5.5 in dialysis bag (MWCO: 3500). Then, the dialysis bag was placed in 40 mL corresponding pH release medium. The resultant dispersions were gently shaken at 37 °C in water bath at 100 rpm. At predetermined time intervals, 4 mL of release medium was taken out and the amount of released DOX was determined by a fluorospectrophotometer (HORIBA, Fluoromax-4, Palaiseau, France). In the meantime, 4 mL of same fresh release medium was replenished.

2.5. In Vitro Photothermal Effect of MWNT-Based Formulations

First, 1 mL of different concentrations free ICG-NH$_2$, MWNT-COOH, and MWNT-HA-ICG/DOX (containing 20 µg/mL ICG-NH$_2$ and 100 µg/mL MWNT-COOH) were dispersed in PBS and placed in 4 mL tubes. Then, different prepared solutions were irradiated by 808 nm laser (1 W/cm^2) for 5 min. The temperatures were recorded every 25 s, and thermographic maps of the dispersion in tubes were taken by thermo imager (FLIR, E64501, Goleta, CA, USA) at the time point of 5 min.

2.6. In Vitro Cytotoxicity Assay

The cytotoxicity of various prepared formulations was evaluated through MTT assay. Briefly, MCF-7 cells were seeded in 96-well plates at a density of 1×10^4 cells/well and incubated for 24 h. For treatment groups without laser irradiation, the culture medium was replaced with 200 µL medium containing ICG-NH$_2$, DOX, MWNT-HA, MWNT-HA/DOX, MWNT-HA-ICG, and MWNT-HA-ICG/DOX at different concentrations and incubated for 24 h. For treatment groups with laser irradiation, the culture medium was replaced with 200 µL medium containing ICG-NH$_2$, MWNT-HA/DOX, MWNT-HA-ICG, MWNT-HA-ICG/DOX, and MWNT-HA and incubated for 4 h, followed by laser irradiation (1 W/cm^2, 5 min); then, laser treatment groups were further incubated for 20 h. After the total 24 h incubation of treatment groups with/without laser irradiation, 20 µL MTT solution (5 mg/mL) was added to each well, and the cells were incubated for 4 h. Afterwards, the medium was removed, and 150 µL DMSO was added. The absorbance measured by microplate reader (Thermo, Multiskan FC) at 570 nm. The cell viability was calculated using the following equation: Cell viability (%) = $(A_{sample} - A_{blank})/(A_{control} - A_{blank}) \times 100\%$.

2.7. Cellular Uptake and Intracellular Trafficking

MCF-7 cells were seeded in 24-well plates and cultured until 80% cell confluence. The medium was discarded, and cells were washed twice with PBS before adding serum-free medium with ICG-NH$_2$, DOX, MWNT-HA-ICG, MWNT-HA/DOX, and MWNT-HA-ICG/DOX at a final concentration of 10 µg/mL ICG-NH$_2$ and/or 10 µg/mL DOX for cellular uptake for 1, 2, and 4 h. For competition assay, cells were pretreated with free HA (5 mg/mL) for 4 h before adding MWNT-HA-ICG/DOX. The cellular uptake of different formulations was determined by flow cytometry (BD FACS Calibur, San Jose, CA, USA) quantitatively according to the fluorescent property of ICG and DOX. The intracellular fluorescence of the above formulations was observed under a confocal laser scanning microscope (CLSM, Leica TCS SP5, Wetzlar, Germany) after staining cells with DAPI.

For intracellular trafficking study, cells were cultured, incubated with different formulations, washed, and fixed by 4% paraformaldehyde for 15 min. After discarding the paraformaldehyde, DAPI was added to stain the cell nucleus for 10 min. Finally, distribution of different formulations was observed by CLSM.

2.8. Cell Apoptosis Assessment

The in vitro antitumor efficacy was carried out by an apoptosis experiment. MCF-7 cells were seeded in 24-well plates at a density of 5×10^4 cells/well and incubated for 24 h. Cells were co-incubated with control (PBS), free DOX, and MWNT-HA-ICG/DOX (containing 10 μg/mL ICG and 10 μg/mL DOX) for 24 h to evaluate the chemotherapeutic efficacy of constructed nanocomplexes. To study the photothermal therapeutic efficacy, cells were co-incubated with control (PBS), ICG-NH$_2$, MWNT-HA-ICG, MWNT-HA/DOX, and MWNT-HA-ICG/DOX (containing 10 μg/mL ICG and 10 μg/mL DOX) for 4 h and then irradiated for 5 min (1 W/cm^2), followed by another 20 h incubation. Afterwards, cells were collected and washed with PBS. Next, cells were suspended in binding buffer and stained with Annexin V-FITC/PI apoptosis detection kit in dark. Finally, the apoptotic cells were detected by flow cytometry.

2.9. In Vivo Imaging Study

Mice were randomly divided into three groups and injected with ICG-NH$_2$, MWNT-HA-ICG, and MWNT-HA-ICG/DOX at a dose of 10 μg ICG-NH$_2$/mouse via tail vein, respectively. Due to the fluorescent characteristic of ICG, the in vivo distribution and targeting efficiency could be evaluated using an in vivo imaging system (FX PRO, Kodak, Rochester, NY, USA) at predetermined time points. After 12 h post-injection, mice were sacrificed to obtain the tumor and major organs for ex vivo imaging quantitative analysis using the same imaging system. The temperature changes of tumors during the irradiation (5 min, 1 W/cm^2) were monitored by a thermo-imager (FLIR, E64501) after 6 h post-injection at the time points of 0 min and 5 min.

2.10. In Vivo Antitumor Efficacy

The in vivo antitumor efficacy was evaluated using MCF-7 tumor bearing xenograft nude mice with average tumor volume around 100 mm^3. All the mice were weighed and randomly divided into nine groups. Of the nine groups, four groups were administrated with PBS, DOX, ICG-NH$_2$, and MWNT-HA/DOX intravenously without irradiation, while the rest groups were treated with PBS, ICG-NH$_2$, MWNT-HA-ICG, MWNT-HA/DOX, and MWNT-HA-ICG/DOX with laser irradiation. All the formulations were administrated at a dose of 10 μg ICG-NH$_2$/mouse and/or 0.5 mg/kg DOX per mouse every 2 days. For laser groups, tumors on mice were irradiated by 808 nm laser (1 W/cm^2) for 5 min at 6 h after injection with different formulations. In the meantime, the tumor volumes and body weights of mice were recorded every 2 days until day 14.

2.11. Statistical Analysis

The data were expressed as mean $\pm$ S.D. from triplicate experiments conducted parallelly, unless otherwise noted. Statistical analysis was carried out through one-way analysis of variance (ANOVA) test for comparison of multiple groups. Statistical significance was regarded as * $p < 0.05$, ** $p < 0.01$, or *** $p < 0.001$.

3. Results

3.1. Synthesis of MWNT-HA-ICG/DOX

Pristine MWNT was long and covered by impurities, which hindered the direct application of MWNT as a drug delivery vehicle. Therefore, mixed acid (H$_2$SO$_4$/HNO$_3$ v/v = 3:1) was used to remove the impurities, shorten the length of MWNT, and enable it to be modified with carboxyl group for further reaction. Moreover, MWNT is highly hydrophobic, which restricts its application in vivo. Thus, hydrophilic polymers are essential to be conjugated on the surface of MWNT to improve its dispersity. Hyaluronic acid (HA) is a negative polysaccharide with high molecular weight and consists of repeated D-glucuronic acid and *N*-acetyl-D-glucosamine disaccharide units [30]. HA has great water solubility and active hydroxyl and carboxyl groups, making it a prospective connection between MWNT-COOH and ICG-NH$_2$ [31]. As the synthesis route shown in Figure 1, ICG-NH$_2$ and HA were firstly conjugated through amide reaction, the resultant HA-ICG improved the water solubility of ICG-NH$_2$ and was used as the reactant in the second step. Next, hydroxyl group on HA-ICG and carboxyl group on MWNT-COOH were connected through ester bond. By virtue of excellent water solubility of HA, the dispersity of MWNT-HA-ICG complex was greatly improved. Moreover, both the photosensitizer ICG and the drug vector MWNT have excellent optical features, which enabled ICG and MWNT to realize synergistic PTT effect. To further improve the antitumor efficacy, chemotherapy drug DOX was conjugated onto the wall of MWNT via a π–π bond; thus, the final MWNT-HA-ICG/DOX nanocomplexes could be obtained.

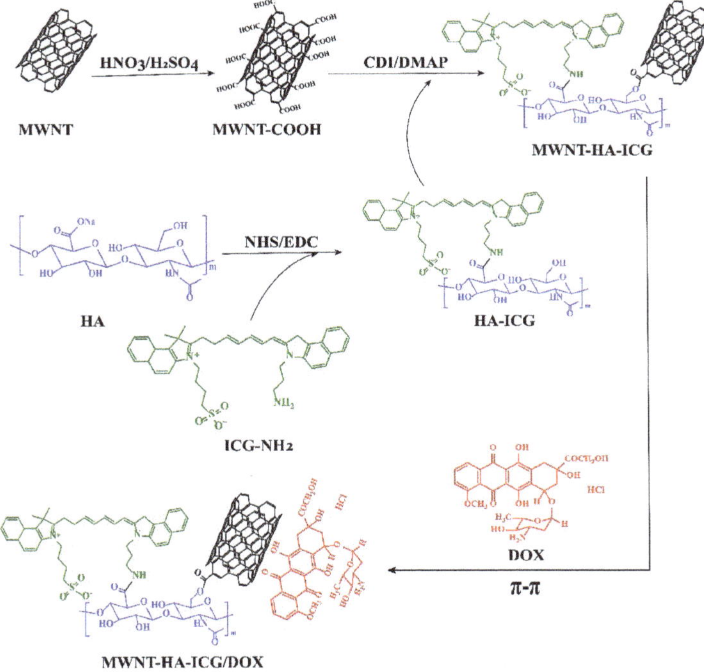

Figure 1. Synthesis route of MWNT-HA-ICG/DOX.

3.2. Characterization of MWNT-HA-ICG/DOX

The successful synthesis of MWNT-HA-ICG/DOX was characterized by Raman, ^{1}H-NMR, UV, fluorescence spectra, and thermogravimetric analysis (TGA). As shown in Figure 2A, both Raman spectra of MWNT and MWNT-COOH displayed two characteristic peaks. Tangential G band (~1590 cm^{-1}) represented the in-plane vibration of C–C bond, and the D band (~1350 cm^{-1}) reflected the disorder in the carbon system. I$_D$/I$_G$ value of raw MWNT was 1.13, and the value of MWNT-COOH increased to 1.31. The increased value of I$_D$/I$_G$ could reflect the elevated degree of deficiency of MWNT, probably due to the carboxylation of MWNT. The conjugation between carboxyl groups on HA and amine groups of ICG-NH$_2$ was verified by ^{1}H-NMR. The solvent for HA and HA-ICG was D$_2$O, and the solvent for ICG-NH$_2$ was DMSO-d_6. As shown in Figure 2B, the typical chemical shift of -CH- in HA was identified at 4.4~4.6 ppm, and the characteristic peaks of N-acetyl group and hydroxyl groups in HA were identified at 2.0 ppm and 4.0 ppm, respectively [32]. In addition, the evident peak of Ar-H due to proton resonance and the characteristic peak of -CH=CH- in ICG-NH$_2$ were identified at 7.5–8.3 ppm and 6.3–6.7 ppm, respectively. In the ^{1}H-NMR spectrum of HA-ICG, the characteristic peaks of HA and ICG-NH$_2$ were both identified, manifesting the successful conjugation of ICG-NH$_2$ onto HA. Moreover, as the UV spectrum was illustrated in Figure 2C, free ICG-NH$_2$, DOX, and MWNT-COOH had absorbance peaks at 780 nm, 496 nm, and 253 nm, respectively, while modified MWNT-HA-ICG showed absorbance at wavelengths of 785 nm and 253 nm, and the final MWNT-HA-ICG/DOX displayed absorbance peaks at 785 nm, 496 nm, and 253 nm, indicating that HA-ICG was connected onto the wall of MWNT-COOH, and MWNT-HA-ICG/DOX was successfully synthesized. Similarly, as the fluorescence spectra shown in Figure 2D, free ICG-NH$_2$ and DOX could be excited by wavelengths of 760 nm and 490 nm, which showed emission peaks at 803 nm and 590 nm, respectively. Meanwhile, MWNT-HA-ICG/DOX also had emission peaks at the wavelength of 792 nm and 590 nm, confirming the successful synthesis of MWNT-HA-ICG/DOX.

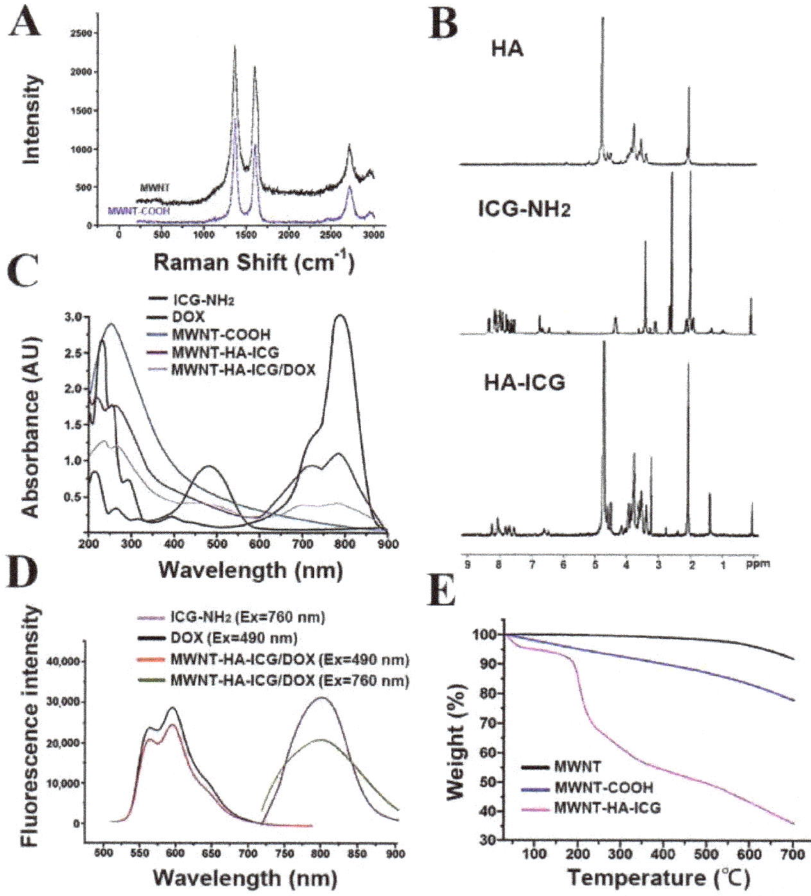

Figure 2. (**A**) Raman spectra of MWNT and MWNT-COOH. (**B**) ^{1}H-NMR spectra of HA, ICG-NH$_2$, and HA-ICG. (**C**) UV spectra of ICG-NH$_2$, DOX, MWNT-COOH, MWNT-HA-ICG, and MWNT-HA-ICG/DOX. (**D**) Fluorescence spectra of ICG-NH$_2$, DOX, and MWNT-HA-ICG/DOX. (**E**) TGA curves of MWNT, MWNT-COOH, and MWNT-HA-ICG.

TGA is a common way to determine the contents of grafting substance on MWNT by measuring the weight loss from 30 °C to 700 °C. As shown in Figure 2E, pristine MWNT was quite stable when the temperature was below 500 °C. In the range of 500 °C to 700 °C, there was around 9.84% weight loss of MWNT, mainly due to the impurities on pristine MWNT. With the temperature rising, the carboxyl groups on MWNT-COOH decomposed. Additionally, at a temperature of 700 °C, the weight loss of MWNT-COOH was approximately 22.37%, which implied that the content of carboxyl groups in MWNT-COOH was about 12.53%. For MWNT-HA-ICG, its weight loss accompanied with temperature increment also reflected the amount of HA-ICG grafted on MWNT-COOH, which was approximately 43.25%, demonstrating that HA-ICG could be effectively conjugated on the wall of MWNT-COOH.

3.3. Particle Size, Zeta Potential, and Morphology

The particle size, zeta potential, and polydispersity index (PDI) were measured to investigate the variation of MWNT after modified with different molecules. As is shown in Figure 3A, the particle size of MWNT-HA-ICG/DOX was around 190 nm and was larger than that of MWNT-COOH, which was due to the functionalization of ICG-conjugated HA and the non-covalent loading of DOX. In addition, the negative-charged nanocomplexes tended to be more stable in vivo due to their avoidance of the interaction with the negative-charged components in plasma [33,34]. The morphology of MWNT, MWNT-COOH, and MWNT-HA-ICG/DOX were observed by transmission electron microscopy (TEM). As shown in Figure 3B, raw MWNT was long, twined, and aggregated together with impurities on its surface. After acidification, MWNT-COOH was much shorter than raw MWNT, with a particle size of approximately 160 nm, which was in accordance with the result measured by dynamic light scattering (DLS). Moreover, the surface of MWNT-COOH was very smooth and no longer woven together, demonstrating that the mixed acid could efficiently shorten the raw MWNT and remove the impurities. After conjugation with HA-ICG and loading with DOX, the thickness of the MWNT was obviously increased and the length of functionalized MWNT was a little longer, demonstrating that HA-ICG and DOX were successfully attached to the surface of MWNT-COOH.

3.4. In Vitro Drug Release and Photothermal Effect of MWNT-Based Formulations

To investigate the in vitro drug release profile of DOX from MWNT-HA-ICG/DOX, three different pH phosphate buffers were used to simulate physiological pH, tumor pH, and lysosomal pH, respectively. As illustrated in Figure 3C, the cumulative released amount of DOX was pH-dependent. At pH 7.4, approximately 20% DOX was released from MWNT-HA-ICG/DOX after 48 h, implying that DOX would not release too much under normal physiological environment because the π–π stacking interaction between DOX and MWNT was stable. In contrast, approximately 30% and 50% of the loaded DOX was released from MWNT-HA-ICG/DOX at pH 6.5 and pH 5.5 after 48 h, respectively, suggesting the accelerated release profile of DOX in acidic tumor sites after internalization inside the tumor cells through receptor-mediated endocytosis. Altogether, the data above demonstrated that DOX could be released from MWNT-HA-ICG/DOX in a sustained manner in tumor sites and the constructed nanocomplexes were stable under normal physiological condition, which was attributed to attenuated π–π stacking interaction between DOX and MWNT due to the amino protonation of DOX under acidic PH conditions. To determine the in vitro PTT efficiency of ICG-NH$_2$, MWNT-COOH, and the synergistic effect of ICG-NH$_2$ and MWNT-COOH, a series of concentrations of the aforementioned three solutions were irradiated for 5 min (808 nm laser, 1.0 W/cm^2) and the corresponding temperatures were recorded at the designed time points. As shown in Figure 3D,E, the photothermal performance of both free ICG-NH$_2$ and MWNT-COOH exhibited a concentration-dependent and time-dependent profile. The temperature of MWNT-COOH was almost proportional to time, while the temperature of ICG-NH$_2$ was not that case. When the concentration of ICG-NH$_2$ was relatively low, the temperature increased significantly with the increment of its concentration and time. However, when the concentration of ICG-NH$_2$ was over 50 µg/mL, the temperature did not show an obvious increase with the increment of concentration, which was probably due to the aggregation of ICG-NH$_2$ that blunted its photothermal properties. As illustrated in Figure 3F, the temperature of MWNT-HA-ICG/DOX that contained 20 µg/mL ICG-NH$_2$ and 100 µg/mL MWNT-COOH was higher than that of free ICG-NH$_2$ and MWNT-COOH alone, demonstrating the synergistic photothermal effect of ICG-NH$_2$ and MWNT-COOH. Moreover, as shown in Figure 3G, the thermographic maps taken at 5 min after laser irradiation displayed an increased temperature in MWNT-HA-ICG/DOX group compared with the others, which was in accordance with the results above.

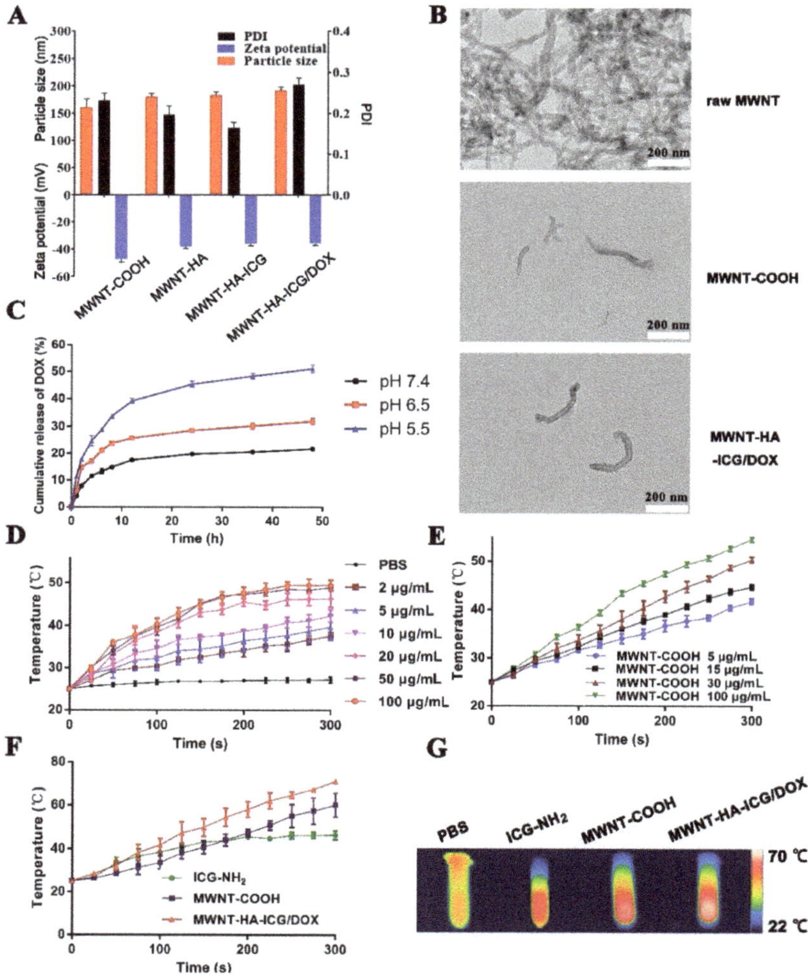

Figure 3. (**A**) Hydrodynamic diameter, zeta potential, and PDI of different MWNT-based formulations. (**B**) TEM images of raw MWNT, MWNT-COOH and MWNT-HA-ICG/DOX. (**C**) Cumulative release curves of DOX from MWNT-HA-ICG/DOX in three phosphate buffers with different pH. In vitro temperature curves of (**D**) free ICG-NH$_2$, (**E**) MWNT-COOH at various concentrations, and (**F**) free ICG-NH$_2$ (20 µg/mL), MWNT-COOH (100 µg/mL) and MWNT-HA-ICG/DOX after laser irradiation for 5 min (808 nm, 1.0 W/cm^2). (**G**) Infrared thermographic maps of free ICG-NH$_2$, MWNT-COOH, and MWNT-HA-ICG/DOX determined at 5 min after continuous laser irradiation.

3.5. In Vitro Cytotoxicity Studies

In vitro cell viability of free ICG-NH$_2$, free DOX, and different MWNT-based formulations with or without laser irradiation was evaluated using MCF-7 cells by MTT assay. Since the amount of ICG-NH$_2$ and DOX in constructed nanocomplexes was the same, and their concentrations were consistent with the concentration increment of MWNT-COOH, cell viability assays were carried out according to the concentrations of therapeutic agents (ICG-NH$_2$/DOX and MWNT-COOH). Figure 4A showed the cell viability of different treatment groups without laser irradiation, which illustrated that the cell viability of free ICG-NH$_2$ group was relatively high, implying the hypotoxicity of ICG-NH$_2$. Meanwhile, the cell viability after MWNT-HA and MWNT-HA-ICG treatment decreased as the concentration increased, indicating that the cytotoxicity of MWNT-HA and MWNT-HA-ICG was concentration-dependent. However, the overall cytotoxicity of this nanocarrier was relatively low, even at the highest concentration, which was attributed to the improved biocompatibility and reduced cytotoxicity of MWNT through HA modification. Furthermore, the cytotoxicity of MWNT-HA/DOX and MWNT-HA-ICG/DOX was significantly higher than that of MWNT-HA and MWNT-HA-ICG, which was mainly due to the cytotoxic effect of DOX. For the laser treatment groups shown in Figure 4B, MCF-7 cell viabilities upon laser irradiation (808 nm, 1.0 W/cm^2) for 5 min declined than those of unirradiated groups in Figure 4A. The higher the concentration of ICG-NH$_2$ and MWNT-COOH, the better the photothermal therapeutic effect. In addition, from all the treatment groups in Figure 4B, cells incubated with MWNT-HA-ICG/DOX plus laser irradiation showed the lowest viability, indicating the synergistic therapeutic effect through chemo-phototherapy of the constructed nanocomplexes.

3.6. In Vitro Cellular Uptake Studies

The evaluation of in vitro cellular uptake of ICG-NH$_2$, DOX, and different MWNT-based formulations was carried out using flow cytometry. ICG-NH$_2$ and DOX were used as fluorescent probes to quantitatively indicate the amount of MWNT-HA-ICG/DOX internalized by MCF-7 cells. In this experiment, ICG-NH$_2$, DOX, and their corresponding MWNT-based formulations were divided into two groups to detect their fluorescent intensity in two channels according to the emission wavelength. As shown in Figure 4C,E, the fluorescent intensity of free DOX was the lowest compared to other groups, while the fluorescent intensities of MWNT-HA/DOX and MWNT-HA-ICG/DOX were significantly stronger than free DOX. Moreover, after pretreatment with HA, the cellular uptake of MWNT-HA-ICG/DOX was significantly reduced, implying that HA-modified MWNT-based formulations could enter the cells through CD44 receptor-mediated endocytosis [32]. All the results above demonstrated that MWNT-based formulations were more likely to enter the cells due to the targeting property of HA. Moreover, the results of cellular uptake by ICG-NH$_2$ fluorescence shown in Figure 4D,F also confirmed the same cellular uptake mechanism of MWNT-HA-ICG/DOX, which were in accordance with the results obtained in Figure 4C,E.

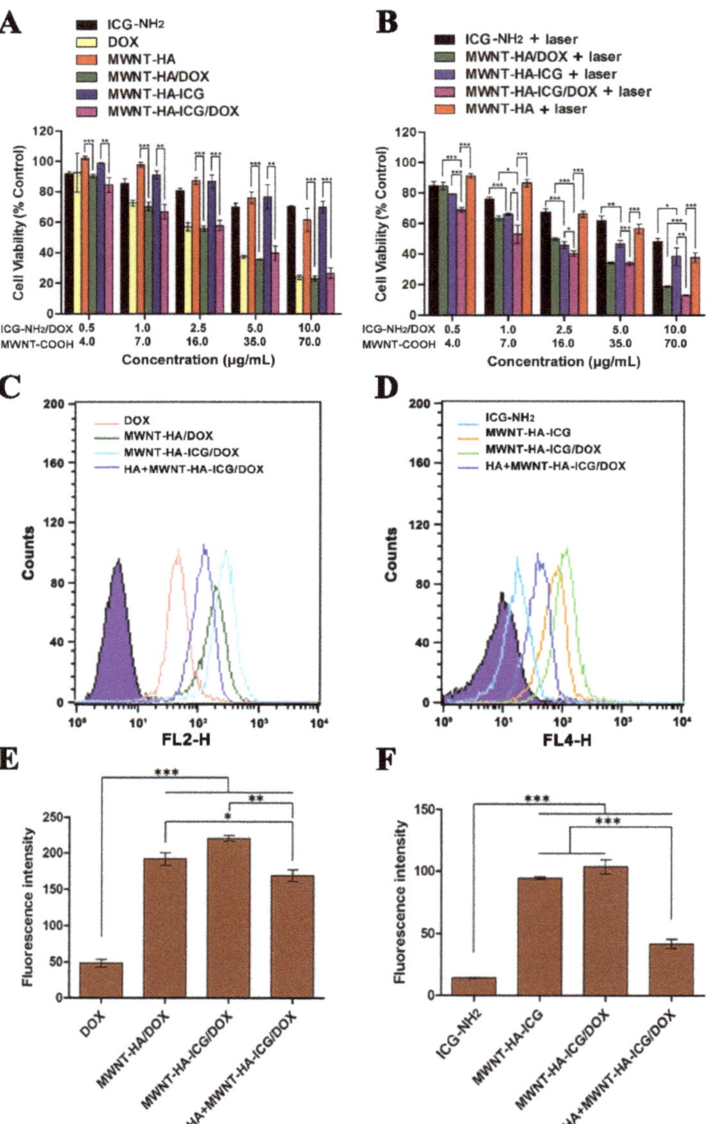

Figure 4. Cell viabilities of MCF-7 cells treated with different formulations at a series of concentrations of ICG-NH$_2$/DOX and MWNT-COOH for 48 h (**A**) without laser irradiation and (**B**) with laser irradiation (ICG-NH$_2$/DOX means ICG-NH$_2$ and/or DOX). Flow cytometric profiles and fluorescence intensities of different formulations of (**C,E**) 10 µg/mL DOX, (**D,F**) 10 µg/mL ICG-NH$_2$ at 4 h after treatment. Data were expressed as mean ± S.D. ($n = 3$). * $p < 0.05$, ** $p < 0.01$, and *** $p < 0.001$.

3.7. Intracellular Distribution and Cell Apoptosis Studies

CLSM was utilized to evaluate the intracellular distribution of MWNT-based formulations. DAPI was employed to dye the cell nucleus (blue). As shown in Figure 5A, the fluorescence of ICG-NH$_2$ (green) and DOX (red) appeared in cytoplasm and nucleus, respectively. The merged fluorescence appeared light purple due to the fluorescent overlay of DOX (red) and DAPI (blue). More fluorescence of ICG-NH$_2$ and DOX could be observed in MWNT-HA/DOX, MWNT-HA-ICG, and MWNT-HA-ICG/DOX groups, indicating that the functionalization of MWNT through HA enabled more ICG-NH$_2$ and DOX to enter into tumor cells due to the targetability of HA. After pretreatment with HA, the cellular uptake of MWNT-HA-ICG/DOX obviously decreased, which was verified by the decreased ICG-NH$_2$ and DOX fluorescence. In summary, the observation of CLSM and the results of flow cytometry were consistent. Cell apoptosis assay was carried out using flow cytometry to quantitatively evaluate the apoptosis-inducing efficacy of different formulations. The double staining of FITC labeled Annexin V and propidium iodide (PI) was used to discriminate live/early apoptotic cells and dead/late apoptotic cells [16]. As demonstrated in Figure 5B, only about 2% and 3% apoptotic cells were examined in the groups of control and control with laser irradiation, respectively, indicating that treatment using 808 nm laser was safe and would not cause damage to normal cells. In contrast, all the treatment groups using MWNT-based formulations exhibited obvious cell apoptosis. Notably, the combination therapy through MWNT-HA-ICG/DOX upon laser irradiation displayed the highest cell apoptotic rate of 98.18%, illustrating the synergistic chemo-photothermal therapeutic effect through the integration of DOX, ICG-NH$_2$, and MWNT.

3.8. In Vivo Targeting Study

A drug carrier that is able to target tumor sites plays a significant role in achieving elevated therapeutic outcome and reducing systemic side effects. Therefore, in vivo tumor targeting performance of constructed nanocomplexes were studied on MCF-7 xenograft tumor in nude mice using an in vivo fluorescence imaging system. As is shown in Figure 6A, mice were administered with ICG-NH$_2$, MWNT-HA-ICG, and MWNT-HA-ICG/DOX and then photographed at the time points of 1 h, 4 h, 6 h, and 12 h. Mice injected with ICG-NH$_2$ exhibited a primary liver accumulation with weak fluorescence at the tumor site. At 4 h post-injection, the fluorescence intensity of ICG-NH$_2$ in the liver became weak, partly due to the quenching aggregation and instability of free ICG [35,36]. In contrast, mice administered with MWNT-based formulations exhibited strong fluorescent signal at the tumor sites, even at 12 h after injection, implying that nanocomplexes conjugated with HA were able to produce selective accumulation and retention at tumor sites due to the specific affinity of HA to overexpressed CD44 receptors on MCF-7 cells [37]. As illustrated in Figure 6C, after 12 h injection, tumors and major organs were isolated for ex vivo imaging, which demonstrated that ICG-NH$_2$ was non-specifically accumulated in liver, spleen, and lung with no accumulation in tumor sites. In contrast, fluorescent images of mice treated with MWNT-HA-ICG and MWNT-HA-ICG/DOX showed favorable accumulation at tumor sites, which was attributed to the EPR effect and CD44-guided tumor targetability [38]. In the quantitative analysis shown in Figure 6D, the accumulation of nanocomplexes in tumor and major organs was consistent with that of ex vivo imaging above, which confirmed the excellent tumor targeting ability of the constructed nanocomplexes. In order to assess the PTT efficacy of the constructed formulations, the local temperatures of tumors upon laser irradiation (808 nm, 1.0 W/cm^2) after 6 h treatment with PBS, free ICG-NH$_2$, MWNT-HA, and MWNT-HA-ICG/DOX were recorded. As is illustrated in Figure 6B, the temperature of mice injected with ICG-NH$_2$ increased from 30.9 °C to 37.6 °C after 5 min laser irradiation, which did not show apparent temperature enhancement compared to that in the mice injected with PBS solution, indicating that free ICG-NH$_2$ could not specifically target tumor sites with a slow temperature increment. In contrast, due to the photothermal property of MWNT and targetability of HA, mice administered MWNT-HA showed an improved PTT performance over free ICG-NH$_2$. Notably, the temperature of mice injected with MWNT-HA-ICG/DOX could rise to 55.6 °C after irradiation for 5 min, which was high enough for tumor ablation [39]. All the data above confirmed that the constructed nanocomplexes could contribute to a synergistic PTT effect due to the combination of MWNT and ICG-NH$_2$ as well as the enhanced tumor targetability mediated by HA.

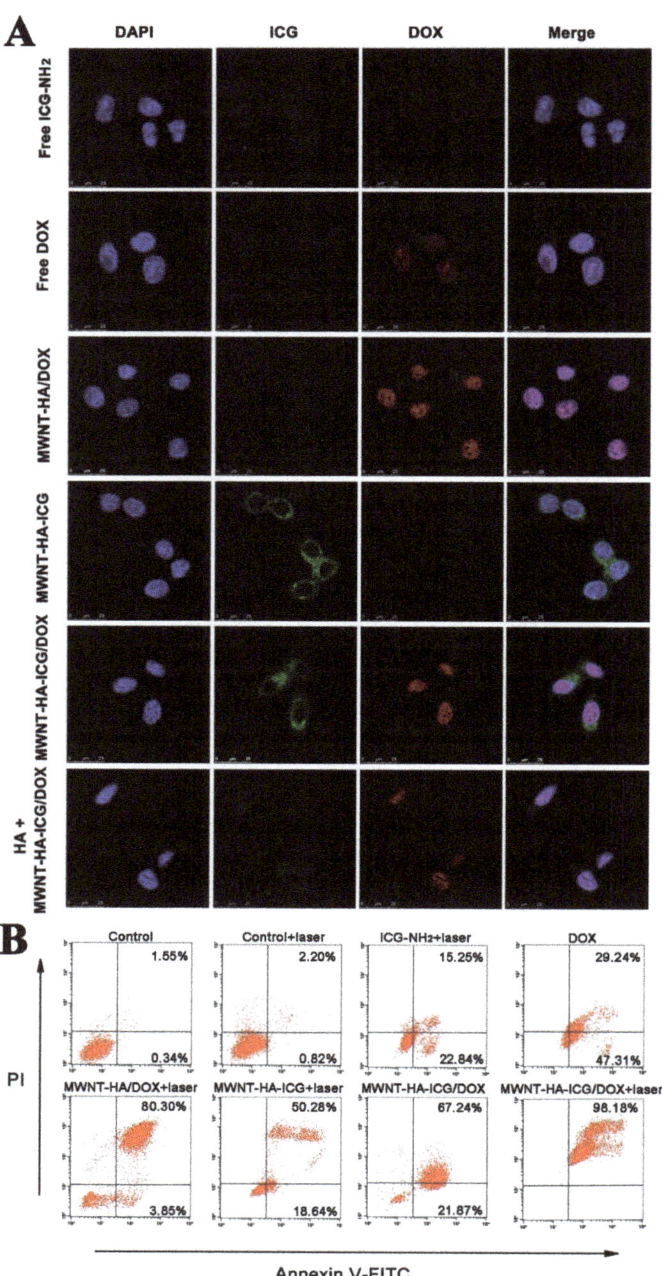

Figure 5. (**A**) Representative confocal images of intracellular trafficking of ICG-NH₂, DOX, MWNT-HA/DOX, MWNT-HA-ICG, MWNT-HA-ICG/DOX, and MWNT-HA-ICG/DOX pretreated with HA in MCF-7 cells. Nucleus stained by DAPI showed blue fluorescence, ICG showed green fluorescence, and DOX showed red fluorescence. (**B**) Cell apoptosis induced by different formulations with and without laser irradiation using flow cytometry analysis. Cells treated with PBS were used as control.

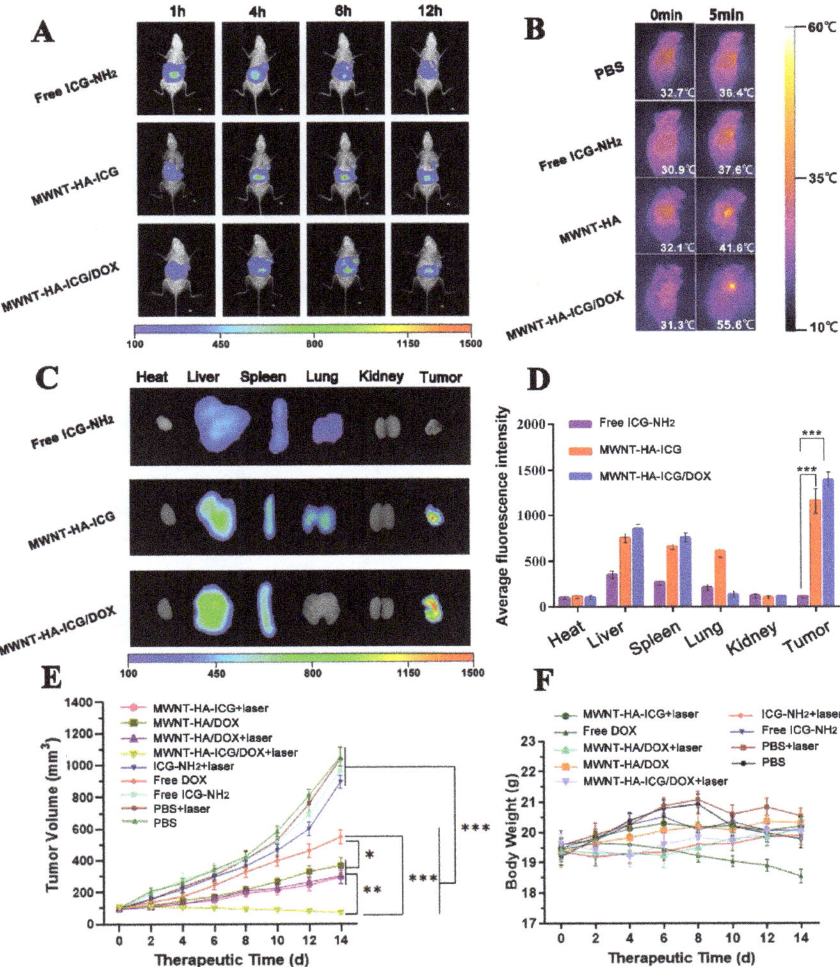

Figure 6. (**A**) Representative time-lapse in vivo imaging and biodistribution of MCF-7-tumor-bearing nude mice intravenously injected with ICG-NH$_2$, MWNT-HA-ICG, and MWNT-HA-ICG/DOX. (**B**) Infrared thermographic maps of mice upon laser irradiation (808 nm, 1.0 W/cm^2) at 6 h after IV injection with PBS, ICG-NH$_2$, MWNT-HA, and MWNT-HA-ICG/DOX. (**C**) Representative ex vivo NIR imaging of tumors and major organs excised from mice at 12 h post-injection. (**D**) Quantitative analysis of the fluorescence intensity in tumors and major organs at 12 h post-injection. (**E**) Tumor growth and (**F**) body weight curves of mice after IV administered with different formulations. Data were expressed as mean ± S.D. (*n* = 3). * *p* < 0.05, ** *p* < 0.01 and *** *p* < 0.001.

3.9. In Vivo Antitumor Efficacy Study

MCF-7 xenograft tumor model was established to investigate the synergistic antitumor efficacy of chemo-photothermal therapy during 14-day treatment. Mice were *i.v.* administered with different formulations with comparable amounts of therapeutics. As illustrated in Figure 6E, tumor volumes of mice treated with PBS, PBS plus laser, free ICG-NH$_2$, and ICG-NH$_2$ plus laser rapidly increased with the time, indicating that single treatment through PTT could not result in obvious inhibitory effect on tumor growth and ICG-NH$_2$ plus laser treatment could not cause considerable antitumor effect either, which was mainly due to the lack of targetability towards tumor sites [40]. Meanwhile, mice injected with free DOX, MWNT-HA-ICG plus laser, MWNT-HA/DOX, and MWNT-HA/DOX plus laser showed a significantly slight tumor volume increment. Moreover, mice treated with MWNT-based formulations exhibited remarkably slower tumor growth compared to the free DOX group, which indicated that formulations based on MWNT-HA exhibited better tumor targetability, thus causing enhanced antitumor efficacy. In contrast, mice injected with the final MWNT-HA-ICG/DOX plus laser showed a significant tumor volume reduction compared to the other groups above, confirming the synergistic therapeutic efficacy of the chemo (DOX)-photothermal (MWNT and ICG-NH$_2$) strategy in MWNT-HA-ICG/DOX nanocomplexes. Meanwhile, as Figure 6F illustrated, the body weight of mice treated with PBS and PBS plus laser increased in the first 8 days due to the growth of tumor and then decreased in the following days, which might be the result of the enlarged tumor that influenced the health of mice. In addition, the body weight of mice treated with DOX continuously reduced due to the cytotoxicity effect of DOX. Moreover, due to the non-specificity of DOX, treatment with free DOX also caused toxic effects on normal cells, which affected the life quality of mice and contributed to their weight reduction [41]. In contrast, the body weight of mice treated with MWNT-based formulations showed an increased tendency, indicating the reduced toxicity and increased therapeutic efficacy of constructed nanocomplexes that improved the life quality of mice, which was due to the elevated tumor-targeting ability and good biocompatibility of MWNT-based formulations after the modification of HA on MWNT.

4. Conclusions

In summary, we successfully fabricated a nano-based drug delivery system with synergistic PTT and chemotherapy effect for efficient tumor elimination. The integration of two photothermal agents ICG-NH$_2$ and MWNT through the connection of HA could not only elevate the photothermal performance compared to the single treatment modality, but also improve the targetability of the whole nanocomplexes due to the specific binding of HA and CD44 receptors overexpressed in tumor cells. Moreover, a simultaneous therapeutic effect could be achieved after involving DOX in this drug delivery system. In vitro results showed that MWNT-HA-ICG/DOX plus laser irradiation could lead to significant cytotoxic effects towards MCF-7 cells. In vivo experiments demonstrated that the combinational treatment strategy through PTT and chemotherapy could result in a favorable inhibitory effect on MCF-7 tumor growth. Therefore, MWNT-HA-ICG/DOX provided a promising therapeutic opportunity based on synergistic strategies for cancer treatment.

Author Contributions: Conceptualization, W.W.; Methodology, L.T., A.Z., Y.M., X.X. and Q.X.; Software and Data Analysis, L.T., X.X. and A.Z.; Original draft preparation, L.T., A.Z., Y.M., Q.X. and X.X.; Review and editing, L.T. and W.W.; Supervision, W.W.; Project administration, W.W.; Funding acquisition, W.W. All authors have read and agreed to the published version of the manuscript.

Funding: This work was funded by National Nature Science Foundation of China (Nos. 31872756 and 32071387), National Major Scientific and Technological Special Project for 'Significant New Drugs Development' (No. 2016ZX09101031), Six Talent Peaks Project in Jiangsu Province (JY-079).

Institutional Review Board Statement: The animal study in this work was conducted with the approval of the Ethics Committee of China Pharmaceutical University (Ethics Code: 2021-12-002).

Informed Consent Statement: Not applicable.

Data Availability Statement: The datasets used and/or analyzed during the current study are available from the corresponding author on reasonable request.

Conflicts of Interest: The authors declare no conflict of interest.

References

1. Chen, K.; Cao, X.; Li, M.; Su, Y.; Li, H.; Xie, M.; Zhang, Z.; Gao, H.; Xu, X.; Han, Y.; et al. A TRAIL-Delivered Lipoprotein-Bioinspired Nanovector Engineering Stem Cell-Based Platform for Inhibition of Lung Metastasis of Melanoma. *Theranostics* **2019**, *9*, 2984–2998. [CrossRef]
2. Su, Y.J.; Wang, T.T.; Su, Y.N.; Li, M.; Zhou, J.P.; Zhang, W.; Wang, W. A neutrophil membrane-functionalized black phosphorus riding inflammatory signal for positive feedback and multimode cancer therapy. *Mater. Horiz.* **2020**, *7*, 574–585. [CrossRef]
3. Li, M.; Su, Y.; Zhang, F.; Chen, K.; Xu, X.; Xu, L.; Zhou, J.; Wang, W. A dual-targeting reconstituted high density lipoprotein leveraging the synergy of sorafenib and antimiRNA21 for enhanced hepatocellular carcinoma therapy. *Acta Biomater.* **2018**, *75*, 413–426. [CrossRef]
4. Jain, R.K.; Stylianopoulos, T. Delivering nanomedicine to solid tumors. *Nat. Rev. Clin. Oncol.* **2010**, *7*, 653–664. [CrossRef]
5. Tang, L.; Li, J.; Zhao, Q.; Pan, T.; Zhong, H.; Wang, W. Advanced and Innovative Nano-Systems for Anticancer Targeted Drug Delivery. *Pharmaceutics* **2021**, *13*, 1151. [CrossRef]
6. Tang, L.; He, S.; Yin, Y.; Liu, H.; Hu, J.; Cheng, J.; Wang, W. Combination of Nanomaterials in Cell-Based Drug Delivery Systems for Cancer Treatment. *Pharmaceutics* **2021**, *13*, 1888. [CrossRef]
7. Ding, Y.; Wang, Y.; Zhou, J.; Gu, X.; Wang, W.; Liu, C.; Bao, X.; Wang, C.; Li, Y.; Zhang, Q. Direct cytosolic siRNA delivery by reconstituted high density lipoprotein for target-specific therapy of tumor angiogenesis. *Biomaterials* **2014**, *35*, 7214–7227. [CrossRef]
8. Sheikhpour, M.; Naghinejad, M.; Kasaeian, A.; Lohrasbi, A.; Shahraeini, S.S.; Zomorodbakhsh, S. The Applications of Carbon Nanotubes in the Diagnosis and Treatment of Lung Cancer: A Critical Review. *Int. J. Nanomed.* **2020**, *15*, 7063–7078. [CrossRef]
9. Liu, X.; Ying, Y.; Ping, J. Structure, synthesis, and sensing applications of single-walled carbon nanohorns. *Biosens. Bioelectron.* **2020**, *167*, 112495. [CrossRef]
10. van Zandwijk, N.; Frank, A.L. Awareness: Potential toxicities of carbon nanotubes. *Transl. Lung Cancer Res.* **2019**, *8* (Suppl. 4), S471–S472. [CrossRef]
11. Deshmukh, M.A.; Jeon, J.Y.; Ha, T.J. Carbon nanotubes: An effective platform for biomedical electronics. *Biosens. Bioelectron.* **2020**, *150*, 111919. [CrossRef]
12. Costa, P.M.; Bourgognon, M.; Wang, J.T.; Al-Jamal, K.T. Functionalised carbon nanotubes: From intracellular uptake and cell-related toxicity to systemic brain delivery. *J. Control. Release* **2016**, *241*, 200–219. [CrossRef] [PubMed]
13. Garriga, R.; Herrero-Continente, T.; Palos, M.; Cebolla, V.L.; Osada, J.; Munoz, E.; Rodriguez-Yoldi, M.J. Toxicity of Carbon Nanomaterials and Their Potential Application as Drug Delivery Systems: In Vitro Studies in Caco-2 and MCF-7 Cell Lines. *Nanomaterials* **2020**, *10*, 1617. [CrossRef]
14. Corletto, A.; Shapter, J.G. Nanoscale Patterning of Carbon Nanotubes: Techniques, Applications, and Future. *Adv. Sci. (Weinh)* **2020**, *8*, 2001778. [CrossRef]
15. Yang, X.; Wang, D.; Shi, Y.; Zou, J.; Zhao, Q.; Zhang, Q.; Huang, W.; Shao, J.; Xie, X.; Dong, X. Black Phosphorus Nanosheets Immobilizing Ce6 for Imaging-Guided Photothermal/Photodynamic Cancer Therapy. *ACS Appl. Mater. Interfaces* **2018**, *10*, 12431–12440. [CrossRef] [PubMed]
16. Su, Y.; Liu, Y.; Xu, X.; Zhou, J.; Xu, L.; Xu, X.; Wang, D.; Li, M.; Chen, K.; Wang, W. On-Demand Versatile Prodrug Nanomicelle for Tumor-Specific Bioimaging and Photothermal-Chemo Synergistic Cancer Therapy. *ACS Appl. Mater. Interfaces* **2018**, *10*, 38700–38714. [CrossRef]
17. Xu, X.; Lu, H.; Lee, R. Near Infrared Light Triggered Photo/Immuno-Therapy Toward Cancers. *Front. Bioeng. Biotechnol.* **2020**, *8*, 488. [CrossRef]
18. Li, X.; Lovell, J.F.; Yoon, J.; Chen, X. Clinical development and potential of photothermal and photodynamic therapies for cancer. *Nat. Rev. Clin. Oncol.* **2020**, *17*, 657–674. [CrossRef] [PubMed]
19. Chen, B.; Mei, L.; Fan, R.; Wang, Y.; Nie, C.; Tong, A.; Guo, G. Facile construction of targeted pH-responsive DNA-conjugated gold nanoparticles for synergistic photothermal-chemotherapy. *Chin. Chem. Lett.* **2021**, *32*, 1775–1779. [CrossRef]
20. Shang, T.; Yu, X.; Han, S.; Yang, B. Nanomedicine-based tumor photothermal therapy synergized immunotherapy. *Biomater. Sci.* **2020**, *8*, 5241–5259. [CrossRef]
21. Jia, Y.; Wang, X.; Hu, D.; Wang, P.; Liu, Q.; Zhang, X.; Jiang, J.; Liu, X.; Sheng, Z.; Liu, B.; et al. Phototheranostics: Active Targeting of Orthotopic Glioma Using Biomimetic Proteolipid Nanoparticles. *ACS Nano* **2019**, *13*, 386–398. [CrossRef]
22. Ding, K.; Zheng, C.; Sun, L.; Liu, X.; Yin, Y.; Wang, L. NIR light-induced tumor phototherapy using ICG delivery system based on platelet-membrane-camouflaged hollow bismuth selenide nanoparticles. *Chin. Chem. Lett.* **2020**, *31*, 1168–1172. [CrossRef]
23. Porcu, E.P.; Salis, A.; Gavini, E.; Rassu, G.; Maestri, M.; Giunchedi, P. Indocyanine green delivery systems for tumour detection and treatments. *Biotechnol. Adv.* **2016**, *34*, 768–789. [CrossRef] [PubMed]
24. Su, Y.; Hu, Y.; Wang, Y.; Xu, X.; Yuan, Y.; Li, Y.; Wang, Z.; Chen, K.; Zhang, F.; Ding, X.; et al. A precision-guided MWNT mediated reawakening the sunk synergy in RAS for anti-angiogenesis lung cancer therapy. *Biomaterials* **2017**, *139*, 75–90. [CrossRef] [PubMed]
25. Li, M.; Sun, J.; Zhang, W.; Zhao, Y.; Zhang, S.; Zhang, S. Drug delivery systems based on CD44-targeted glycosaminoglycans for cancer therapy. *Carbohydr. Polym.* **2021**, *251*, 117103. [CrossRef]
26. Luo, Z.; Dai, Y.; Gao, H. Development and application of hyaluronic acid in tumor targeting drug delivery. *Acta Pharm. Sin. B* **2019**, *9*, 1099–1112. [CrossRef]

27. Li, W.; Zhou, C.; Fu, Y.; Chen, T.; Liu, X.; Zhang, Z.; Gong, T. Targeted delivery of hyaluronic acid nanomicelles to hepatic stellate cells in hepatic fibrosis rats. *Acta Pharm. Sin. B* **2020**, *10*, 693–710. [CrossRef] [PubMed]
28. Shafei, A.; El-Bakly, W.; Sobhy, A.; Wagdy, O.; Reda, A.; Aboelenin, O.; Marzouk, A.; El Habak, K.; Mostafa, R.; Ali, M.A.; et al. A review on the efficacy and toxicity of different doxorubicin nanoparticles for targeted therapy in metastatic breast cancer. *Biomed. Pharmacother.* **2017**, *95*, 1209–1218. [CrossRef] [PubMed]
29. Mohammadi, M.; Arabi, L.; Alibolandi, M. Doxorubicin-loaded composite nanogels for cancer treatment. *J. Control. Release* **2020**, *328*, 171–191. [CrossRef]
30. Huang, G.; Chen, J. Preparation and applications of hyaluronic acid and its derivatives. *Int. J. Biol. Macromol.* **2019**, *125*, 478–484. [CrossRef]
31. Singh, P.; Wu, L.; Ren, X.; Zhang, W.; Tang, Y.; Chen, Y.; Carrier, A.; Zhang, X.; Zhang, J. Hyaluronic-acid-based beta-cyclodextrin grafted copolymers as biocompatible supramolecular hosts to enhance the water solubility of tocopherol. *Int. J. Pharm.* **2020**, *586*, 119542. [CrossRef] [PubMed]
32. Wang, W.; Li, M.; Zhang, Z.; Cui, C.; Zhou, J.; Yin, L.; Lv, H. Design, synthesis and evaluation of multi-functional tLyP-1-hyaluronic acid-paclitaxel conjugate endowed with broad anticancer scope. *Carbohydr. Polym.* **2017**, *156*, 97–107. [CrossRef]
33. Wang, Y.; Wang, C.; Ding, Y.; Li, J.; Li, M.; Liang, X.; Zhou, J.; Wang, W. Biomimetic HDL nanoparticle mediated tumor targeted delivery of indocyanine green for enhanced photodynamic therapy. *Colloids Surf. B Biointerfaces* **2016**, *148*, 533–540. [CrossRef]
34. Toth, D.; Hailegnaw, B.; Richheimer, F.; Castro, F.A.; Kienberger, F.; Scharber, M.C.; Wood, S.; Gramse, G. Nanoscale Charge Accumulation and Its Effect on Carrier Dynamics in Tri-cation Perovskite Structures. *ACS Appl. Mater. Interfaces* **2020**, *12*, 48057–48066. [CrossRef] [PubMed]
35. Li, D.H.; Smith, B.D. Deuterated Indocyanine Green (ICG) with Extended Aqueous Storage Shelf-Life: Chemical and Clinical Implications. *Chemistry* **2021**, *27*, 14535–14542. [CrossRef]
36. Della Pelle, G.; Delgado Lopez, A.; Salord Fiol, M.; Kostevsek, N. Cyanine Dyes for Photo-Thermal Therapy: A Comparison of Synthetic Liposomes and Natural Erythrocyte-Based Carriers. *Int. J. Mol. Sci* **2021**, *22*, 6914. [CrossRef] [PubMed]
37. Amorim, S.; Reis, C.A.; Reis, R.L.; Pires, R.A. Extracellular Matrix Mimics Using Hyaluronan-Based Biomaterials. *Trends Biotechnol.* **2021**, *39*, 90–104. [CrossRef] [PubMed]
38. Kalyane, D.; Raval, N.; Maheshwari, R.; Tambe, V.; Kalia, K.; Tekade, R.K. Employment of enhanced permeability and retention effect (EPR): Nanoparticle-based precision tools for targeting of therapeutic and diagnostic agent in cancer. *Mater. Sci. Eng. C Mater. Biol. Appl.* **2019**, *98*, 1252–1276. [CrossRef]
39. Park, T.; Lee, S.; Amatya, R.; Cheong, H.; Moon, C.; Kwak, H.D.; Min, K.A.; Shin, M.C. ICG-Loaded PEGylated BSA-Silver Nanoparticles for Effective Photothermal Cancer Therapy. *Int. J. Nanomed.* **2020**, *15*, 5459–5471. [CrossRef]
40. Zheng, M.; Yue, C.; Ma, Y.; Gong, P.; Zhao, P.; Zheng, C.; Sheng, Z.; Zhang, P.; Wang, Z.; Cai, L. Single-step assembly of DOX/ICG loaded lipid–polymer nanoparticles for highly effective chemo-photothermal combination therapy. *ACS Nano* **2013**, *7*, 2056–2067. [CrossRef]
41. Li, C.; Guan, H.; Li, Z.; Wang, F.; Wu, J.; Zhang, B. Study on different particle sizes of DOX-loaded mixed micelles for cancer therapy. *Colloids Surf. B Biointerfaces* **2020**, *196*, 111303. [CrossRef] [PubMed]

pharmaceutics

Article

Dendrimer-Coated Gold Nanoparticles for Efficient Folate-Targeted mRNA Delivery In Vitro

Londiwe Simphiwe Mbatha, Fiona Maiyo, Aliscia Daniels and Moganavelli Singh *

Nano-Gene and Drug Delivery Group, Discipline of Biochemistry, School of Life Sciences, University of KwaZulu-Natal, Private Bag X54001, Durban 4000, South Africa; londiwem3@dut.ac.za (L.S.M.); fcmaiyo@kabarak.ac.ke (F.M.); DanielsA@ukzn.ac.za (A.D.)
* Correspondence: singhm1@ukzn.ac.za; Tel.: +27-31-2607170

Abstract: Messenger RNA (mRNA) is not an attractive candidate for gene therapy due to its instability and has therefore received little attention. Recent studies show the advantage of mRNA over DNA, especially in cancer immunotherapy and vaccine development. This study aimed to formulate folic-acid-(FA)-modified, poly-amidoamine-generation-5 (PAMAM G5D)-grafted gold nanoparticles (AuNPs) and to evaluate their cytotoxicity and transgene expression using the luciferase reporter gene (F*Luc*-mRNA) in vitro. Nanocomplexes were spherical and of favorable size. Nanocomplexes at optimum nanoparticle:mRNA (w/w) binding ratios showed good protection of the bound mRNA against nucleases and were well tolerated in all cell lines. Transgene expression was significantly ($p < 0.0001$) higher with FA-targeted, dendrimer-grafted AuNPs (Au:G5D:FA) in FA receptors overexpressing MCF-7 and KB cells compared to the G5D and G5D:FA NPs, decreasing significantly ($p < 0.01$) in the presence of excess competing FA ligand, which confirmed nanocomplex uptake via receptor mediation. Overall, transgene expression of the Au:G5D and Au:G5D:FA nanocomplexes exceeded that of G5D and G5D:FA nanocomplexes, indicating the pivotal role played by the inclusion of the AuNP delivery system. The favorable properties imparted by the AuNPs potentiated an increased level of luciferase gene expression.

Keywords: gold nanoparticles; PAMAM dendrimers; folic acid; mRNA; gene expression

Citation: Mbatha, L.S.; Maiyo, F.; Daniels, A.; Singh, M. Dendrimer-Coated Gold Nanoparticles for Efficient Folate-Targeted mRNA Delivery In Vitro. *Pharmaceutics* **2021**, *13*, 900. https://doi.org/10.3390/pharmaceutics13060900

Academic Editors: Francisco José Ostos, José Antonio Lebrón and Pilar López-Cornejo

Received: 16 April 2021
Accepted: 20 May 2021
Published: 17 June 2021

Publisher's Note: MDPI stays neutral with regard to jurisdictional claims in published maps and institutional affiliations.

1. Introduction

Over the years, non-viral gene delivery modalities based on plasmid DNA (pDNA) were extensively evaluated in vitro as potential treatments of inherited diseases [1]. However, their failure to demonstrate potency at a clinical level due to their inability to bypass hurdles posed by the nuclear membrane of non-dividing cells and immunogenic responses of cytosine-phosphate-guanine (CpG) motifs contained by unmethylated DNA has aroused interest in using mRNA instead of pDNA [2,3].

Since an early study conducted by Malone and co-workers, the use of mRNA in gene therapy was limited by the belief that mRNA is too unstable when transfected into cells [4,5]. Recently, researchers have disproved that notion by successfully demonstrating the feasibility of mRNA-based modalities in several therapeutic applications, including tumor vaccination [6] and cancer immunotherapy. The feasibility and non-toxicity of naked mRNA and mRNA complexed with protamine were demonstrated in human patients via intradermal injections, resulting in promising immunological responses [7,8].

The recent interest in mRNA-based systems is due to the pharmaceutical safety advantages demonstrated over their pDNA-based counterparts. These include, first, the ease of mRNA to be formulated into an efficient therapeutic agent since it does not require the incorporation of promoters and terminators such as pDNA. It lacks immunogenic CpG motifs, which are present in pDNA, and does not need to traverse the nuclear membrane to elicit expression, as it is delivered into the cytoplasm, resulting in early and improved transfection activities [9]. Lastly, mRNA can transfect non-dividing cells, and its inability to integrate into the host genome eliminates insertional mutagenesis, making it safer to deliver than pDNA [10]. However, few studies have explored mRNA transfection over the years, and consequently, knowledge regarding mRNA transfection is limited, as the application of mRNA is still restricted by the need for improved delivery systems [11]. Thus far, the general consensus is that the use of cationic non-viral mRNA-based delivery systems, particularly cationic polymers (e.g., dendrimers), results in significantly improved transgene activity compared to that elicited by pDNA-based delivery systems [5], with some researchers recently using lipid nanoparticles (LNPs) for mRNA delivery [12]. Dendrimers, particularly PAMAM, are shown to elicit high transfection activities in vitro due to their hyperbranched, well-defined, three-dimensional (3D) structure with multiple surface functionalities, extreme buffering capacity, and ability to be protonated at physiological pH for efficient nucleic acid binding [13–16]. However, their high cytotoxic profiles induced by an excess of the surface amines (tertiary, 3° internal and peripheral primary, 1°) amines, especially at higher generations (>5), have tarnished their use in drug/gene delivery in the past [17]. Many reports, however, have shown that modifying these surface amines via pegylation, methylation, alkylation, acetylation, and conjugation with vitamins or amino acids significantly reduced this cytotoxicity [18–20].

Recently, several studies have exploited the remarkable properties of dendrimers as stabilizers of metal nanoparticles (NPs) [14–16,21–23]. This strategy combines the unique properties of metal NPs with those of cationic dendrimers to produce safe and highly efficient non-viral gene delivery systems. Gold nanoparticles are among the most commonly used metallic NPs to date due to their facile synthesis, biocompatibility, favorable surface-to-volume ratio, ability to be modified, and low cytotoxicity [24,25]. To the best of our knowledge, the transfection of mRNA using PAMAM dendrimer-grafted gold nanoparticles (AuNPs) was never explored. For that reason, this proof of principle study focused on designing FA-modified PAMAM-grafted AuNPs and PAMAM-grafted AuNPs and evaluating their cytotoxicity profiles and capacity to efficiently deliver F*Luc*-mRNA in vitro. FA-modified PAMAM nano-conjugates and PAMAM nano-conjugates were also evaluated for comparison purposes.

2. Materials and Methods

2.1. Materials

Starburst PAMAM dendrimer, generation five (PAMAM G5D), (*Mw* of 28,826, 128 surface amino groups), bicinchoninic acid (BCA), folic acid, 1-(3-dimethylaminopropyl)-3-ethyl carbodiimide (EDC), dimethylformamide (DMF), sodium dodecyl sulfate (SDS), dialysis tubing (*MWCO*, 12,000 Daltons), and ribonuclease A (RNase A) were supplied by Sigma-Aldrich (St. Louis, MO, USA). Ultra-pure DNA-grade agarose was acquired from Bio-Rad Laboratories (Richmond, VA, USA). Tris (hydroxymethyl)-aminomethane hydrochloride (Tris-HCl), 3-(4,5-dimethylthiazol-2-yl)-2,5-diphenyltetrazolium bromide (MTT), 2-[4-(2-hydroxyethyl)-1-piperazinyl] ethane sulphonic acid (HEPES), Dimethyl sulphoxide (DMSO), ethidium bromide (ETB), and gold (III) chloride trihydrate 99% (HAuCl4) were purchased from Merck (Darmstadt, Germany). F*Luc*-mRNA (5-methylcytidine and pseudouridine modified) was purchased from TriLink BioTechnologies, Inc (San Diego, CA, USA). Minimum essential medium (EMEM) containing Earle's salts and L-glutamine, penicillin (500 units/mL)/streptomycin (5000 µg/mL), and trypsin-versene were purchased from Lonza-BioWhittaker (Walkersville, MD, USA). Fetal bovine serum (FBS) was purchased from Highveld Biological (Lyndhurst, South Africa). Human embryonic kidney (HEK293), hepatocellular carcinoma (HepG2), breast adenocarcinoma (MCF-7), cervical adenocarcinoma cells (KB), and colorectal adenocarcinoma (Caco-2) cells were originally obtained from the American Type Culture Collection (Manassas, VA, USA).

2.2. Synthesis of Gold Nanoparticles (AuNPs)

An adaptation of the Turkevich method was followed to synthesize the AuNPs [26]. Briefly, HAuCl4 (0.03 M, 0.1 mL) was dissolved in 25 mL of 18 MOhm water, stirred vigorously, and heated for 15 min until boiling. This was followed by the slow addition of 1 mL of 1% trisodium citrate ($Na_3C_6H_5O_7$) with stirring until a red color change was produced. The mixture was then removed from the heat and stirred until it cooled to room temperature.

2.3. Modification of PAMAM G5D with Folic Acid (FA)

PAMAM G5D (dried under nitrogen) was dissolved in 18 MOhm water and conjugated to folic acid (FA) via carbodiimide chemistry as described previously by the authors [15,16]. FA (2.8 µmol in 3 mL of DMF) was reacted with 38.2 µmol EDC for 45 min with constant stirring under nitrogen. The activated FA was then added slowly with stirring into the dendrimer (3 µmol, 100 µL) solution, and the pH maintained at 9.5. The solution was stirred for 3 days under nitrogen, followed by the removal of unreacted by-products by dialysis (*MWCO* 12 000 Da) against 18 MOhm water for 24 h.

2.4. Formulation of Dendrimer-Coated AuNPs (Au:G5D NPs, and Folic-Acid-Targeted, Dendrimer-Coated AuNPs (Au:G5D:FA NPs)

The G5D and previously synthesized G5D:FA (Section 2.3) were conjugated to the citrate-reduced AuNP solution as previously described by the authors [15,16] to produce Au:G5D and Au:G5D:FA NPs in a 25:1 gold/dendrimer molar ratio. NPs were dialyzed as in Section 2.3.

2.5. Ultra-Violet (UV) and Proton Nuclear Magnetic Resonance (^1H NMR) Spectroscopy

Successful functionalization of the G5D and AuNPs was monitored by UV-vis spectroscopy (UV-1650PC, Shimadzu, Japan) using a wavelength range of 200–800 nm. Further confirmation of NP synthesis was achieved using ^1H NMR spectroscopy (Bruker DRX 400) with deuterated (D_2O) water as a solvent.

2.6. Transmission Electron Microscopy (TEM) and Nanoparticle Tracking Analysis (NTA)

The ultrastructural morphology of the NPs and their mRNA nanocomplexes at optimum binding ratios (w/w) were determined by cryo-TEM, using a Jeol JEM-1010 transmission electron microscope containing a Soft Imaging System (SIS) fitted with a MegaView III digital camera with iTEM UIP software, operating at an acceleration voltage of 200 kV (Tokyo, Japan). The z-average hydrodynamic diameters and zeta (ζ) potentials were determined by nanoparticle tracking analysis (NTA, NanoSight NS500; Malvern Instruments, Worcestershire, UK) at 25 °C. NPs (1 mL) were diluted 1:100 in 18 MOhm and sonicated before analysis. Although the characterization of these NPs was reported previously by the authors [15,16], the mRNA-based nanocomplexes are reported here for the first time.

2.7. Nanocomplex Preparation and Binding Studies

Nanocomplexes for mRNA binding, cell viability, and transfection studies contained a constant amount of F*Luc*-mRNA (0.05 µg) together with increasing amounts of G5D, Au:G5D, G5D:FA, and Au:G5D:FA NPs. Nanocomplexes were briefly mixed and incubated at room temperature for 60 min.

2.7.1. Band Shift Assay

Band shift assays [27] were utilized to determine the binding of mRNA to the NPs. Nanocomplexes prepared as in Section 2.7 were subjected to electrophoresis on 1% (w/v) agarose gels containing ethidium bromide (ETB) (1 µg/mL) in a Bio-Rad mini-sub electrophoresis apparatus containing 1× electrophoresis buffer (36 mM Tris-HCl, 30 mM, sodium phosphate (NaH_2PO_4), 10 mM ethylenediamine tetra-acetic acid (EDTA), pH 7.5), for 45 min at 50 Volts. Gels were viewed and images captured using a Vacutec Syngene G: Box BioImaging system (Syngene, Cambridge, UK).

2.7.2. Ethidium Bromide Displacement Assay

The compaction of the nanocomplexes was assessed using a dye displacement assay [27]. ETB solution (24 µL, 100 µg/mL) and HBS (100 µL) were initially added to a 96-well FluorTrac flat-bottom black plate, and fluorescence read in a Glomax®-Multi + detection system (Promega, Sunnyvale, CA, USA) at an excitation wavelength of 520 nm and an emission wavelength of 600 nm. This measurement was set as 0% relative fluorescence (RF). The 100% RF was obtained after the addition of 0.05 µg F*Luc*-mRNA. Thereafter, 1 µL aliquots of the respective NPs were added, and fluorescence was measured until a plateau in fluorescence was achieved.

2.7.3. RNase A Protection Assay

The stability of the nanocomplexes and the protection afforded to the mRNA in the presence of degrading enzymes were evaluated by an RNase protection assay adapted from [27]. NP:mRNA nanocomplexes prepared at the sub-optimum, optimum, and supra-optimum ratios (obtained from Section 2.7.1) were exposed to 10% RNase A for 2 h at 37 °C. This was followed by the addition of 10 mM EDTA to halt the reaction and 0.5% SDS to release the nucleic acid from the nanocomplex. Samples were subsequently incubated at 55 °C for 20 min, followed by electrophoresis as described previously (Section 2.7.1).

2.8. Cell Culture-Based Assays

All cells were maintained and propagated at 37 °C and 5% CO_2 in 25 cm^2 flasks containing sterile EMEM, FBS (10%, *v/v*), penicillin G (100 U/mL), and streptomycin sulfate (100 µg/mL). The cells were split upon confluency into desired ratios when necessary and the medium changed routinely.

2.8.1. MTT Cell Viability Assay

The MTT assay was used to determine the viability of the cells after treatment with the respective nanocomplexes as described previously [28,29]. All cells were seeded into 48-well plates at densities of 2.5×10^5 cells/well, and incubated for 24 h at 37 °C. Thereafter, nanocomplexes at selected ratios were added in triplicate, and cells were incubated for 48 h at 37 °C. Cells containing no nanocomplexes were used as the positive control (100% cell viability). Following the 48 h incubation, a fresh medium containing the 10% MTT reagent (5 mg/mL in PBS) was added, followed by a 4 h incubation at 37 °C. The medium MTT mixture was then aspirated, cells washed with PBS (2×0.3 mL), and 0.3 mL of DMSO was added to solubilize the resulting formazan crystals. Absorbance was then measured at 570 nm in a Mindray MR-96A microplate reader (Vacutec, Hamburg, Germany) using DMSO as the blank. The percentage cell viability was calculated against the positive control (100%).

2.8.2. Apoptosis Assay

To determine if apoptosis was instrumental in the cell death recorded, a fluorescent dual-stain apoptosis assay was conducted as previously described [30]. Cells (2.9×10^5 cells/well) were plated into 12-well plates and incubated for 24 h at 37 °C. Following the addition of nanocomplexes at optimum binding ratios, the cells were incubated for 48 h at 37 °C. Thereafter, cells were washed with PBS, and 10 µL of AO/ETB (AO/ETB, 1:1 *v/v*, 100 µg/mL) was added. Cells were viewed for structural and morphological changes under an Olympus fluorescent microscope ($\times 200$ magnification), fitted with a CC12 fluorescent camera (Olympus Co., Tokyo, Japan). Apoptosis was quantified by calculating the apoptotic index (AI) as below:

Apoptotic Index = Number of apoptotic cells/Total number of cells

2.8.3. Transfection and Competition Assays

The transfection and competition assays were conducted as previously described [15,16,28,29]. Cells with densities of 2.5×10^5 cells/well were seeded into

48-well plates and incubated for 24 h at 37 °C. The nanocomplexes (ratios as used for the MTT assay) were then added, and the cells were incubated for 48 h at 37 °C. Thereafter, the cells were washed with PBS (2 × 0.5 mL) and lysed with 80 µL/well cell lysis buffer (Promega) for 15 min with shaking at 30 rpm in a Scientific STR 6 platform rocker (Stuart Scientific, Staffordshire, UK). Cell suspensions were then centrifuged at 12,000× *g* for 1 min. The cell-free extract (20 µL) was added to 100 µL luciferase assay reagent, mixed, and luminescence recorded in relative light units (RLU) in a Glomax®-Multi+Detection System (Promega Biosystem, Sunnyvale, CA, USA). The standard BCA assay was used to determine the protein concentrations of the cell-free extracts as described previously [29,31]. The luminescence recorded was normalized against the protein concentration, and luciferase activity was expressed as RLU/mg protein.

For the competition assay, cells were seeded and treated as for the normal transfection, but FA (250 µg) was incubated with folate receptor-positive cells (MCF-7 and KB cells) for 20 min at 37 °C before the addition of the targeted nanocomplexes. Luciferase activity was then determined as described above.

2.9. Statistical Analysis

Cell viability and transfection studies were performed in triplicate and results expressed as means ± standard deviation (SD). The experimental data was analyzed by a two-way ANOVA and *t*-test using GraphPad Prism 6.0 and statistically significant values are indicated by * $p < 0.05$, ** $p < 0.01$, *** $p < 0.001$, and **** $p < 0.0001$, # $p > 0.05$.

3. Results

3.1. UV-Visible and ^1H NMR Spectroscopy

The attachment of G5D and FA on the AuNPs was first confirmed by UV-vis spectroscopy (Figure 1).

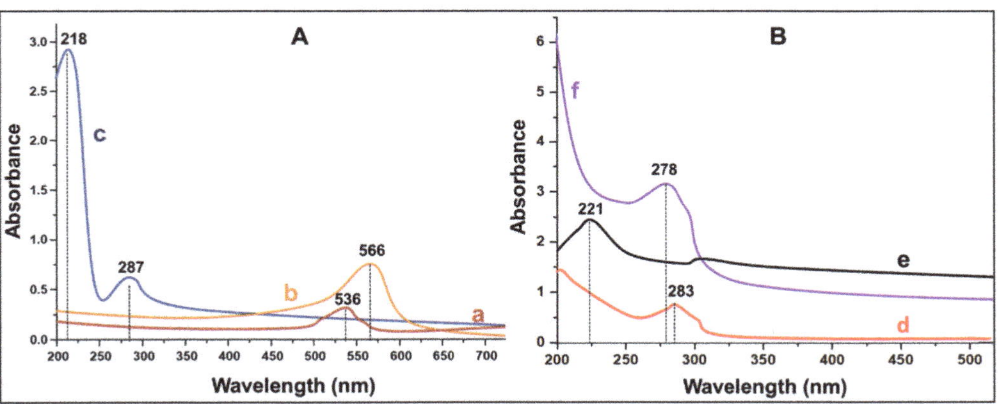

Figure 1. (**A**) UV spectra of (**a**) AuNPs, (**b**) Au:G5D NPs, (**c**) Au:G5D:FA NPs; and (**B**) UV-spectra of (**d**) G5D, (**e**) G5D:FA NPs, and (**f**) FA.

The absorption band at 536 nm confirmed the formation of AuNPs, since the known absorption band of AuNPs range between 520 and 550 nm [32]. The band shift from 536 nm to 566 nm confirmed the attachment of G5D on the surface of the AuNPs [33]. Furthermore, the covalent attachment of the FA onto the surface of NPs is known by its absorption maxima at 280 nm with a saddle point at 360 nm [34,35] (Figure 1A), corresponding to the absorption peak of Au:G5D:FA observed at 287 nm. Figure 1B shows the λ max for G5D and FA which caused the changes in the surface plasmon resonance of the AuNPs upon functionalization. The UV-vis absorbances were further utilized to estimate the amount of bound G5D and FA, which were 53.8% and 60.6%, respectively.

The formation of Au:G5D and Au:G5D:FA NPs was also verified by ^{1}H NMR spectroscopy (Figure 2). Significant differences in the chemical shift of protons related to Au:G5D (D), Au:G5D:FA (B), G5D:FA (A) were observed when compared to G5D(C). The ^{1}H NMR of the G5D shows six broad peaks (Figure 2C, peaks 1–6) as indicated by a chemical shift ranging from 2.25 to 3.34 ppm, representing the protons of the amino (NH_2) and methylene groups (CH_2). These findings correlated with those reported [36,37]. Moreover, the three peaks between 6.50 and 8.63 ppm observed in Figure 2A,B indicate the attachment of FA protons (H-Ar (7 and 13), NH (18)). The formation of Au:G5D nanocomplexes resulted in the downfield shift of protons 4, 5, and 6 of G5D, which indicated the interaction of the surface of the AuNPs with the internal amines of the dendrimer. These findings correlate to that in literature [38].

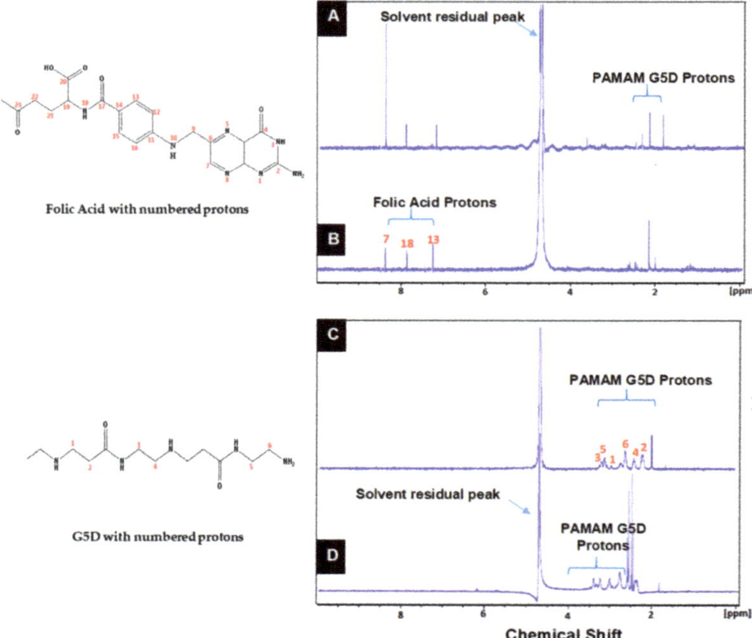

Figure 2. The ^{1}H NMR spectra of PAMAM dendrimer (G5D) and folic acid-functionalized gold nanoparticles in D_2O. (**A**) G5D:FA, (**B**) Au:G5D:FA, (**C**) G5D, (**D**) Au:G5D.

3.2. Morphology, Size, and Zeta Potential of Nanoparticles and Nanocomplexes

The NPs appeared spherical (Figure 3A,B,D) with a uniform distribution and hydrodynamic diameters from NTA ranging from 65 nm to 128 nm (Table 1). Nanocomplexes prepared at optimum binding ratios (Figure 3C,E), presented as clusters of smaller particles with hydrodynamic diameters ranging from 101 nm to 265 nm (Table 1). There was no significant size difference (# $p > 0.05$) between the Au:G5D/Au:G5D:FA and G5D/G5D:FA nanocomplexes (Table 1).

Overall, ζ potentials ranged from 20.9 mV to 87.2 mV for the NPs and from −21.0 mV to −65 mV for the nanocomplexes, indicating good colloidal stability (Table 1). Au:G5D and Au:G5D:FA nanocomplexes had the highest ζ potentials of −37.3 mV and −65.7 mV, respectively. The polydispersity indices (PDI) revealed that all the NPs and nanocomplexes are highly monodisperse and uniform in size with PDI values below 0.2 (Table 1), suggesting that these NPs and nanocomplexes have a lower tendency to agglomerate [39].

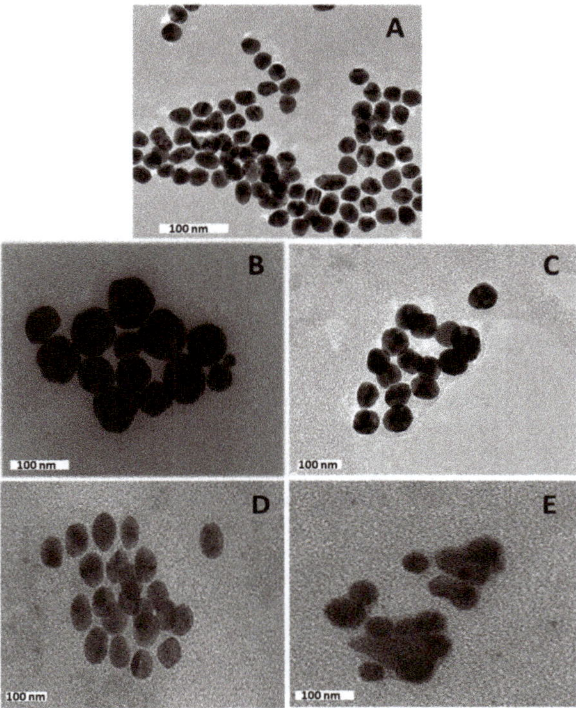

Figure 3. TEM micrograph of (**A**) AuNPs, (**B**) Au:G5D, (**C**) Au:G5D-mRNA nanocomplex, (**D**) Au:G5D:FA, and (**E**) Au:G5D:FA-mRNA nanocomplex. Nanocomplexes were prepared at optimum binding ratios of 3:1 (*w/w*) for Au:G5D-mRNA and 4:1 (*w/w*) for Au:G5D:FA-mRNA, respectively.

Table 1. Hydrodynamic size, ζ potential measurements, and polydispersity indices of nanoparticles and nanocomplexes. Data are presented as mean diameter $\pm$ standard deviation (SD) ($n = 3$).

Nanoparticles/ Nanocomplexes	NP:mRNA (*w/w*) Ratio	Mean Diameter (nm) $\pm$ SD	ζ Potential (mV) $\pm$ SD	Polydispersity Index
Au [16]	-	65.9 $\pm$ 9.8	-7.3 ± 1.6	0.022
G5D [16]	-	161.3 $\pm$ 11.9	$+87.2 \pm 2.4$	0.005
Au:G5D [16]	-	100.5 $\pm$ 44.1	$+20.9 \pm 2.2$	0.193
G5D:FA [16]	-	128.0 $\pm$ 1.20	$+71.2 \pm 3.4$	0.00009
Au:G5D:FA [16]	-	77.7 $\pm$ 12.5	$+29.0 \pm 0.5$	0.026
Au:G5D-mRNA	3:1	207.2 $\pm$ 35.5 #	-37.3 ± 0.1 ***	0.029
Au:G5D:FA-mRNA	4:1	101.8 $\pm$ 36.9 #	-65.7 ± 1.4 ***	0.131
G5D-mRNA	2:1	118.0 $\pm$ 6.20 #	-21.0 ± 0.5 ***	0.028
G5D:FA-mRNA	4:1	265.2 $\pm$ 51.6 #	-25.8 ± 0.0 ***	0.038

$p > 0.05$, *** $p < 0.001$, when dendrimer-only-based nanocomplexes are compared to gold–dendrimer nanocomplexes.

3.3. The Band Shift Assay

The binding of mRNA to the prepared NPs can be seen in Figure 4.

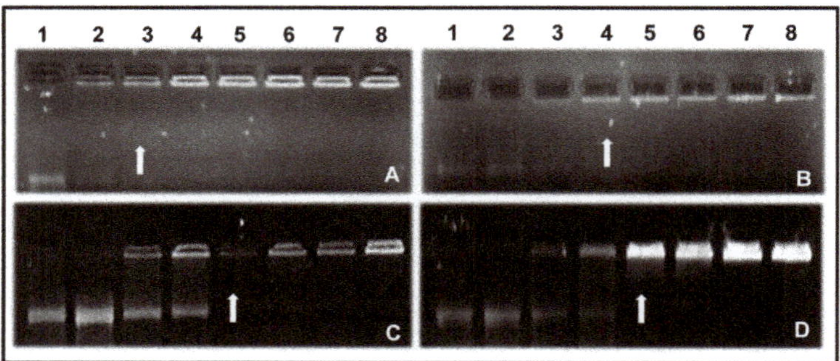

Figure 4. Band shift assay of the interaction between (**A**) G5D, (**B**) Au:G5D, (**C**) G5D:FA, (**D**) Au:G5D:FA, and mRNA. Incubation mixtures (20 μL) in HBS contained varying amounts of the nanoparticle preparation and 0.05 μg FLuc-mRNA corresponding to *w/w* ratios of 1:1, 2:1, 3:1, 4:1, 5:1, 6:1, 7:1, and 8:1 in lanes 2–8, respectively (**A–D**). Lane 1: naked mRNA control. Arrows indicate endpoint ratios.

All prepared NPs were able to bind and complex with the mRNA. This can be credited to the ability of G5D to become protonated at physiological pH [11]. G5D and Au:G5D NPs completely retarded the mRNA at ratios of 2:1 and 3:1 (*w/w*), respectively, while both G5D:FA and Au:G5D:FA NPs completely retarded mRNA at a ratio of 4:1 (*w/w*).

3.4. Ethidium Bromide Dye Displacement Assay

All NPs displaced ethidium bromide (ETB), indicating a significant degree of mRNA compaction, which bodes well for their stability and protection under physiological conditions. The degree of mRNA compaction by the G5D and Au:G5D NPs ranged from 50 to 80%, while that of G5D:FA and Au:G5D:FA NPs ranged from 40 to 70% (Figure 5).

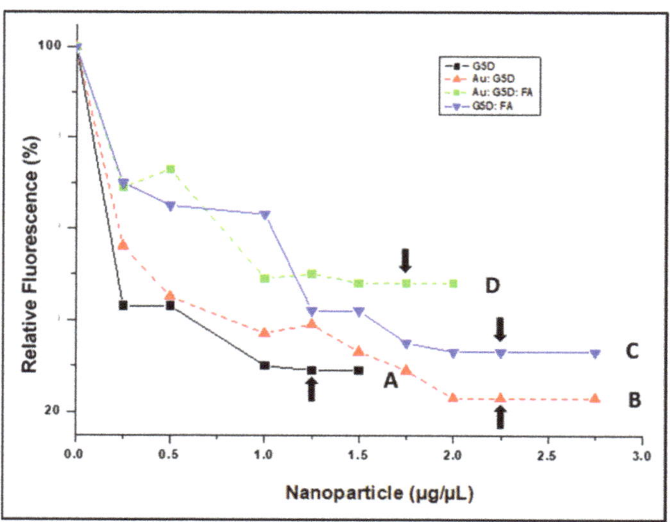

Figure 5. Ethidium bromide displacement assay of (**A**) G5D, (**B**) Au:G5D, (**C**) G5D:FA, and (**D**) Au:G5D:FA NPs. Arrows indicate a point of complexation.

3.5. RNase A Digestion Assay

To assess the ability of the NPs to protect the mRNA cargo against nucleases, which would be encountered in circulation in an in vivo system, an RNase A digestion assay was conducted.

Figure 6 clearly shows the exceptional ability of all NPs to fully protect mRNA following treatment with 10% RNase A, as depicted by the presence of undigested bands in all tested ratios. By contrast, the treatment of naked mRNA with RNase A showed complete degradation (negative control), as illustrated in Lane 2.

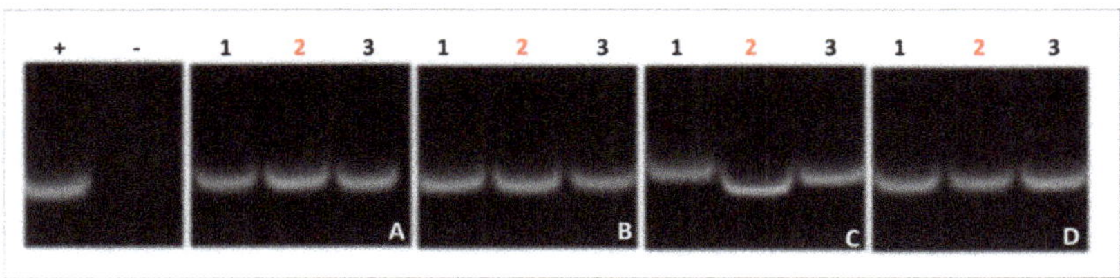

Figure 6. RNase A digestion assay of nanocomplexes. (**A**) G5D, (**B**) Au:G5D, (**C**) G5D: FA, (**D**) Au:G5D:FA. Control: naked mRNA in the absence (+ = positive control) or presence (− = negative control) of RNase A. Lanes 1–3 contain nanocomplexes at sub-optimum, optimum, and supra-optimum nanoparticle: mRNA ratios. (**A**) 1:1, 2:1, 3:1; (**B**) 2:1, 3:1, 4:1; (**C**) 3:1, 4:1, 5:1; (**D**) 3:1, 4:1, 5:1 (w/w). Red-colored numbers indicate the optimum binding ratios.

3.6. The MTT Assay

To monitor cell viability after treatment with prepared nanocomplexes in selected cell lines, the MTT assay was conducted. This assay uses the MTT reagent, which enters the cells and passes into the mitochondria, where it is reduced to an insoluble, purple-colored formazan product that can be quantified spectroscopically and used as an indication of metabolically active cells. No significant ($p > 0.05$) change in cell viability was observed following treatment with all nanocomplexes. Higher cell viabilities (80–97%) were observed in all cell lines for the Au:G5D:mRNA and Au:G5D:FA:mRNA nanocomplexes, compared to the G5D:mRNA and G5D:FA:mRNA nanocomplexes (68–78%) (Figure 7A,B).

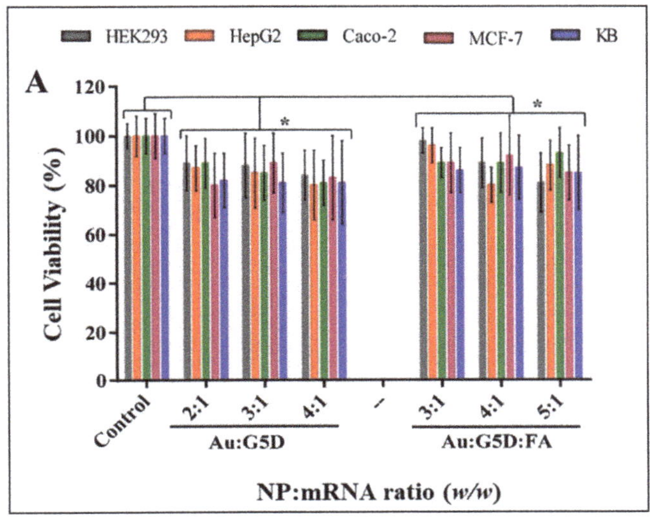

Figure 7. *Cont.*

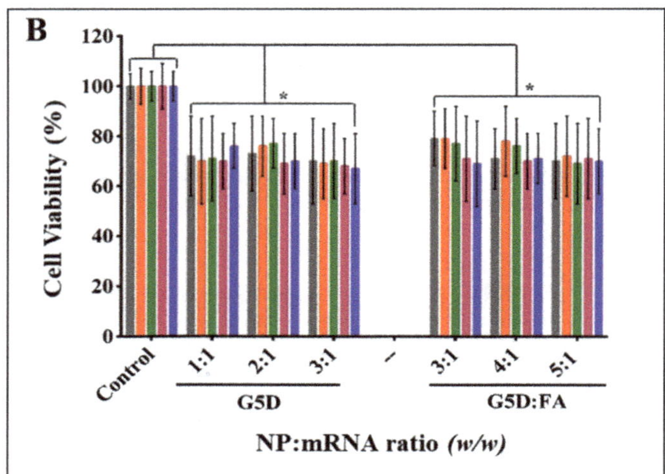

Figure 7. Cell viability assay of nanocomplexes containing (**A**) Au:G5D and Au:G5D:FA; and (**B**) G5D and G5D:FA, in HEK293, HepG2, Caco-2, MCF-7, and KB cells. Cells were incubated with nanocomplexes containing 0.05 µg F*Luc*-mRNA at indicated ratios (w/w). Nanocomplexes were prepared at sub-optimum, optimum, and supra-optimum ratios. Data are presented as means $\pm$ SD ($n = 3$). Control = untreated cells. * $p > 0.05$.

Noticeably, all FA-targeted nanocomplexes showed higher cell viability than their untargeted nanocomplex counterparts (average cell viability of 88% for Au:G5D:FA and 72% for G5D:FA).

3.7. Apoptosis Assay

Cell death was also studied by evaluating the ability of NPs to induce apoptosis in selected cell lines. All nanocomplexes induced little or no apoptosis in the cells, as evidenced by very few apoptotic (yellow-orange/red) cells visible and low apoptotic indices (AI) (Figure 8 and Table 2). Noticeably, the AI values of Au:G5D:mRNA and Au:G5D:FA:mRNA nanocomplexes were significantly ($p < 0.0001$) lower than those of the G5D:mRNA and G5D:FA:mRNA nanocomplexes particularly, in all cell lines (Table 2).

Table 2. Apoptotic indices of nanocomplexes in selected cell lines.

Cell Lines	Apoptotic Indices				
	Cell Control	Nanocomplexes			
		Au:G5D	Au:G5D:FA	G5D	G5D:FA
HEK293	0.0	0.03 ± 0.0001	0.04 ± 0.0004	0.07 ± 0.0010	0.08 ± 0.0020
HepG2	0.0	0.06 ± 0.0015	0.04 ± 0.0018	0.08 ± 0.0012	0.09 ± 0.0011
Caco-2	0.0	0.05 ± 0.0010	0.04 ± 0.0011	0.13 ± 0.0015	0.11 ± 0.0030
MCF-7	0.0	0.04 ± 0.0011	0.06 ± 0.0003	0.25 ± 0.0030	0.23 ± 0.0010
KB	0.0	0.05 ± 0.0021	0.06 ± 0.0003	0.19 ± 0.0015	0.20 ± 0.0012

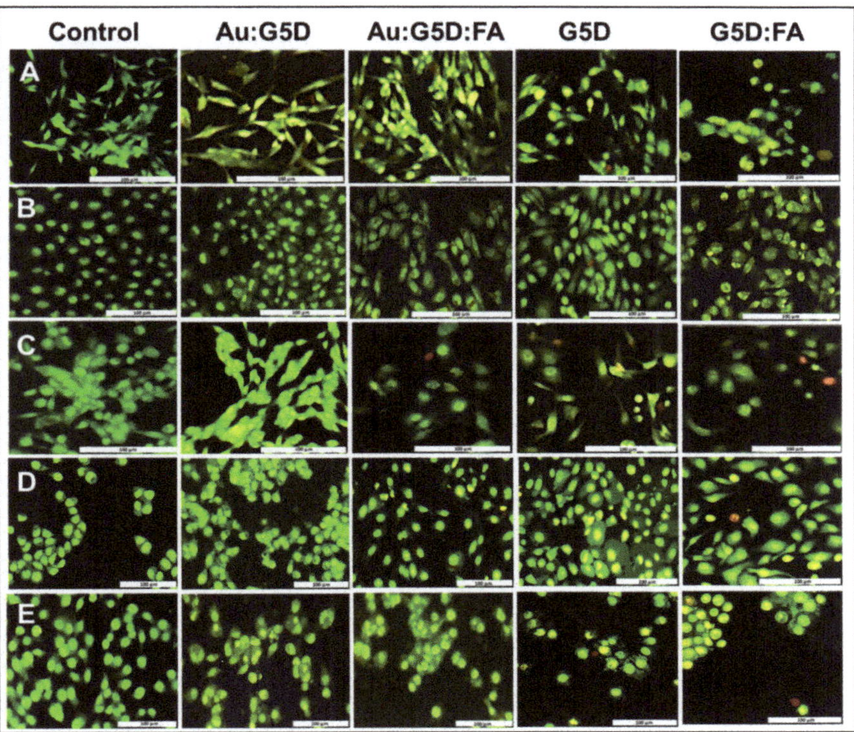

Figure 8. Fluorescence images of (**A**) HEK293, (**B**) HepG2, (C) Caco-2, (**D**) MCF-7, and (**E**) KB cells treated with test and control nanocomplexes prepared at sub-optimum ratios for 24 h, showing induction of apoptosis. Green = live (L), light orange = early apoptotic (EA), and dark orange = late apoptotic (LA) cells. Scale = 100 μm.

3.8. Transfection and Competition Assays

The ability of the NPs to deliver mRNA was evaluated in folate receptor-negative cell lines, HEK293, Caco-2, and folate receptor-positive cell lines HepG2, MCF-7, and KB (KB > MCF-7 > HepG2), with KB cells often being used as a model for folate receptors (FRs) [40]. The transfection efficacy of the nanocomplexes was assessed as a function of weight ratios (sub-optimum, optimum, and supra-optimum). The transfection activity of the Au:G5D:mRNA and Au:G5D:FA:mRNA nanocomplexes (Figure 9A,B) was much higher than that of the naked mRNA (control). Moreover, the transfection levels in HEK293 and Caco-2 cells were significantly ($p < 0.001$) lower than those elicited in the receptor-positive cells.

All nanocomplexes showed excellent transfection activity, with Au:G5D:mRNA and Au:G5D:FA:mRNA nanocomplexes (Figure 9A) showing higher transfection efficiencies ranging from 5×10^7–6×10^8 RLU/mg protein. On the other hand, G5D:mRNA and G5D:FA:mRNA nanocomplexes (4×10^7–3×10^8 RLU/mg protein) produced decreased transfection activity (Figure 9B). Noticeably, the Au:G5D:FA:mRNA nanocomplexes showed a four-fold increase in transfection activity (6×10^8 RLU/mg protein), compared to Au:G5D:mRNA nanocomplexes (2×10^8 RLU/mg protein) at the optimum ratios in the FR positive cell line, MCF-7.

To confirm the mechanism of the cellular uptake of the nanocomplexes, a competition assay was conducted. This involved flooding the cells with excess free FA (250 μg) before exposure to the FA-targeted nanocomplexes (Au:G5D:FA:mRNA and G5D:FA:mRNA). The assay was conducted in the cell lines with overall higher targeted transgene expression, viz. MCF-7 and KB cell. A significant ($p < 0.01$) drop of approximately 30% in transgene activity was observed as depicted in Figure 10, which suggests that a large portion of these nanocomplexes were taken up by receptor-mediated endocytosis [41], confirming that FA receptor mediation was a key player in the high transgene expression obtained.

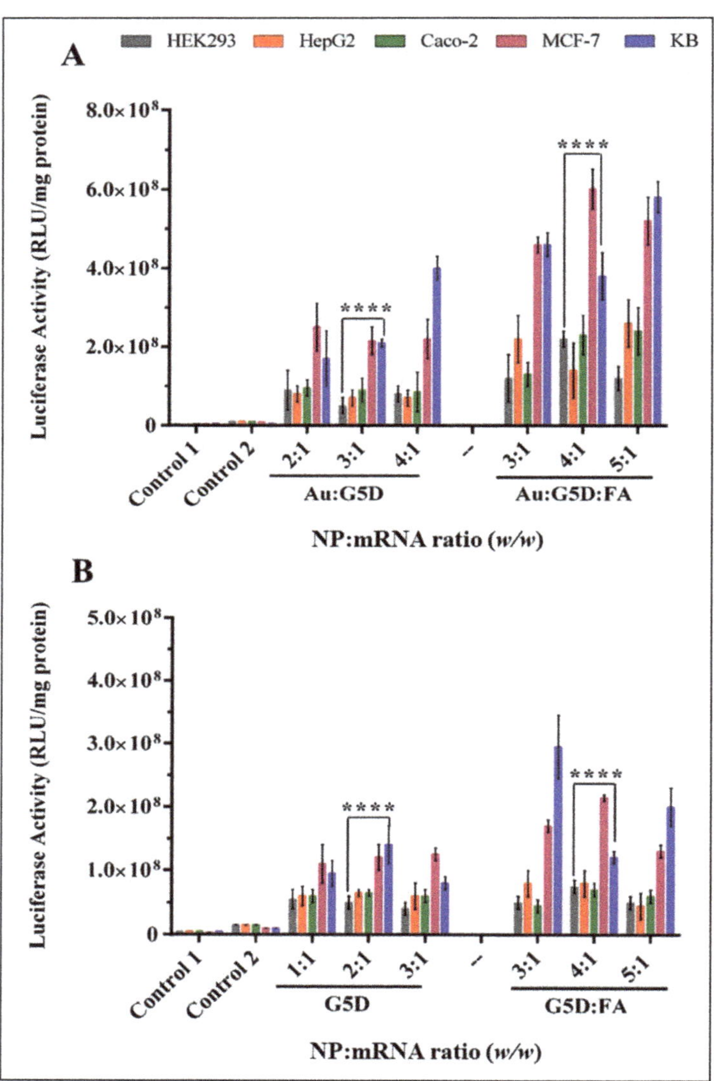

Figure 9. Transgene expression for (**A**) Au:G5D and Au:G5D:FA nanocomplexes, and (**B**) G5D and G5D:FA nanocomplexes in HEK293, HepG2, Caco-2, MCF-7, and KB cells. Nanocomplexes contained 0.05 µg mRNA with varying amounts of nanoparticles to constitute the sub-optimum, optimum, and supra-optimum (w/w) ratios. Control 1 = untreated cells. Control 2 = cells treated with naked F*Luc*-mRNA. The transgene expression is reported as RLU/mg protein. Data are presented as means ± SD ($n = 3$). **** $p < 0.0001$ for optimum ratios.

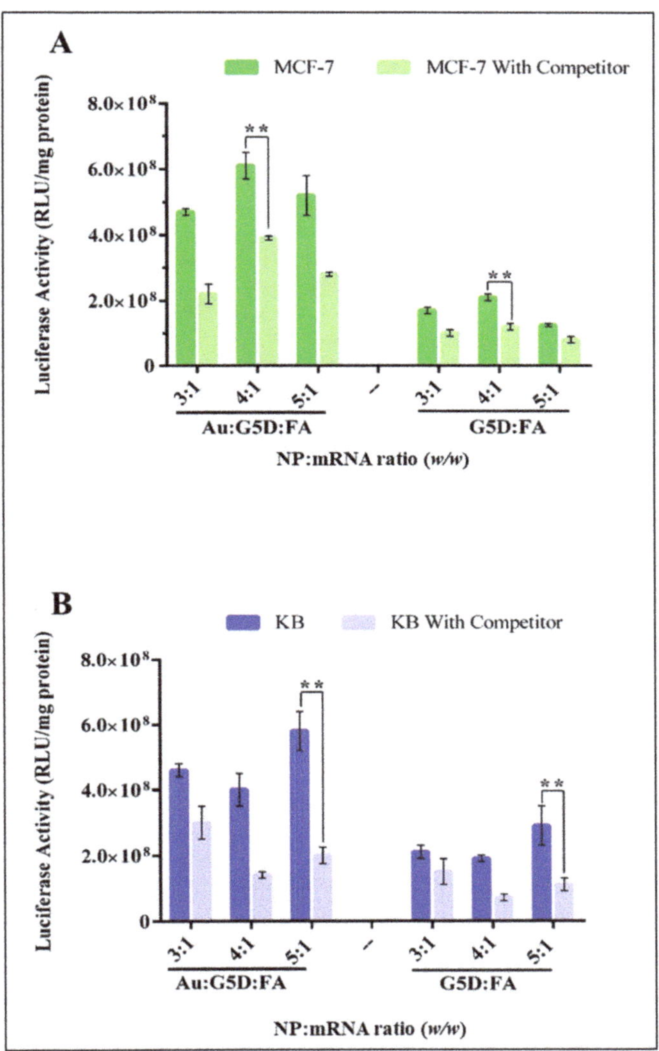

Figure 10. Competition studies of FA-targeted mRNA nanocomplexes in (**A**) MCF-7 and (**B**) KB cells. Cells were first exposed to excess folic acid (250 µg) then treated with FA-targeted nanocomplexes at selected ratios. Transgene expression is reported as RLU/mg protein. Data are presented as means ± SD (n = 3). ** $p < 0.01$.

4. Discussion

All NPs were successfully synthesized to produce spherical, monodispersed NPs. NP synthesis was confirmed by UV-vis and NMR spectroscopy, which also confirmed that the G5D polymer and FA moiety were successfully conjugated to the AuNPs. The Au:G5D NPs produced a redshift in the spectrum, whereas the Au:G5D:FA NPs produced a blue shift. The G5D generally has a very weak peak between 280 and 285 nm [14], especially at higher or at physiological pH due to the presence of the protonated amine groups of the G5D [14,42]. In this study, a small peak was noted at 283 nm. However, this peak often seems to disappear at lower pH. In NMR, the formation of Au:G5D NPs resulted in the downfield shift of protons 4, 5, and 6 of G5D, which indicated the interaction of the surface of the AuNPs with the internal amines of the dendrimers [43].

Favorably sized NPs (<200 nm) with the most zeta potentials, except for the AuNPs on their own being above 20 mV, were produced. All nanocomplexes, with the exception of the G5D:FA nanocomplexes (265.2 nm) fell within the ideal size range (100–200 nm) required for gene delivery via non-specific or receptor-specific uptake [44–46]. Zeta potential measurements greater than +25 mV or less than −25 mV are reported to be associated with good colloidal stability [47]. These AuNPs alone showed poor stability (−7.3 mV), but upon G5D and FA functionalization, the stability improved immensely to +20.9 mV for Au:G5D and +29 mV for Au:G5D:FA. This confirms that the functionalization of the NPs with the dendrimers improved their stability, as seen in a recent study where dendrimer was used to functionalize selenium NPs [14]. The improved stability achieved with the targeted NPs could be due to the partial shielding effect imparted by FA, in addition to the repulsive cationic amine groups on the dendrimers, which prevents particle aggregation [14,48]. From these findings, it can be predicted that these nanocomplexes may be efficient in delivering mRNA.

The differences observed in the binding efficiency between the FA-targeted and untargeted NPs could be due to the possible shielding of the cationic charges of the dendrimers on the targeted NPs by the FA moiety, which meant that more positive charges and more NPs were required to fully neutralize the negative charges on the mRNA [49]. Overall, the NP:mRNA nanocomplex formation occurred at very low ratios (w/w), which could be accredited to the single-stranded nature of the mRNA, which is quickly embedded by the highly cationic G5D. The G5D and Au:G5D showed greater quenching of the ethidium bromide fluorescence, which could be attributed to more amine groups being available to bind the mRNA [14]. The compaction was seen for the targeted nanocomplexes, further suggested a weaker binding of the mRNA, which could translate into easy dissociation of the mRNA from the nanocomplexes during transfection, hence avoiding degradation by the lysosomal compartment, and in turn, enhancing gene-transfection efficiency [50]. Overall, all NPs were able to efficiently bind and compact mRNA to varying degrees.

The integrity of the nanocomplexes may be compromised by degrading nuclease enzymes such as RNase A, leading to a reduced transgene expression [51]. The good nuclease protection afforded by the NPs in this study could be due to the highly organized globular structures that formed as a result of the electrostatic interaction between the negatively charged single-stranded mRNA and the highly cationic G5D-containing NPs [52]. The use of the RNase enzyme was a stringent test for these NPs due to its specificity for RNA molecules and was reported previously [53,54]. Various studies have used the less-specific fetal bovine serum containing nucleases to determine the integrity of RNA-based nanocomplexes [55,56] to achieve similar results. In the circulatory system, it is possible that the nanoparticles may encounter less-specific enzymes and possibly at lower concentrations as well. However, this assay confirmed that all NPs afforded exceptional protection to the mRNA cargo, boding well for future in vivo studies.

The first step towards understanding the biocompatibility of a delivery system often involves the use of cell-culture-based studies, commencing with the assessment of cytotoxicity. The gold-containing NPs achieved higher cell viability, which may be due to the presence of the gold in the NP and partly to the reduction of the cationic charges of the 1° amines of the G5D, some of which are responsible for stabilizing the entrapped AuNPs [21]. Furthermore, unmodified AuNPs have been shown to have little or no impact on cytotoxicity in non-cancer HEK293 and cervical cancer (HeLa) cells [24], which could be attributed to their inherent biocompatibility and favorable physicochemical properties that have been widely mentioned. Noticeably, all FA-targeted nanocomplexes showed higher cell viabilities compared to their untargeted nanocomplex counterparts, which could be as a result of the shielding effect of FA, which may have covered a portion of the positive charges on the surface of G5D, hence reducing the strong electrostatic interaction between the cells and the NPs [17]. Overall, more than 80% of cells were still viable after being exposed to the gold-containing nanocomplexes at the selected ratios, suggesting that these nanocomplexes were superior and well-tolerated in all tested cell lines, and therefore relatively safe to use. Apoptosis studies corroborated these results, confirming that the Au:G5D:mRNA and Au:G5D:FA:mRNA nanocomplexes were safe and stable and did not induce any significant apoptotic effects.

The introduction of naked mRNA into cells is known to be associated with poor transgene expression, mainly due to enzymatic degradation [57], as evidenced in the RNase A digestion assay. All nanocomplexes displayed significant transfection in the cell lines tested. The Au:G5D:mRNA and Au:G5D:FA:mRNA nanocomplexes produced the highest luciferase activity, which could be due to three reasons. First, since the translation of mRNA occurred in the cytoplasm—and the major limiting step, which is the nuclear pore entry, was avoided—resulting in an increased transgene expression. Second, the transfection studies were conducted over a duration of 48 h, and more protein may have expressed, considering that mRNA may have a limited half-life [5]. Lastly, the efficient encapsulation of the mRNA by the dendrimer and its exceptional buffering capacity could have helped protect the mRNA from degradation and facilitated the endosomal escape of the nanocomplexes [58].

HepG2 cells exhibited lower luciferase expression, possibly due to fewer receptors on their cell surface compared to MCF-7 and KB cells. The low targeted expression is associated with a lack of specific transcription factors and cell-surface receptors [35]. The higher transfection efficiencies of Au:G5D:mRNA and Au:G5D:FA:mRNA nanocomplexes can be accredited to the entrapment of AuNPs within the 1° amines of the dendrimers, which helped preserve the structural integrity of the dendrimers, allowing for efficient interaction between the dendrimers and the mRNA [21]. This could lead to favorable cellular uptake and high gene expression. The decreased transfection activities of the G5D:mRNA and G5D:FA:mRNA nanocomplexes could be due to their higher cytotoxicity compared to their gold-containing counterparts and the poor dissociation between the mRNA and the G5D due to their strong binding affinity. The mRNA may have been entrapped by a network formed by the branches of the dendrimer. Earlier studies have demonstrated a direct correlation between the binding affinity of the single-stranded mRNA to cationic polymers and transgene expression [59].

The Au:G5D:FA:mRNA nanocomplexes showed a superior transfection activity to the Au:G5D:mRNA nanocomplexes, most likely due to ligand–receptor interaction that occurred between the FA and the FRs abundantly, decorating the surface of the MCF-7 and KB cells [60]. It is generally known that FA has a high affinity for FRs overexpressed by a majority of cancer cells [35], with KB cells generally regarded as models for the folate receptor, as previously mentioned [40]. The significant ($p < 0.01$) drop in transgene activity in the competition assay suggested that a large portion of these nanocomplexes were taken up by receptor-mediated endocytosis, confirming that FA receptor-mediation was a key player in the high transgene expression obtained.

5. Conclusions

Both Au:G5D and Au:G5D:FA NPs were highly efficient in *FLuc*-mRNA binding and delivery. They formed stable nanocomplexes and afforded excellent protection to the mRNA against RNases. Furthermore, more than 80% cell viability was observed, suggesting that these nanocomplexes were well tolerated by all cells. This was also demonstrated in their superior transfection efficiency, indicating the significant and synergistic roles played by both the dendrimer and the AuNPs in their formulation. This study further confirmed that folate-receptor-mediated delivery was the main route of entry into the receptor-positive cells, as evidenced by the transfection levels in the FA receptor negative cell lines, being significantly lower than that in FA receptor positive cell lines. Since this proof in principle study has shown potential, future studies would encompass the NP optimization for in vivo delivery using a therapeutic mRNA molecule.

Author Contributions: Conceptualization, L.S.M. and M.S.; methodology, L.S.M. and F.M., software, L.S.M., F.M., and A.D.; validation, M.S. and A.D.; data curation, L.S.M., F.M., and A.D.; resources, M.S.; writing—original draft preparation, L.S.M., F.M., and A.D.; writing—review and editing, M.S.; supervision, M.S.; project administration, M.S.; funding acquisition, M.S. All authors have read and agreed to the published version of the manuscript.

Funding: This research was funded by the National Research Foundation of South Africa, grant numbers 120455 and 129263.

Institutional Review Board Statement: Not applicable.

Informed Consent Statement: Not applicable.

Data Availability Statement: The data and contributions presented in the study are included in the article. Further inquiries can be directed to the corresponding author.

Acknowledgments: The authors acknowledge members of the Nano-Gene and Drug Delivery group for advice and technical support.

Conflicts of Interest: The authors declare no conflict of interest.

References

1. McCrudden, C.M.; McCarthy, H.O. Cancer Gene Therapy-Key Biological Concepts in the Design of Multifunctional Non-Viral Delivery Systems. In *Gene Therapy—Tools and Potential Applications*, 1st ed.; Martin, F., Ed.; InTechOpen: London, UK, 2013; pp. 213–235.
2. Kreig, A.M. CpG motifs in Bacterial DNA and Their Immune Effects. *Annu. Rev. Immunol.* **2002**, *20*, 709–760. [CrossRef]
3. Su, X.; Fricke, J.; Kavanagh, D.; Irvine, D.J. In vitro and in Vivo Mrna Delivery Using Lipid-Enveloped Ph-Responsive Polymer Nanoparticles. *Mol. Pharm.* **2011**, *8*, 774–787. [CrossRef] [PubMed]
4. Malone, R.W.; Felgner, P.L.; Verma, I.M. Cationic liposome-mediated RNA transfection. *Proc. Natl. Acad. Sci. USA* **1989**, *86*, 6077–6081. [CrossRef] [PubMed]
5. Tavernier, G.; Andries, O.; Demeester, J.; Sanders, N.N.; De Smedt, S.C.; Rejman, J. mRNA as Gene Therapeutic: How to Control Protein Expression. *J. Control. Release* **2011**, *150*, 238–247. [CrossRef]
6. Saenz-Badillos, J.; Amin, S.; Granstein, R. RNA as a Tumor Vaccine: A Review of the Literature. *Exp. Dermatol.* **2001**, *10*, 143–154. [CrossRef] [PubMed]
7. Weide, B.; Carralot, J.-P.; Reese, A.; Scheel, B.; Eigentler, T.K.; Hoerr, I.; Rammensee, H.-G.; Garbe, C.; Pascolo, S. Results of the First Phase I/II Clinical Vaccination Trial with Direct Injection of Mrna. *J. Immunother.* **2008**, *31*, 180–188. [CrossRef]
8. Weide, B.; Pascolo, S.; Scheel, B.; Derhovanessian, E.; Pflugfelder, A.; Eigentler, T.K.; Pawelec, G.; Hoerr, I.; Rammensee, H.-G.; Garbe, C. Direct Injection of Protamine-Protected Mrna: Results of a Phase 1/2 Vaccination Trial in Metastatic Melanoma Patients. *J. Immunother.* **2009**, *32*, 498–507. [CrossRef]
9. Mockey, M.; Gonçalves, C.; Dupuy, F.P.; Lemoine, F.M.; Pichon, C.; Midoux, P. Mrna Transfection of Dendritic Cells: Synergistic Effect of ARCA Mrna Capping with Poly (A) Chains in Cis and in Trans for a High Protein Expression Level. *Biochem. Biophys. Res. Commun.* **2006**, *340*, 1062–1068. [CrossRef]
10. Van Tendeloo, V.F.; Ponsaerts, P.; Berneman, Z.N. mRNA-Based Gene Transfer as a Tool for Gene and Cell Therapy. *Curr. Opin. Mol. Ther.* **2007**, *9*, 423–431.
11. Kowalski, P.S.; Rudra, A.; Miao, L.; Anderson, D.G. Delivering the Messenger: Advances in Technologies for Therapeutic mRNA Delivery. *Mol. Ther.* **2019**, *27*, 710–728. [CrossRef]
12. Xionga, H.; Liua, S.; Weia, T.; Chenga, Q.; Siegwarta, D.J. The Theranostic Dendrimer-Based Lipid Nanoparticles Containing Pegylated BODIPY Dyes for Tumor Imaging and Systemic Mrna Delivery in Vivo. *J. Control. Release* **2020**, *325*, 198–205. [CrossRef] [PubMed]
13. Chaplot, S.P.; Rupenthal, I.D. Dendrimers for Gene Delivery–A Potential Approach for Ocular Therapy? *J. Pharm. Pharmacol.* **2014**, *66*, 542–556. [CrossRef] [PubMed]
14. Pillay, N.S.; Daniels, A.; Singh, M. Folate-Targeted Transgenic Activity of Dendrimer Functionalized Selenium Nanoparticles in vitro. *Int. J. Mol. Sci.* **2020**, *21*, 7177. [CrossRef]
15. Mbatha, L.S.; Singh, M. Starburst Poly(amidoamine) Dendrimer Grafted Gold Nanoparticles as a Scaffold for Folic Acid-Targeted Plasmid DNA Delivery in vitro. *J. Nanosci. Nanotechnol.* **2019**, *19*, 1959–1970. [CrossRef] [PubMed]
16. Mbatha, L.S.; Maiyo, F.C.; Singh, M. Dendrimer Functionalized Folate-Targeted Gold Nanoparticles for Luciferase Gene Silencing in vitro: A Proof of Principle Study. *Acta Pharm.* **2019**, *69*, 49–61. [CrossRef] [PubMed]
17. Xiao, T.; Hou, W.; Cao, X.; Wen, S.; Shen, M.; Shi, X. Dendrimer-Entrapped Gold Nanoparticles Modified with Folic Acid for Targeted Gene Delivery Applications. *Biomater. Sci.* **2013**, *1*, 1172–1180. [CrossRef]
18. Luo, D.; Haverstick, K.; Belcheva, N.; Han, E.; Saltzman, W.M. Poly (ethylene glycol)-Conjugated PAMAM Dendrimer for Biocompatible, High-Efficiency DNA Delivery. *Macromolecules* **2002**, *35*, 3456–3462. [CrossRef]

19. Lee, J.H.; Lim, Y.-B.; Choi, J.S.; Lee, Y.; Kim, T.-I.; Kim, H.J.; Yoon, J.K.; Kim, K.; Park, J.-S. Polyplexes Assembled with Internally Quaternized PAMAM-OH Dendrimer and Plasmid DNA Have a Neutral Surface and Gene Delivery Potency. *Bioconjug. Chem.* **2003**, *14*, 1214–1221. [CrossRef]

20. Kim, T.-I.; Seo, H.J.; Choi, J.S.; Jang, H.-S.; Baek, J.-U.; Kim, K.; Park, J.-S. PAMAM-PEG-PAMAM: Novel Triblock Copolymer as a Biocompatible and Efficient Gene Delivery Carrier. *Biomacromolecules* **2004**, *5*, 2487–2492. [CrossRef]

21. Shan, Y.; Luo, T.; Peng, C.; Sheng, R.; Cao, A.; Coa, X.; Shen, M.; Guo, R.; Tomas, H.; Shi, X. Gene Delivery Using Dendrimer-Entrapped Gold Nanoparticles as Nonviral Vectors. *Biomaterials* **2012**, *33*, 3025–3035. [CrossRef]

22. Yuan, X.; Wen, S.; Shen, M.; Shi, X. Dendrimer-Stabilized Silver Nanoparticles Enable Efficient Colorimetric Sensing of Mercury Ions in Aqueous Solution. *Anal. Methods* **2013**, *5*, 5486–5492. [CrossRef]

23. Figueroa, E.R.; Lin, A.Y.; Yan, J.; Luo, L.; Foster, A.E.; Drezek, R.A. Optimization of PAMAM-Gold Nanoparticle Conjugation for Gene Therapy. *Biomaterials* **2014**, *35*, 1725–1734. [CrossRef]

24. Oladimeji, O.; Akinyelu, A.; Singh, M. Co-polymer Functionalised Gold Nanoparticles Show Efficient Mitochondrial Targeted Drug Delivery in Cervical Carcinoma Cells. *J. Biomed. Nanotechnol.* **2020**, *16*, 853–866. [CrossRef] [PubMed]

25. Akinyelu, A.; Oladimeji, O.; Singh, M. Lactobionic Acid-Chitosan Functionalized Gold Coated Poly(lactide-co-glycolide) Nanoparticles for Hepatocyte Targeted Gene Delivery. *Adv. Nat. Sci. Nanosci. Nanotechnol.* **2020**, *11*, 045017. [CrossRef]

26. Turkevich, J.; Stevenson, P.C.; Hillier, J. A Study of the Nucleation and Growth Processes in the Synthesis of Colloidal Gold. *Discuss. Faraday Soc.* **1951**, *11*, 55–75. [CrossRef]

27. Singh, M. Assessing Nucleic acid: Cationic Nanoparticle Interaction for Gene Delivery. In *Bio-Carrier Vectors*; Narayan, K., Ed.; Methods in Molecular Biology Series; Springer-Nature: New York, NY, USA, 2021; Volume 2211, pp. 43–55.

28. Maiyo, F.; Singh, M. Folate-Targeted mRNA Delivery Using Chitosan Functionalized Selenium Nanoparticles: Potential in Cancer Immunotherapy. *Pharmaceuticals* **2019**, *12*, 164. [CrossRef]

29. Singh, D.; Singh, M. Hepatocellular-Targeted mRNA Delivery Using Functionalized Selenium Nanoparticles in Vitro. *Pharmaceutics* **2021**, *13*, 298. [CrossRef]

30. Maiyo, F.; Moodley, R.; Singh, M. Cytotoxicity, Antioxidant and Apoptosis Studies of Quercetin-3-O-Glucoside and 4-(B-D-Glucopyranosyl-1→4-A-L-Rhamnopyranosyloxy)-Benzyl Isothiocyanate from Moringa Oleifera. *Anti-Cancer Agents Med. Chem.* **2016**, *16*, 648–656. [CrossRef] [PubMed]

31. Smith, P.K.; Krohn, R.I.; Hermanson, G.; Mallia, A.; Gartner, F.; Provenzano, M.; Fujimoto, E.; Goeke, N.; Olson, B.; Klenk, D. Measurement of Protein Using Bicinchoninic Acid. *Anal. Biochem.* **1985**, *150*, 76–85. [CrossRef]

32. Haiss, W.; Thanh, N.T.; Aveyard, J.; Fernig, G.D.G. Determination of Size and Concentration of Gold Nanoparticles from UV—Vis Spectra. *Anal. Chem.* **2007**, *79*, 4215–4221. [CrossRef] [PubMed]

33. Pan, D.; Turner, J.L.; Wooley, K.L. Folic Acid-Conjugated Nanostructured Materials Designed for Cancer Cell Targeting. *Chem. Commun.* **2003**, *19*, 2400–2401. [CrossRef]

34. Zhang, Z.; Zhou, F.; Lavernia, E. On the Analysis of Grain Size in Bulk Nanocrystalline Materials via X-ray Diffraction. *Metall. Mater. Trans. A* **2003**, *34*, 1349–1355. [CrossRef]

35. Mansoori, G.A.; Brandenburg, K.S.; Shakeri-Zadeh, A. A Comparative Study of Two Folate-Conjugated Gold Nanoparticles for Cancer Nanotechnology Applications. *Cancers* **2010**, *2*, 1911–1928. [CrossRef] [PubMed]

36. Zhang, G.Y.; Thomas, T.P.; Desai, A.; Zong, H.; Leroueil, P.R.; Majoros, I.J.; Baker, J.R. Targeted Dendrimeric Anticancer Prodrug: A Methotrexate-Folic Acid-Poly (Amidoamine) Conjugate and a Novel, Rapid, "One Pot" Synthetic Approach. *Bioconjug. Chem.* **2010**, *21*, 489–495. [CrossRef] [PubMed]

37. Santos, J.L.; Oliviera, H.; Pandita, D.; Rodrigues, J.; Pêgo, A.P.; Granja, P.L.; Tomas, H. Functionalization of Poly(Amidoamine) Dendrimers with Hydrophobic Chains for Improved Gene Delivery in Mesenchymal Stem Cells. *J. Control. Release* **2010**, *144*, 55–64. [CrossRef]

38. Chang, Y.; Liu, N.; Chen, L.; Meng, X.; Liu, Y.; Li, Y.; Wang, J. Synthesis and Characterization of DOX-Conjugated Dendrimer-Modified Magnetic Iron Oxide Conjugates for Magnetic Resonance Imaging, Targeting, and Drug Delivery. *J. Mater. Chem.* **2012**, *22*, 9594–9601. [CrossRef]

39. Danaei, M.; Dehghankhold, M.; Ataei, S.; Hasanzadeh Davarani, F.; Javanmard, R.; Dokhani, A.; Khorasani, S.; Mozafari, M.R. Impact of Particle Size and Polydispersity Index on the Clinical Applications of Lipidic Nanocarrier Systems. *Pharmaceutics* **2018**, *10*, 57. [CrossRef] [PubMed]

40. Siwowska, K.; Schmid, R.M.; Cohrs, S.; Schibli, R.; Müller, C. Folate Receptor-Positive Gynecological Cancer Cells: In Vitro and in Vivo Characterization. *Pharmaceuticals* **2017**, *10*, 72. [CrossRef] [PubMed]

41. Zhang, S.; Gao, H.; Bao, G. Physical Principles of Nanoparticle Cellular Endocytosis. *ACS Nano* **2015**, *9*, 8655–8671. [CrossRef]

42. Boroumand, S.; Safari, M.; Shaabani, E.; Shirzad, M.; Faridi-Majidi, R. Selenium Nanoparticles: Synthesis, Characterization and Study of Their Cytotoxicity, Antioxidant and Antibacterial Activity. *Mater. Res. Express* **2019**, *6*, 0850d8. [CrossRef]

43. Shi, X.; Sun, K.; Baker, J.R. Spontaneous Formation of Functionalized Dendrimer-Stabilized Gold Nanoparticles. *J. Phys. Chem. C* **2008**, *112*, 8251–8258. [CrossRef]

44. Azzam, T.; Domb, A.J. Current Developments in Gene Transfection Agents. *Curr. Drug Deliv.* **2004**, *1*, 165–193. [CrossRef]

45. Rejman, J.; Oberle, V.; Zuhorn, I.S.; Hoekstra, D. Size-Dependent Internalization of Particles Via the Pathways of Clathrin-and Caveolae-Mediated Endocytosis. *Biochem. J.* **2004**, *377*, 159–169. [CrossRef] [PubMed]

46. Grosse, S.; Aron, Y.; Thévenot, G.; François, D.; Monsigny, M.; Fajac, I. Potocytosis and Cellular Exit of Complexes as Cellular Pathways for Gene Delivery by Polycations. *J. Gene Med.* **2005**, *7*, 1275–1286. [CrossRef] [PubMed]
47. Honary, S.; Zahir, F. Effect of Zeta Potential on the Properties of Nano-Drug Delivery Systems—A Review (Part 2). *Trop. J. Pharm. Res.* **2013**, *12*, 265–273.
48. Fratila, R.M.; Mitchell, S.G.; Del Pino, P.; Grazu, V.; De La Fuente, J.S.M. Strategies for the Biofunctionalization of Gold and Iron Oxide Nanoparticles. *Langmuir* **2014**, *30*, 15057–15071. [CrossRef]
49. Mbatha, L.; Chakravorty, S.; de Koning, C.B.; van Otterlo, W.A.; Arbuthnot, P.; Ariatti, M.; Singh, M. Spacer Length: A Determining Factor in the Design of Galactosyl Ligands for Hepatoma Cell-Specific Liposomal Gene Delivery. *Curr. Drug Deliv.* **2016**, *13*, 935–945. [CrossRef] [PubMed]
50. Chuang, C.-C.; Chang, C.-W. Complexation of Bioreducible Cationic Polymers with Gold Nanoparticles for Improving Stability in Serum and Application on Nonviral Gene Delivery. *ACS Appl. Mater. Interfaces* **2015**, *7*, 7724–7731. [CrossRef]
51. Obata, Y.; Saito, S.; Takeda, N.; Takeoka, S. Plasmid DNA-Encapsulating Liposomes: Effect of a Spacer between the Cationic Head Group and Hydrophobic Moieties of the Lipids on Gene Expression Efficiency. *Biochim. Biophys. Acta* **2009**, *1788*, 1148–1158. [CrossRef]
52. Pitard, B. Supramolecular Assemblies of DNA Delivery Systems. *Somat. Cell Mol. Genet.* **2002**, *27*, 5–15. [CrossRef]
53. Daniels, A.; Singh, M.; Ariatti, M. Pegylated and Non-Pegylated siRNA Lipoplexes Formulated with Cholesteryl Cytofectins Promote Efficient Luciferase Knockdown in Hela Tat Luc Cells. *Nucleos Nucleot Nucleic* **2013**, *32*, 206–220. [CrossRef]
54. Daniels, A.N.; Singh, M. Sterically Stabilized siRNA: Gold Nanocomplexes Enhance c-MYC Silencing in a Breast Cancer Cell Model. *Nanomedicine* **2019**, *14*, 1387–1401. [CrossRef] [PubMed]
55. Nundkumar, N.; Singh, S.; Singh, M. Amino Acid Functionalized Hydrotalcites for Gene Silencing. *J. Nanosci. Nanotechnol.* **2020**, *20*, 3387–3397. [CrossRef] [PubMed]
56. Habib, S.; Daniels, A.; Ariatti, M.; Singh, M. Anti-c-MYC Cholesterol based Lipoplexes as Onco-Nanotherapeutic Agents in vitro. *F1000 Res.* **2020**, *9*, 770. [CrossRef]
57. Elsabahy, M.; Nazarali, A.; Foldvari, M. Non-Viral Nucleic Acid Delivery: Key Challenges and Future Directions. *Curr. Drug Deliv.* **2011**, *8*, 235–244. [CrossRef] [PubMed]
58. Kambhampati, S.P.; Clunies-Ross, A.J.; Bhutto, I.; Mishra, M.K.; Edwards, M.; McLeod, D.S.; Kannan, R.M.; Lutty, G. Systemic and Intravitreal Delivery of Dendrimers to Activated Microglia/Macrophage in Ischemia/Reperfusion Mouse Retina. *Investig. Ophthalmol. Vis. Sci.* **2015**, *56*, 4413–4424. [CrossRef] [PubMed]
59. Bettinger, T.; Carlisle, R.C.; Read, M.L.; Ogris, M.; Seymour, L.W. Peptide-Mediated RNA Delivery: A Novel Approach for Enhanced Transfection of Primary and Post-Mitotic Cells. *Nucleic Acids Res.* **2001**, *29*, 3882–3891. [CrossRef]
60. Srinivasarao, M.; Galliford, C.V.; Low, P.S. Principles in the Design of Ligand-Targeted Cancer Therapeutics and Imaging Agents. *Nat. Rev. Drug Discov.* **2015**, *14*, 203–219. [CrossRef]

 *pharmaceutics*

Article

Engineered Human Heavy-Chain Ferritin with Half-Life Extension and Tumor Targeting by PAS and RGDK Peptide Functionalization

Shuang Yin [1], Yan Wang [2], Bingyang Zhang [1], Yiran Qu [1], Yongdong Liu [3], Sheng Dai [4], Yao Zhang [3], Yingli Wang [2,*] and Jingxiu Bi [1,*]

[1]   School of Chemical Engineering and Advanced Materials, The University of Adelaide, Adelaide SA5005, Australia; shuang.yin@adelaide.edu.au (S.Y.); bingyang.zhang@adelaide.edu.au (B.Z.); yiran.qu@adelaide.edu.au (Y.Q.)
[2]   School of Chinese Medicine and Food Engineering, Shanxi University of Traditional Chinese Medicine, Jinzhong 030619, China; wangyan81823@aliyun.com
[3]   State Key Laboratory of Biochemistry Engineering, Institute of Process Engineering, Chinese Academy of Sciences, Beijing 100190, China; ydliu@ipe.ac.cn (Y.L.); zhangyao@ipe.ac.cn (Y.Z.)
[4]   Department of Chemical Engineering, Brunel University London, Uxbridge UB8 3PH, UK; sheng.dai@brunel.ac.uk
*   Correspondence: wyl@sxtcm.edu.cn (Y.W.); jingxiu.bi@adelaide.edu.au (J.B.)

Citation: Yin, S.; Wang, Y.; Zhang, B.; Qu, Y.; Liu, Y.; Dai, S.; Zhang, Y.; Wang, Y.; Bi, J. Engineered Human Heavy-Chain Ferritin with Half-Life Extension and Tumor Targeting by PAS and RGDK Peptide Functionalization. *Pharmaceutics* **2021**, *13*, 521. https://doi.org/10.3390/pharmaceutics13040521

Academic Editors: Francisco José Ostos, José Antonio Lebrón, Pilar López-Cornejo and Patrick J. Sinko

Received: 24 February 2021
Accepted: 6 April 2021
Published: 9 April 2021

Publisher's Note: MDPI stays neutral with regard to jurisdictional claims in published maps and institutional affiliations.

Abstract: Ferritin, one of the most investigated protein nanocages, is considered as a promising drug carrier because of its advantageous stability and safety. However, its short half-life and undesirable tumor targeting ability has limited its usage in tumor treatment. In this work, two types of functional peptides, half-life extension peptide PAS, and tumor targeting peptide RGDK (Arg-Gly-Asp-Lys), are inserted to human heavy-chain ferritin (HFn) at C-terminal through flexible linkers with two distinct enzyme cleavable sites. Structural characterizations show both HFn and engineered HFns can assemble into nanoparticles but with different apparent hydrodynamic volumes and molecular weights. RGDK peptide enhanced the internalization efficiency of HFn and showed a significant increase of growth inhibition against 4T1 cell line in vitro. Pharmacokinetic study in vivo demonstrates PAS peptides extended ferritin half-life about 4.9 times in Sprague Dawley rats. RGDK peptides greatly enhanced drug accumulation in the tumor site rather than in other organs in biodistribution analysis. Drug loaded PAS-RGDK functionalized HFns curbed tumor growth with significantly greater efficacies in comparison with drug loaded HFn.

Keywords: ferritin; drug delivery; tumor targeting; half-life extension

1. Introduction

Ferritin is one of the most attractive protein nanocages for drug delivery, due to its extraordinary thermal and chemical stability. In mammals, ferritin is a 12 nm sphere with an 8 nm cavity, made up of 24 subunits [1]. Two types of ferritin subunits exist in mammal tissues, called heavy-chain (H-chain) and light-chain (L-chain) (21 kDa and 19 kDa), respectively. Both two types of subunits consist of five α-helices (helices A-E), one long loop connecting helix B and C (BC loop) and three turns connecting helices. Exposed BC loop of Human H-chain ferritin (HFn) has a binding site of human transferrin receptor 1 (TfR1) and gives rise to an intrinsic tumor active targeting ability [2]. Researchers have loaded various chemotherapeutics into H-chain ferritin and explored its anti-tumor efficacy. For example, 5-fluorouracil attached Au nanoparticles inside ferritin decreased IC_{50} against HepG2 cells by 15 times [3]. A single dose of doxorubicin (DOX) loaded HFn (HFn/DOX) successfully inhibited TfR1 overexpressed HT-29 human colon cancer cells growth in mice [4]. Neuronal drugs carbachol and atropine loaded ferritin is proven to be able to regulate pancreatic cancer progression [5].

133

In spite of ferritin's multiple advantages, it is still facing challenges as a drug nanocarrier. It has a half-life in circulation of approximate 2 h in rats, shorter than the majority of other drug nanocarriers because of its relatively small particle size. Wang fused albumin binding domain (ABD) to increase ferritin half-life to 17.2 h [6]. In addition, the innate tumor targeting ability of HFn cannot be guaranteed in all tumors. The expression level of its receptor, human TfR1, varies in different tumor cell lines and in different stages of tumor progression [7,8]. Head and neck cancer, colorectal cancer and cervical cancer tissues have the highest expression level of human TfR1, whilst no human TfR1 was detected in carcinoid, prostate and testicular tumor tissues [9]. Human TfR1 is also ubiquitously expressed in healthy human tissues, such as bone marrow, lung, colon and liver, to import iron into cells, so the usage of HFn has the risk of undesired drug accumulation in healthy tissues.

To address (1) the short half-life and (2) the limited tumor targeting ability of HFn, two functional peptides were fused to HFn subunit C-terminal to construct three functionalized HFns (HFn-PAS, HFn-GFLG-PAS-RGDK and HFn-PLGLAG-PAS-RGDK). One peptide, PAS peptide, comprises repetitive P, A and S residues. It was designed by Schlapschy, and aimed to mimic poly ethylene glycol (PEG) [10]. In three previous studies, Falvo et al. have fused 40 aa and 75 aa PAS peptides to human ferritin subunit at N-terminal to increase half-life in circulation [11–13]. Another peptide is a tetrapeptide named as RGDK. It belongs to tumor penetration peptide (TPP) and possesses two functions. RGDK enhances drug tumor delivery, and drug distribution inside whole tumor tissue instead of only tumor cells alongside tumor vessels [14]. It specifically binds to two receptors, integrin $\alpha v\beta 3/5$ and neuropilin-1, both overexpressed in a wide range of tumor cells [15]. Integrin $\alpha v\beta 5$ is highly expressed in cancers such as gliomas and urothelial cancer, and neoropilin-1 expression is upregulated in ovarian cancer, colorectal cancer and stomach cancer [9]. Therefore, the addition of RGDK peptide can improve HFn tumor targeting ability and broaden HFn application. The GFLG (Gly-Phe-Leu-Gly) and PLGLAG (Pro-Leu-Gly-leu-Ala-Gly) in HFn-GFLG-PAS-RGDK and HFn-PLGLAG-PAS-RGDK are enzyme cleavable sites responding to cathepsin B and matrix metalloproteinase-2/9 (MMP-2, MMP-9), respectively [16,17]. Both enzymes are overexpressed in tumors but Cathepsin B is located inside cell lysosome and MMP-2 is secreted outside tumor cells [18,19]. The PAS-RGDK functional moiety in these two dually-functionalized HFns is theoretically to be cleaved from HFn before and after cell internalization, respectively. In total, four HFn-based proteins were compared with each other to investigate the impacts of PAS and RGDK on HFn performance as a drug nanocarrier.

A total of four HFn-based proteins were expressed in *Escherichia coli* (*E. coli*), and purified. High-Performance Size Exclusion Chromatography coupled with Multiple Angle Laser Light Scattering (HPSEC-MALLS) was used to characterize protein structures. In vitro and in vivo tests were designed to compare anti-tumor drug delivery performance of four HFn-based proteins in tumors lacking overexpressed human TfR1. Therefore, 4T1, a BALB/c mice breast tumor cell line was selected; 4T1 does not express human TfR1 and overexpresses integrin $\alpha v\beta 3/5$ and neuropilin-1 [20,21]. Cellular uptake assay investigated RGDK functionalization impact on 4T1 cellular internalization efficiency. Intracellular distribution monitored if drug can be released from proteins and enter nucleus for killing tumor cells after internalization. Cytotoxicity assay compared IC_{50} values of drug carried by four HFn-based proteins. Pharmacokinetic study mainly assessed PAS impact on half-life in circulation. Biodistribution study assessed tumor targeting ability of four HFn-based proteins. In vivo anti-tumor test was conducted to compare the tumor growth inhibition efficacy of DOX carried HFn and functionalized HFns.

2. Materials and Methods

2.1. Materials

A total of four recombinant HFn-based proteins (HFn, HFn-PAS, HFn-GFLG-PAS-RGDK and HFn-PLGLAG-PAS-RGDK) were expressed in *Escherichia coli* (*E. coli*) BL21 (DE3). HFn-PAS was constructed by inserting a 15 aa flexible linker (GGGSGGGTGGGSGGG), an enzyme-cleavable site GFLG, a 40 aa PAS peptide (ASPAAPAPAPAAPAPSAPAASPAA-PAPASPAAPAPSAPA) together with another 5 aa flexible liner (GGSGG) to HFn Subunit C-terminus. HFn-GFLG-PAS-RGDK was constructed by adding RGDK tretapeptide to HFn-PAS C-terminus. HFn-PLGLAG-PAS-RGDK, was designed by substitution of enzyme-cleavable site GFLG in HFn-GFLG-PAS-RGDK by a six residue MMP-2 cleavable site PLGLAG. Proteins were purified using a two-step pathway. Briefly, HFn was purified by heat-acidic precipitation at 60 °C, pH 4.5 5 min followed by butyl fast flow hydrophobic interaction chromatography (GE Healthcare, Waukesha, WI, USA). The other three functionalized HFns were purified by heat-acidic precipitation at 60 °C, pH 4.5 5 min followed by mono Q ion-exchange chromatography (GE Healthcare, Waukesha, WI, USA).

Doxorubicin hydrochloride (DOX) was purchased from Dalian Meilun Biotechnology (Dalian, China). 4T1 cells were purchased from Cellbank (Sydney, NSW, Australia). RPMI-1640 medium, penicillin-streptomycin solution (100 ×), fetal bovine serum (FBS), 0.25% trypsin-EDTA (1 ×) solution, Hoechst 33258 reagent and MTT reagent were purchased from Invitrogen (Thermo Scientific, Adelaide, SA, Australia). Propidium iodide and trypan blue solution were bought from Sigma-Aldrich (Sydney, NSW, Australia). All of the other reagents were of analytical reagent quality. Mili Q water was utilized throughout the whole procedure, produced by Merck Mili Q direct (Melbourne, VIC, Australia).

2.2. HPSEC-MALLS Characterization of Purified Proteins and DOX Loading

The four protein purities were analyzed by reducing 12% SDS-PAGE (Bio-Rad, Gladesville, NSW, Australia). Sizes and molecular weights (Mws) were measured by HPSEC-MALLS. In HPSEC-MALLS analysis, TSK G4000 SWxl column (Tosoh bioscience, Tokyo, Japan) was connected to HPLC (Shimadzu, Melbourne, VIC, Australia) coupled with DAWN MALLS and Optilab refractive index (RI) detector (Wyatt, Santa Barbara, CA, USA). Equilibration buffer was 20 mM phosphate buffer (PB), 0.1 M Na_2SO_4, pH 7.0. Flow rate was 0.8 mL min^{-1}. Absorbance of fractions at 280 nm was monitored. Sample loading volume was 50 μL.

In DOX loading, briefly, 1 mg mL^{-1} HFn-based protein in 20 mM phosphate buffer, 5 mM guanidinium chloride, pH 7.5 was heated at 50 °C for 6 h with 0.2 mg mL^{-1} DOX. Excessive DOX was separated from DOX loaded protein (protein/DOX) by desalting on Hitrap G25 desalting column (GE Healthcare, Waukesha, WI, USA) using AKTA PURE (GE Healthcare, Waukesha, WI, USA). Collected protein/DOX underwent measurement of OD280 and OD480. DOX has absorbance at both 280 and 480 nm, and protein only has absorbance at 280 nm. Therefore, two (2) assumptions were made: (1) $OD480_{protein/DOX} = OD480_{DOX}$; (2) $OD280_{protein/DOX} = OD280_{DOX} + OD280_{protein}$. Standard OD vs. C linear curves of DOX and HFn-based proteins were determined by serial concentrations of DOX (1–40 μg mL^{-1}) and proteins (0.1–1.2 mg mL^{-1}). Standard curves were used to calculate the concentration of DOX (C_{DOX}) and the concentration of proteins ($C_{protein}$) in protein/DOX. Consequently, the calculation of loading ratio in protein/DOX was as follows:

$$\text{Loading ratio} = \frac{\text{number of DOX}}{\text{number of protein}} = \frac{C_{DOX} \bullet Mw_{protein}}{C_{protein} \bullet Mw_{DOX}} \tag{1}$$

2.3. Cellular Uptake Test

The 4T1 cell line was cultured in RPMI-1640 medium supplemented with 10% FBS at 37 °C in a 5% CO_2 atmosphere. Cellular uptake test procedure was modified from a previous paper [15]. For each protein/DOX group, three different treatments were conducted to obtain three fluorescence intensities, total fluorescence, internalized fluorescence, fluorescence after RGDK peptide pre-incubation.

The procedure was as follows: (1) Cell seeding. 4T1 Cells in the exponential growth phase were seeded in 24-well plates at a density of 1×10^5 cells per well and cultured for 48 h for attachment. (2) RGDK peptide pre-incubation in wells for fluorescence after RGDK peptide pre-incubation determination. To investigate the impact of fused RGDK on cellular uptake characteristics, 500 μM free RGDK peptide was pre-incubated with the cell for 1 h at 37 °C to saturate RGDK specific receptors. (3) Drug incubation. The media with or without RGDK peptide in all wells were discarded and cells were washed with phosphate buffered saline (PBS) three times, prior to adding 100 μL serum-free culture medium containing free DOX or protein/DOX (15 μg mL^{-1} DOX-equivalent). Then the cells were incubated for 90 min at 37 °C and washed three times with PBS to remove drugs. (4) Trypan blue quenching. In wells for internalized fluorescence and fluorescence after RGDK peptide pre-incubation determination in all five groups, cells were incubated with trypan blue (0.25% in 0.85% NaCl) for 5 min at 25 °C, and then washed five times with PBS to remove trypan blue. (5) Detachment of cells for flow cytometry analysis. A total of 400 μL of 0.25% trypsin–0.05% EDTA solution was added to all wells for digestion for 5 min at 37 °C and 2 mL of complete medium was added to stop the digestion. Detached cells were spun at 112 rcf for 3 min at 4 °C and re-suspended in 1 mL PBS. In total, five microliters of propidium iodide (PI) was added to incubate with cells for 10 min at 25 °C for differentiation of alive and dead cells in flow cytometry detection. (6) Flow cytometry analysis. Csampler flow cytometry (Becton Dickinson, San Jose, CA, USA) was employed to determine the mean fluorescence of 5000 cells in each sample. A cell control underwent PI staining but without drug incubation, trypan blue and RGDK peptide treatment was used for gating and parameter setting prior to sample detection. PE channel (excitation laser light: 488 nm, emission: 578 nm) was utilized for DOX fluorescence detection. Mean fluorescence intensity of each sample was recorded.

2.4. Intracellular Distribution Analysis

Intracellular distribution analysis was designed to monitor if DOX carried by HFn-based proteins could reach tumor cell nucleus for disruption of cell division. Exponentially growing 4T1 cells were placed on a 6-well plate at a density of 4×10^5 cells per well and cultured for 24 h. One cover-glass slide was put in each well prior to seeding. The medium was then discarded and cells were treated with fresh media containing protein/DOX or free DOX (20 μM DOX-equivalent) in 2 mL per well for 3 h. Drugs in wells were then removed and cells were washed three times using PBS. Fresh complete medium was added to wells for another 36 h incubation. Subsequently, the cells were washed three times with PBS and fixed with 4% paraformaldehyde for 10 min at 25 °C. Following another three times wash with PBS, cell nucleus were stained with 0.5 μg mL^{-1} Hoechst 33258 at 25 °C for 5 min. A ZOE fluorescence cell imager (Bio-Rad, Gladesville, NSW, Australia) was used to visualize cells. Images of cells under bright filed channel, green channel (Excitation: 480/17 nm, Emission: 517/23 nm) and blue channel (Excitation: 355/40 nm, Emission: 433/36 nm) were captured. Green channel and green channel monitored Hoechst 33258 and DOX signal, respectively.

2.5. Cytotoxicity Study

The cytotoxicity of four protein/DOX and free DOX against 4T1 cell was evaluated by MTT assays. Exponential growth-phase cells were digested by 0.25% trypsin-0.05% EDTA, and cell density was adjusted to 1×10^5 cells per mL by complete medium. 100 μL of cells were seeded in wells of 96-well plates. Then, four wells without cells were adopted as blank control on each plate. After incubation for 24 h, medium was replaced with new complete medium separately containing either free DOX or protein/DOX, whose concentrations ranged from 0 to 30 μg mL^{-1} (equivalent DOX). Four 0 μg mL^{-1} DOX wells on each plate were cell control wells. After incubation for another 60 h, the media were removed and cells were washed three times by PBS. Then, 90 μL of new complete medium with 10 μL of MTT solution was added to each well for another 4 h. A total of 100 μL dimethyl sulfoxide (DMSO) was added to wells to ensure complete solubilization of the formed form-azan crystals. Finally, the absorbance of the solution was measured at 595 nm (background: 630 nm) by a Microplate Reader (Biotek, Winooski, VT, USA). Absorbance of each well (A$_{well}$) was defined as A$_{595}$–A$_{630}$. Cell viability (%) were calculated using Equation (2).

A_{cell} was the A_{well} of cell controls, and A_{blank} was the A_{well} of blank controls. IC_{50} value of each group was calculated using dose-response fitting in origin 9.0 software (Originlab, Northampton, MA, USA).

$$\text{Cell viability (\%)} = (A_{well} - A_{blank})/(A_{cell} - A_{blank}) \times 100 \text{ (\%)} \tag{2}$$

2.6. Pharmacokinetics Study

All animal experiments were performed with the approval of the medical ethics committee of Shanxi University of Chinese Medicine (Approval Number 2019LL137, approval date: 13 June 2019). Specific-pathogen free Sprague Dawley rats (male, 230–250 g, SPF Biotechnology Co., Ltd. Beijing, China) were randomly assigned to six groups (three rats in each group), and administrated with PBS, free DOX and protein/DOX (3.0 mg kg^{-1} DOX equivalent) separately via intravenous injection at tail vein. After injection, blood samples were collected from the retro orbital sinus at fixed time points (10, 30 min, 1, 2, 4, 8, 12, 24, 36, 48 h) and followed by clotting for at least 0.5 h at 37 °C. Serum was obtained by centrifugation at 4032 rcf for 30 min at 4 °C. Finally, 100 μL of serum of each sample was transferred to a 96-well microplate, and the DOX contents were determined using SpectraMax i3x microplate reader (Molecular devices, San Jose, CA USA). Excitation wavelength was set at 480 nm and emission at 580 nm. Meanwhile, the standard curve of the fluorescence intensity with varying concentrations of DOX in rat serum was also measured for quantitative analysis. Half-lives of DOX and protein/DOX were calculated using Drug Analysis System 2.0 software (Drug China, Shanghai, China) by fitting data in single-compartment mode.

2.7. In Vivo Imaging

The four HFn-based proteins were first labelled by Sulfo-cy5 NHS ester (Lumiprobe, Hunt Valley, MD, USA) with a molar ratio of 1:30 (Protein to Cy5) and the uncoupled Cy5 was removed by Hitrap G25 desalting chromatography. As 4T1 is a BALB/c breast tumor cell line, female BALB/c mice were chosen to establish tumor-bearing animal model. 1×10^6 4T1 cells in 100 μL of PBS were injected into right armpit of 8-week old female BALB/c mice (specific-pathogen free, SPF Biotechnology Co., Ltd. Beijing, China) to form mice tumor model. Each group had three mice. When tumor volume reached about 300 mm^3, a 150 μL sample of Cy5 or protein-Cy5 conjugates (0.2 mg kg^{-1} Cy5 equivalent) was intravenously injected into the tumor-bearing mice via tail vein. After treatment, the mice were anesthetized using isoflurane at 2, 4, 6.5, 12, 24 and 52 h and fluorescence images were taken under excitation wavelength of 646 nm and emission wavelength of 662 nm using FX Pro in vivo imaging system (Bruker BioSpin, Carteret, NJ, USA).

2.8. Anti-Tumor Assay

Then, 1×10^6 4T1 cells in 100 μL of PBS were injected into right armpit of 8-week old female BALB/c mice. For in vivo inhibition of tumor progression assessment, female BALB/c mice bearing 4T1 tumors of approximate 250 mm^3 in size were randomly assigned to six groups ($n = 6$ in each group) and treated with protein/DOX (3 mg kg^{-1} DOX equivalent), free DOX (3 mg kg^{-1}), or PBS via 200 μL intravenous injection. The drug injection was carried out every 4 days for two doses. The volumes of tumors were measured every other day. Mice were monitored for up to 17 days post-implantation and then sacrificed. Primary tumors were harvested for ex vivo imaging.

2.9. Statistical Analysis

Data were presented in Mean ± Standard deviation (SD). T-test was applied to evaluate statistical significance of results. p value < 0.05 was considered significant.

3. Results

3.1. Purified HFn-Based Protein Characterizations and Drug Loading

The purity of each protein after purification reached above 90% based on the SDS-PAGE gel (Figure 1A), calculated by density scan using software Image J [22]. The apparent subunit molecular weights of HFn-PAS, HFn-GFLG-PAS-RGDK and HFn-PLGLAG-PAS-RGDK on gel were higher than their theoretical molecular weights (26 kDa, 26.5 kDa and 26.6 kDa), which are due to the hydration of PAS peptides. Bands of two PAS-RGDK functionalized HFns in SDS-PAGE gel were slightly higher than HFn-PAS probably due to the presence of extra residues.

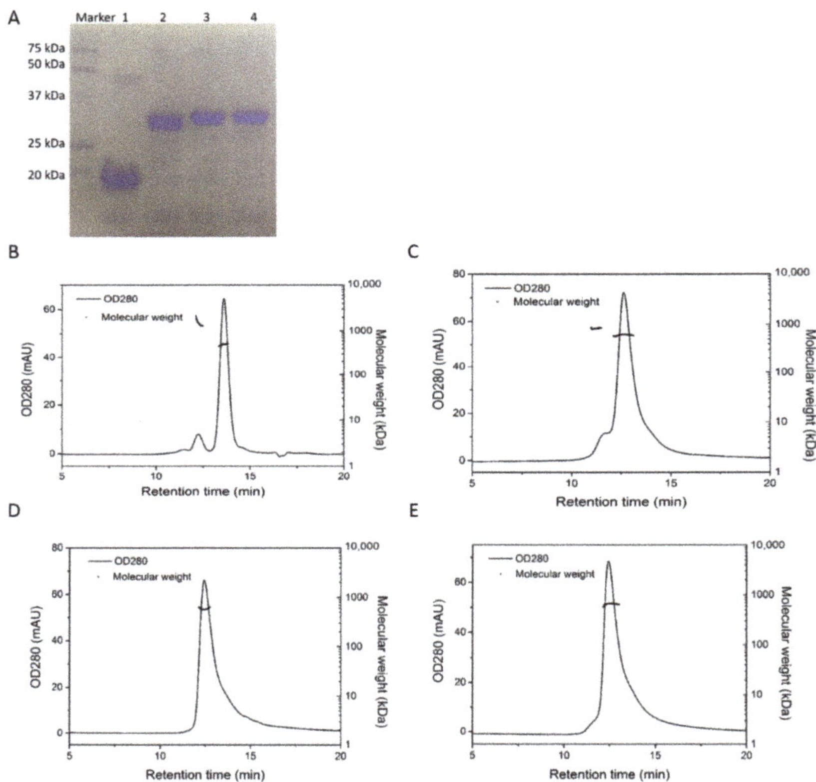

Figure 1. Characterizations of purified HFn-based proteins. (**A**), 12% reducing SDS-PAGE analysis of purified proteins. Lane 1, HFn; 2, HFn-PAS; 3, HFn-GFLG-PAS-RGDK; 4, HFn-PLGLAG-PAS-RGDK. (**B**), HPSEC-MALLS chromatogram of HFn. (**C**), HPSEC-MALLS chromatogram of HFn-PAS. (**D**), HPSEC-MALLS chromatogram of HFn-GFLG-PAS-RGDK. (**E**), HPSEC-MALLS chromatogram of HFn-PLGLAG-PAS-RGDK.

The apparent hydrodynamic radius and molecular weight (Mw) of four HFn-based proteins were further characterized by HPSEC-MALLS analysis. In Figure 1B–E, the horizontal Mw lines of the main peaks show the uniform Mws of all four HFn-based proteins. Table 1 lists the hydrodynamic size and Mw of each protein. Due to PAS peptides, HFn-PAS possessed 1.4 nm higher apparent hydrodynamic radius in contrast to HFn. The adding of RGDK peptide and enzyme-cleavable site into HFn slightly further increased hydrodynamic radius. HFn-GFLG-PAS-RGDK and HFn-PLGLAG-PAS-RGDK were 1.75 nm and 1.91 nm larger than HFn, respectively. MALLS determined Mw order is in accordance with theoretical order: HFn-PLGLAG-PAS-RGDK > HFn-GFLG-PAS-RGDK > HFn-PAS > HFn, and average Mw of all three proteins determined are similar to their theoretical Mw (Table 1).

Table 1. Hydrodynamic radius and molecular weight determined by HPSEC-MALLS.

Protein	Particle Hydrodynamic Radius (nm)	Measured Average Molecular Weight (kDa)	Theoretical Molecular Weight (kDa)
HFn	6.31 (±0.53%)	493.3(±0.07%)	506.0
HFn-PAS	7.72 (±0.54%)	608.6(±2.18%)	625.1
HFn-GFLG-PAS-RGDK	8.06 (±0.55%)	630.8(±0.12%)	636.1
HFn-PLGLAG-PAS-RGDK	8.22 (±0.55%)	638.7(±0.11%)	639.3

After purification, the model drug DOX was loaded by thermally induced passive diffusion. On average, incubation with DOX at 50 °C loaded 33.5, 38.4, 36.9 and 42.1 DOX in one HFn, HFn-PAS, HFn-PLGLAG-PAS-RGDK and HFn-GFLG-PAS-RGDK nanocage, respectively. HFn DOX loading ratio in this study is comparable with previous pH-induced disassembly–reassembly method [23] and 8 M urea method adopting HFn [4].

3.2. Cellular Uptake Efficiency

In cellular uptake test, we investigated the RGDK functionalization impact on cellular uptake efficiency and the mechanism. Figure 2 presents DOX fluorescence intensities of all groups. Total fluorescence of DOX measured in flow cytometry came from two sources, DOX internalized by 4T1 cells and un-specifically bound to cell membranes. Trypan blue treatment quenched the signal from membrane-bound DOX, and, therefore, a lower internalized fluorescence intensity compared with total fluorescence intensity was observed in all groups (Figure 2).

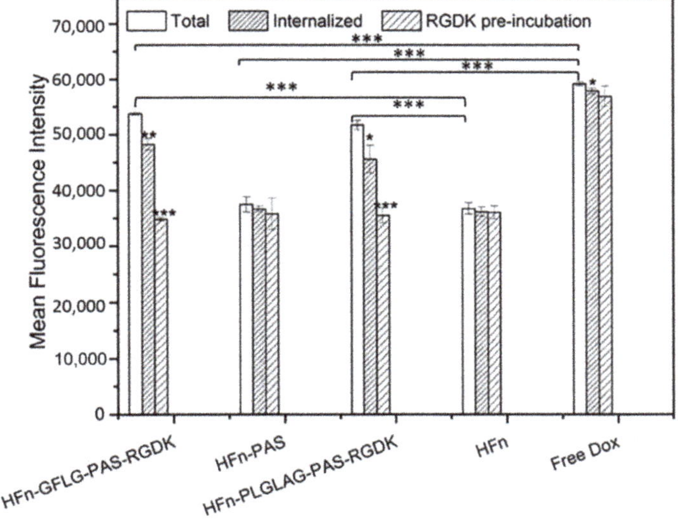

Figure 2. Mean DOX fluorescence intensity in 4T1 cellular uptake test. Data were represented as mean ± standard deviation (n = 3), * $p < 0.05$, ** $p < 0.01$, *** $p < 0.001$. Symbol '*' on top of column represents the significance of p value between this column and the white column (total DOX fluorescence) in the same group.

Free DOX showed significantly greater internalized cellular uptake than others. HFn-GFLG-PAS-RGDK/DOX and HFn-PLGLAG-PAS-RGDK/DOX had the second highest efficiencies and were significantly different from the rest two. This means the insertion of RGDK peptide has significantly enhanced cellular uptake efficiency. HFn-PAS/DOX and HFn/DOX had similar internalized cellular uptake efficiencies. RGDK peptide pre-incubation treatment, with the use of excessive amount of free RGDK, is intended to mask RGDK-specific receptors, integrin $\alpha v \beta 3/5$ and neuropilin-1, on cells to hamper RGDK-related cellular uptake. In Figure 2, the uptake of HFn-GFLG/PLGLAG-PAS-RGDK groups were significantly inhibited by RGDK peptide pre-incubation while in other groups no obvious difference occurred. After the pre-incubation of RGDK peptide, internalized fluorescence intensities of HFn-GFLG-PAS-RGDK/DOX and HFn-PLGLAG-PAS-RGDK/DOX were similar to that of HFn/DOX and HFn-PAS/DOX. This proves that RGDK facilitated 4T1 cells' internalization of HFn-GFLG-PAS-RGDK/DOX and HFn-PLGLAG-PAS-RGDK/DOX by providing RGDK-specific receptor-mediated pathway.

The difference of tumor cell uptake efficiencies lies in various uptake mechanisms. DOX is a small molecule and enters cells via passive diffusion. Passive diffusion is energy-free and concentration gradient-driven. It is quicker compared with all other internalization pathways when directly incubating drugs with cells. As 4T1 does not express human TFR1, HFn/DOX and HFn-PAS/DOX probably enter the cell through non-specific pinocytosis. In contrast with HFn/DOX and HFn-PAS/DOX, HFn-GFLG-PAS-RGDK/DOX and HFn-PLGLAG-PAS-RGDK/DOX have an extra internalization pathway by binding to RGDK recognized receptors, integrin $\alpha v \beta 3/5$ and neuropilin-1.

3.3. Intracellular Distribution

DOX is an anthracycline topoisomerase inhibitor and exerts its function mainly inside the cell nucleus [24]. Free DOX directly diffuses into nucleus and disrupts cell division after internalization, whilst the protein/DOX are supposed to first be broken down by enzymes in lysosome to release loaded DOX and then reach nucleus. Intracellular distribution test was designed to check if the drugs loaded on HFn-based proteins could enter cell nucleus to kill tumor cells. In Figure 3, the blue color indicates where cell nucleus is and green color represents the fluorescence from DOX. Clearly, the majority of DOX has entered and accumulated inside nucleus of 4T1 cells in all groups, as the cyan color is the dominant color in merged images. This shows that the DOX in four protein/DOX groups could accumulate in 4T1 cell nucleus, the same as free DOX.

3.4. Functionalization Effect on Cytotoxicity

In order to test the inhibition of protein/DOX on tumor cell proliferation, we adopted an MTT assay. All DOX loaded HFn-based proteins demonstrate obvious anti-proliferation abilities (Figure 4). Free DOX group had the lowest IC_{50} (Table 2) and this is due to its relatively high cellular internalization efficiency. Inhibition impacts of HFn-GFLG-PAS-RGDK/DOX and HFn-PLGLAG-PAS-RGDK/DOX on tumor cell growth were similar and the second strongest, HFn-PAS/DOX ranked third, and HFn/DOX showed the worst anti-proliferation effect. T-test shows there was significant differences between IC_{50} values of free DOX and HFn-GFLG/PLGLAG-PAS-RGDK/DOX group ($p < 0.05$). Significant differences of IC_{50} values were also found between HFn-GFLG/PLGLAG-PAS-RGDK/DOX and the other two HFn-based protein/DOX groups ($p < 0.05$). That implies the RGDK in HFn-GFLG/PLGLAG-PAS-RGDK/DOX has enhanced HFn/DOX performance in terms of drug cytotoxicity towards tumor cells.

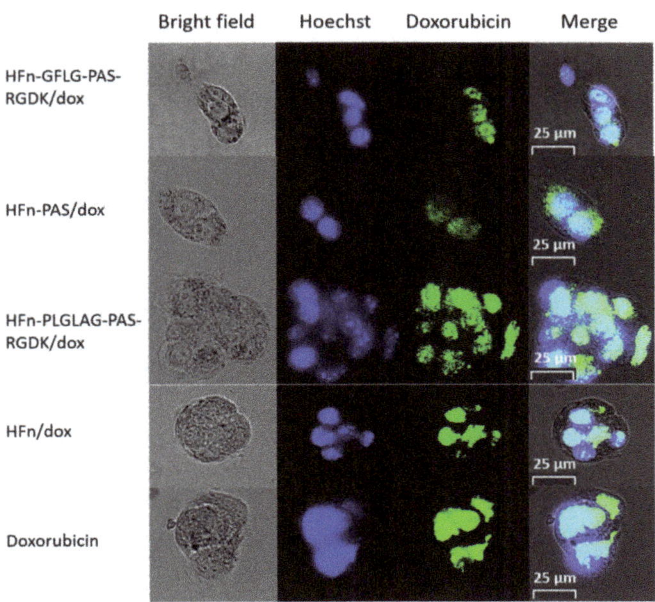

Figure 3. Intracellular distribution of DOX in 4T1 cells. Under the Bio-Rad Zoe cell imager, blue: nucleus after being stained with Hoechst 33258. Green: DOX because of its intrinsic fluorescence. Cyan: merged florescence signal.

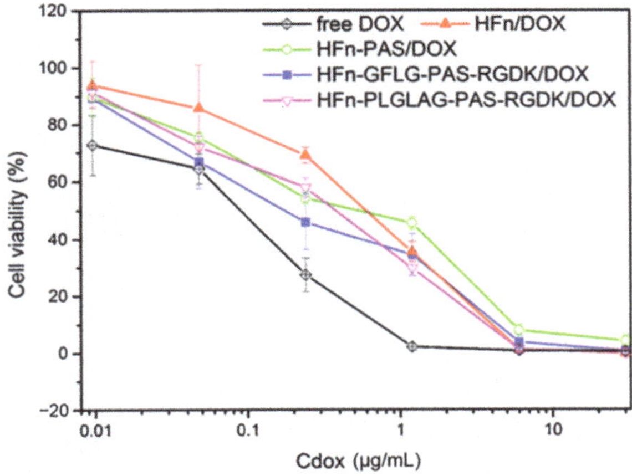

Figure 4. Proliferation inhibition on 4T1 cells. Data were mean ± standard deviation ($n = 4$).

Table 2. IC$_{50}$ values of all groups.

Group	IC$_{50}$ (μg mL^{-1})
DOX	0.08 ± 0.03
HFn/DOX	0.49 ± 0.11
HFn-PAS/DOX	0.38 ± 0.09
HFn-GFLG-PAS-RGDK/DOX	0.17 ± 0.01
HFn-PLGLAG-PAS-RGDK/DOX	0.18 ± 0.04

3.5. Functionalization Effect on Pharmacokinetic Profile

Pharmacokinetic profile of all four protein/DOX and free DOX were obtained through tail vein injection of healthy Sprague Dawley rats. Line chart of DOX concentrations in plasma over time (10 min–48 h) is shown in Figure 5 and half-lives in circulation of all protein/DOX are listed in Table 3. Standard curve of fluorescence intensity-doxorubicin concentration is shown in supplementary material Figure S1. Plasma drug concentrations of free DOX group rats reduced rapidly right after administration (Figure 5). At 10 min, average plasma drug concentration was 15 μg mL^{-1}. Then, 8 h later, almost all free DOX was eliminated from circulation. In terms of HFn/DOX group rats, their plasma drug cleaning out speed ranked the second, with average drug concentration of 21.6 μg mL^{-1} at 10 min and below 5 μg mL^{-1} after 12 h. Average plasma drug concentrations in all three functionalized HFn/DOX group rats (approximate 35 μg mL^{-1}) were more than double of those in free DOX group at 10 min. 48 h after administration, more than 5 μg mL^{-1} drug still remained in plasma of all three functionalized HFn/DOX group rats. Single compartment fitting of the drug concentration-time curve was applied to evaluate drug half-life in circulation. The results show free doxorubicin only had about 25 min of half–life in circulation (Table 3). HFn/DOX half-life, approximate 3 h, was 7.3 times of free DOX. PAS peptide in HFn-PAS/DOX has increased half-life almost 4.9 times (14.96 h) compared with HFn/DOX. The extra RGDK residues in HFn-GFLG-PAS-RGDK/DOX further extended the half-life to 17.61 h. HFn-PLALGA-PAS-RGDK/DOX possessed the longest half-life, 18.93 h. Differences in half-life of all three functionalized HFns/DOX compared with HFn/DOX and free drug were statistically significant in *t*-test ($p < 0.001$).

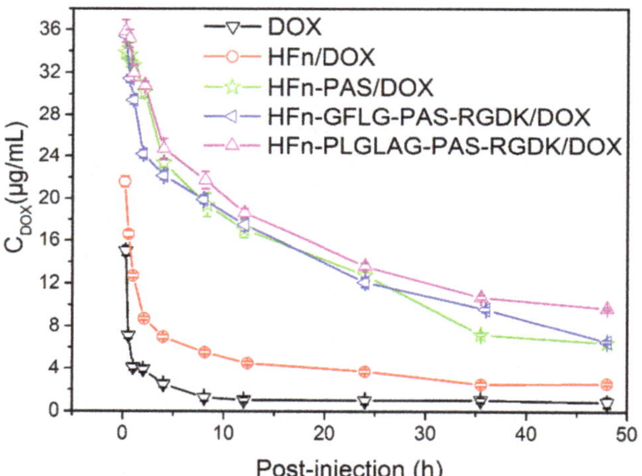

Figure 5. Plasma concentrations of protein/DOX and free DOX in Sprague Dawley rats of different groups. Data were expressed as mean $\pm$ SD ($n = 3$).

Table 3. Half-life of each protein/DOX in Sprague Dawley rats ($n = 3$).

Group	$T_{1/2}$(h)
DOX	0.42 ± 0.03
HFn/DOX	3.07 ± 0.06
HFn-PAS/DOX	14.96 ± 0.29
HFn-GFLG-PAS-RGDK/DOX	17.61 ± 0.39
HFn-PLGLAG-PAS-RGDK/DOX	18.93 ± 0.61

3.6. Functionalization Effect on Protein Biodistribution

To monitor distribution of all HFn-based proteins in tumor-bearing mice after tail vein administration over time, we used in vivo imaging. In this analysis, fluorescence label Cy5 was attached to all employed proteins and free cy5 worked as free drug control. The reagent in use reacts with primary amine group on protein outer surface. On the outer surface of HFn assembly, there are 24 exposed subunit N-terminals –NH_2 groups and 144 Lys (K) residues. Due to the large number of accessible reaction sites, the possibility of cy5 blocking some or all K residues of RGDK in HFn-GFLG/PLGLAG-PAS-RGDK is low. The cy5 conjugation is unlikely to affect RGDK function. Real-time biodistribution of Cy5 attached proteins and free Cy5 were visualized in BALB/c mice with 4T1 tumor in right armpit, and fluorescence intensities of tumor areas were recorded. Two control groups, mice injected with free cy5 and HFn-cy5 were scanned at the same time, and mice in other three groups (HFn-PAS-cy5, HFn-PLALGA-PAS-RGDK-cy5 and HFn-GFLG-PAS-RGDK-cy5) were scanned together.

Figure 6A shows the top half of the mice where there was fluorescence signal captured by camera. 4–12 h after injection, fluorescence of free cy5 in liver was captured by camera (Figure 6A). At 24 and 52 h, fluorescence was barely visible. Free cy5 preferred to accumulate in the liver, perhaps due to the fact that liver is the main organ for metabolism. Theoretically, as a nanoparticle, HFn has passive tumor targeting ability. However, from the results in Figure 6A, signal of HFn-cy5 fluorescence was captured in liver rather than in tumor from 4 to 12 h. It seems that HFn-cy5 did not show desirable tumor targeting ability and it preferred liver. The particle size of HFn is probably still too small to achieve desirable passive tumor targeting ability. No obvious fluorescence was captured at all time points in HFn-PAS-cy5 group (Figure 6A). However, as is shown in Figure 6B, there actually was fluorescence detected in tumor area. Perhaps because of the sharp contrast between signal intensities of HFn-PAS-cy5 and HFn-GFLG/PLGLAG-PAS-RGDK-cy5, lower intensity of HFn-PAS-cy5 failed to be captured by the camera under the same exposure time. In HFn-GFLG/PLGLAG-PAS-RGDK-cy5 groups, fluorescence signal was captured from 4 h to 52 h after injection (Figure 6A). The armpit fluorescence areas at 6.5, 12 and 24 h were larger than the area of armpit lymph node, proving the protein accumulation in tumor tissues. However, it is uncertain that if lymph node accumulation co-existed or not. Figure 6A shows that the tumor area of HFn-GFLG/PLGLAG-PAS-RGDK-cy5 groups had stronger signals than liver at all time points. At 52 h after injection, whilst HFn-cy5 and free cy5 were almost completely eliminated, HFn-GFLG/PLGLAG-PAS-RGDK-cy5 were still detectable in region of tumor site, implying functionalized HFns were retained in tumor by longer and stronger accumulation.

As is presented in Figure 6B, free cy5 and HFn-cy5 had the lowest tumor florescence intensity at all time points. At 2 h, free cy5 had a greater intensity than HFn-cy5 but was surpassed by HFn-cy5 afterwards. Free cy5 tumor area fluorescence intensity peaked at 4 h and decreased rapidly after that, suggesting a fast clearance. HFn-cy5 achieved the highest concentration in tumor at around 4 h after injection (Figure 6B). The difference in free cy5 and HFn-cy5 is likely to be due to a quicker distribution and a shorter half-life of small molecule cy5 than HFn-cy5. HFn-PAS-cy5 demonstrated significantly stronger and longer lasting tumor intensities than HFn-cy5 at all detected time points ($p < 0.001$). As proven in the pharmacokinetic study, the insertion of the PAS peptide could lead to a longer half-life in circulation and probably result in the slower clearance of HFn-PAS-cy5 than HFn-cy5. The best drug targeting delivery results were from HFn-GFLG/PLGLAG-PAS-RGDK-cy5. They had significantly greater signal intensities in tumor area at all times than all the other groups ($p < 0.001$). This shows that the RGDK peptide can significantly improve HFn biodistribution. A previous study of RGDK fused Albumin binding domain has also proven the tumor targeting ability improvement of RGDK peptide *in vivo*. [24] Overall, both PAS and RGDK functionalization, and particularly RGDK functionalization, improved the tumor biodistribution of HFn.

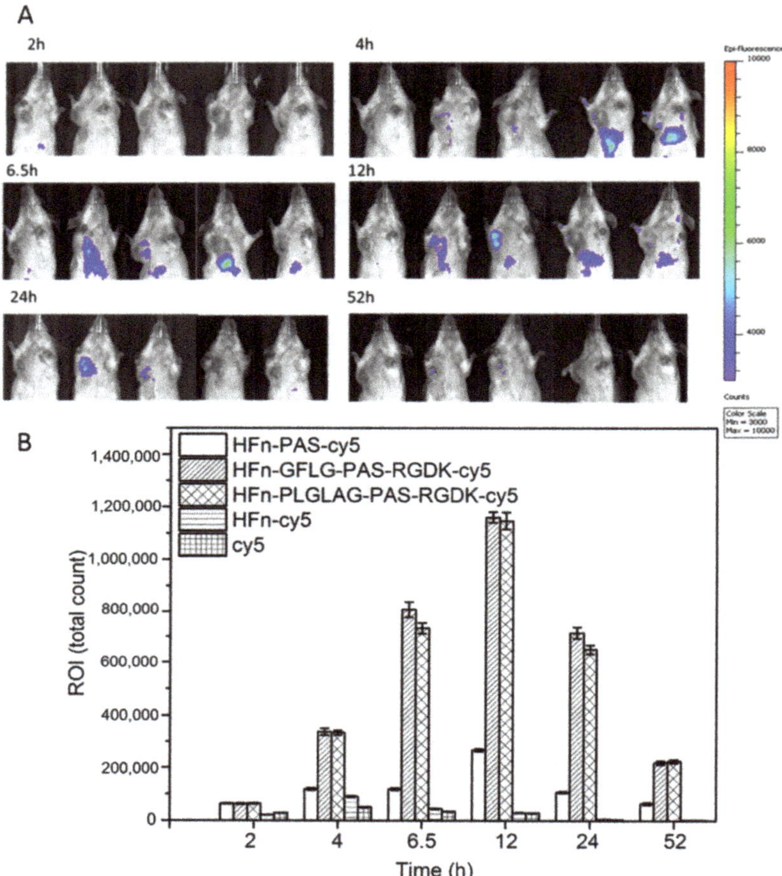

Figure 6. Biodistribution of cy5 and cy5 conjugated with HFn-based proteins. (**A**), in vivo fluorescence imaging of tumor-bearing mice at different time points, from left to right: HFn-PAS-cy5, HFn-GFLG-PAS-RGDK-cy5, HFn-PLGLAG-PAS-RGDK-cy5, HFn-cy5 and free cy5. (**B**), the sum fluorescent intensity of region of interest (ROI, tumor area) at each time point.

3.7. Functionalization Effect on Protein/DOX Anti-Tumor Efficacy

To compare tumor treatment efficacies of all protein/DOX and free DOX, 4T1 tumor bearing BALB/c mice model was built and 36 mice with around 250 mm³ tumor were randomly assigned into six groups. Intravenous injections of four HFn-based protein/DOX, free DOX and PBS were conducted at day 0 and day 5. As is shown in Figure 7A, the fastest mice tumor growth rate was observed in PBS control group rats which underwent no drug treatment. The average tumor volume reached 2030 mm³ after 17 days. The second fastest tumor growth rate was in free DOX group mice and their average group tumor volume were 1667 mm³ at day 17. HFn/DOX showed a better tumor growth inhibition and at day 17, tumor volume grew to 1521 mm³. In HFn-PAS/DOX group, mice tumor volume reached 1432 mm³ in the end. Two PAS-RGDK functionalized HFn/DOX treated group had the strongest tumor-growth inhibition. In spite of just two administrations, average tumor volume of these two group mice at day 17 were just around 1100 mm³, close to half of the volume of the PBS group tumor.

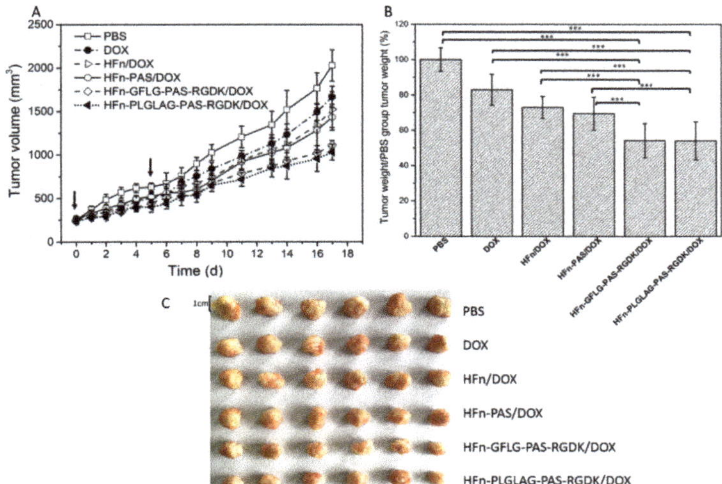

Figure 7. In vivo tumor inhibitory effects on 4T1 tumor-bearing mice. (**A**), tumor volume change over time. (**B**), group tumor weight/ PBS group tumor weight (%) on day 17. *** $p < 0.001$. Symbol '*' represents the significance of p value between groups. (**C**), the photo of excised tumor tissues on day 17. Arrows indicated the injection days; data are mean ± RSD ($n = 6$).

On day 17, the corresponding average tumor weights were measured and the percentages of tumor weights compared with tumor weights of control PBS group mice are presented in Figure 7B. The photo of excised tumor tissues is shown as Figure 7C. Tumor weights on day 17 of HFn-GFLG-PAS-RGDK/DOX and HFn-PLGLAG-PAS-RGDK group mice were 54.2 ± 9.7% and 54.0 ± 10.8% of tumor weights of the control PBS group. HFn-PAS/DOX group mice had 69.5 ± 9.4% and HFn/DOX group mice had 72.99 ± 6.2% of PBS control group mice tumor weights. Free DOX group mice tumor weight was 82.9 ± 8.7% of PBS control group mice. T-test results demonstrate that there were significant statistical differences between final tumor masses of protein/DOX and free DOX group. Both HFn and functionalized HFns had significantly increased DOX anti-tumor efficacy ($p < 0.001$). Compared with HFn/DOX group, HFn-PAS/DOX did not show statistical distinction ($p = 0.468$), showing that PAS functionalization alone was not enough to significantly improve anti-tumor efficacy. Masses of tumors from two PAS-RGDK protein/DOX groups, however, were significantly lower than those of both HFn/DOX group and HFn-PAS/DOX group. This indicates RGDK functionalization primarily accounts for the significant improvement of growth inhibition efficacy of 4T1 tumor. Difference between HFn-GFLG-PAS-RGDK/DOX and HFn-PLGLAG-PAS-RGDK/DOX ($p = 0.977$) was not significant in statistical analysis. Two different enzyme-cleavable sites did not make a statistical difference in anti-tumor efficacy.

4. Discussion

Based on all of the results above, PAS and RGDK functionalization have both improved HFn anti-tumor performance. PAS functionalization impacts HFn mainly by extension of half-life in circulation. In pharmacokinetic study, differences in half-lives in circulation of all protein/DOX mostly stemmed from the insertion of PAS peptide. In biodistribution assay, PAS provided HFn-PAS with a longer retaining time than HFn in tumor area. Constituted by repetitive P, A and S residues, PAS peptide is hydrophilic and uncharged in neutral solutions and plasma. Circular dichroism shows it is a flexible random coil [25]. Its properties are similar to polyethylene glycol (PEG), but it is advantageous in terms of biodegradability and biocompatibility. When it is attached to another molecule, it can attract water molecules to increase molecule hydrodynamic volume, thereby extending half-life in circulation. PAS peptide usually has to be over 200 residues to achieve half-life extension, but because of the repeated and organized presentation manner on ferritin shell, 40 aa and 75 aa PAS peptides have been proven to be long enough when fused onto N-terminal of ferritin subunit [11]. In this study, the enlargement of HFn hydrodynamic volume after PAS insertion was detected in HPSEC-MALLS characterization.

RGDK peptide significantly improved anti-tumor performance of HFn/DOX. In cellular uptake assay, it has increased cellular internalization efficiency through binding to specific receptors. It has also led to the best tumor targeting abilities in biodistribution assay and the greatest anti-tumor efficacy in cytotoxicity and in vivo anti-tumor assay.

Comparing HFn-GFLG-PAS-RGDK and HFn-PLGLAG-PAS-RGDK, the sequence length difference caused the minor differences in hydrodynamic volume in HPSEC-MALLS and half-life in circulation in pharmacokinetic study. However, the two enzyme-cleavable sites, GFLG and PLGLAG, did not make a significant difference in cellular uptake efficiency, and in any other in vitro and in vivo tests. This suggest the PLGLAG enzyme-cleavable site probably was not digested by MMP-2/9 before cell internalization. It could be caused by the insufficient activity of MMP-2/9 in vitro and in vivo and/or the low accessibility of PLGLAG to enzymes. A further detailed investigation of the cleavage of PLGLAG is needed.

Figure 8 illustrates the assumed tumor cell internalization pathways of all groups. Free DOX enters tumor cells via unspecific passive diffusion due to its small size (Figure 8A). The short half-life in circulation and the lack of tumor targeting ability caused a great loss of DOX before it reached tumor cells. As a result, the in vivo anti-tumor efficacy was the lowest. Both HFn/DOX and HFn-PAS/DOX enter cells through non-specific pinocytosis [26] (Figure 8B), because there is no corresponding receptor, human TfR1, on 4T1 cells. Therefore, in in vitro assessments, cellular uptake assay and cytotoxicity assay, there were no statistical differences between HFn/DOX and HFn-PAS/DOX.

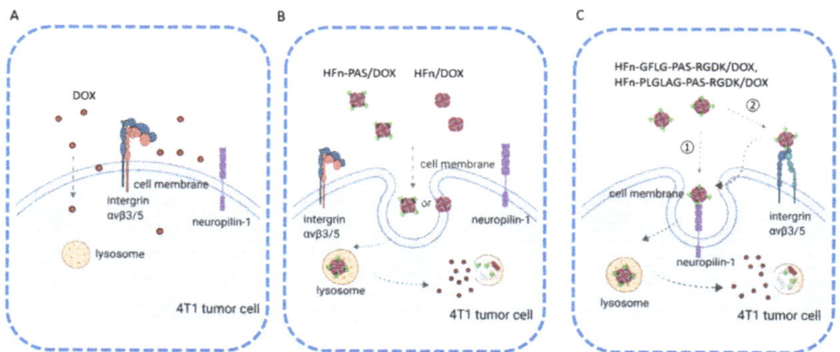

Figure 8. Schematic of different tumor cellular internalization mechanisms of DOX and protein/DOX. (**A**), free DOX passive diffusion pathway. (**B**), HFn/DOX and HFn-PAS/DOX pinocytosis internalization. (**C**), two possible receptor-mediated internalization pathways of HFn-GFLG-PAS-RGDK/DOX and HFn-PLGLAG-PAS-RGDK. (Created with Biorender).

HFn-GFLG-PAS-RGDK/DOX and HFn-PLGLAG-PAS-RGDK/DOX have extra drug internalization mechanisms compared with HFn/DOX and HFn-PAS/DOX (Figure 8C). Since RGDK has RGD motif and an exposed free C-terminal K residue, it can be directly recognized by both integrin αvβ3/5 and neuropilin-1 (NRP1), two kinds of receptors overexpressed on 4T1 cells [14]. The overexpression of these two receptors has led to the increase of internalization efficiency of HFn-GFLG/PLGLAG-PAS-RGDK/DOX. As shown in Figure 8C, when HFn-GFLG/PLGLAG-PAS-RGDK/DOX reaches tumor tissue, there are two possible internalization pathways. RGDK can either directly bind to NRP1 or firstly interact with integrin αvβ3/5 and then be transferred to NRP1, followed by endocytosis.

After that, some of the HFn-GFLG/PLGLAG-PAS-RGDK/DOX inside tumor cells would be digested in lysosome while some will travel to other cells nearby via paracellular pathway or transcytosis [27].

5. Conclusions

All three functionalized HFns expressed in *E. coli* have self-assembled into nanoparticles such as HFn. RGDK peptide has enhanced HFn tumor cell uptake efficiency and improved biodistribution, resulting in a significant improvement in anti-tumor treatment outcome. PAS has expanded HFn hydrodynamic volume and helped ferritin stay longer in circulation, which also has improved anti-tumor efficacy of ferritin. In summary, we successfully prepared and evaluated three new functionalized HFn constructs (HFn-PAS, HFn-GFLG-PAS-RGDK, HFn-PLGLAG-PAS-RGDK), especially two PAS-RGDK fused ones, which hold greater potentials as anti-tumor drug delivery nanoparticles than HFn.

Supplementary Materials: The following are available online at https://www.mdpi.com/article/10.3390/pharmaceutics13040521/s1, Figure S1: standard curve of fluorescence intensity-doxorubicin concentration in SD rat plasma.

Author Contributions: Conceptualization, J.B., Y.W. (Yinli Wang) and Y.L.; methodology, S.Y., Y.W. (Yan Wang) and B.Z.; software, S.Y.; validation, S.Y., Y.Q. and Y.Z.; formal analysis, S.Y.; investigation, S.Y.; resources, Y.L., J.B., Y.W. (Yan Wang) and Y.W. (Yinli Wang); writing—original draft preparation, S.Y.; writing—review and editing, J.B.; supervision, Y.W. (Yinli Wang), S.D., Y.L. and J.B.; project administration, J.B. and Y.W. (Yinli Wang); funding acquisition, Y.L., J.B. and Y.W. (Yinli Wang). All authors have read and agreed to the published version of the manuscript.

Funding: This work was funded by joint PhD Scholarship Scheme of the University of Adelaide and Institute of Process Engineering, Chinese Academy of Sciences, the National Natural Science Foundation of China [Grant No. 21576267], Beijing Natural Science Foundation [Grant Number 2162041], and Shanxi Education Science "1331 project" special research project (Research and Development of Traditional Chinese Medicine Micro-emulsion and New Biological Preparation).

Institutional Review Board Statement: The animal study in this work was conducted with the approval of the medical ethics committee of Shanxi University of Chinese Medicine (Approval Number 2019LL137).

Informed Consent Statement: Not applicable.

Data Availability Statement: The data presented in this study are available on request from the corresponding author.

Acknowledgments: Thanks to Iain Comerford from the University of Adelaide for his help with flow cytometry. Great appreciations to Anton Middelberg from the University of Adelaide for his helpful advice and support on this work.

Conflicts of Interest: The authors declare no conflict of interest.

References

1. He, D.; Marles-Wright, J. Ferritin family proteins and their use in bionanotechnology. *New Biotechnol.* **2015**, *32*, 651–657. [CrossRef] [PubMed]
2. He, J.; Fan, K.; Yan, X. Ferritin drug carrier (FDC) for tumor targeting therapy. *J. Control Release* **2019**, *311–312*, 288–300. [CrossRef]
3. Liu, X.; Wei, W.; Huang, S.; Lin, S.-S.; Zhang, X.; Zhang, C.; Du, Y.; Ma, G.; Li, M.; Mann, S.; et al. Bio-inspired protein–gold nanoconstruct with core–void–shell structure: Beyond a chemo drug carrier. *J. Mater. Chem. B* **2013**, *1*, 3136–3143. [CrossRef] [PubMed]
4. Liang, M.; Fan, K.; Zhou, M.; Duan, D.; Zheng, J.; Yan, X. H-ferritin–nanocaged doxorubicin nanoparticles specifically target and kill tumors with a single-dose injection. *Proc. Natl. Acad. Sci. USA* **2014**, *111*, 14900–14905. [CrossRef] [PubMed]
5. Lei, Y.; Hamada, Y.; Li, J.; Cong, L.; Wang, N.; Li, Y.; Zheng, W.; Jiang, X. Targeted tumor delivery and controlled release of neuronal drugs with ferritin nanoparticles to regulate pancreatic cancer progression. *J. Control Release* **2016**, *232*, 131–142. [CrossRef]
6. Wang, C.; Zhang, C.; Li, Z.; Yin, S.; Wang, Q.; Guo, F.; Zhang, Y.; Yu, R.; Liu, Y.; Su, Z. Extending half-life of H-ferritin nanoparticle by fusing albumin binding domain for doxorubicin encapsulation. *Biomacromolecules* **2018**, *19*, 773–781. [CrossRef] [PubMed]

7. Corte-Rodriguez, M.; Blanco-González, E.; Bettmer, J.; Montes-Bayon, M. Quantitative analysis of transferrin receptor 1 (TfR1) in individual breast cancer cells by means of labeled antibodies and elemental (ICP-MS) detection. *Anal. Chem.* **2019**, *91*, 15532–15538. [CrossRef]

8. Prutki, M.; Poljak-Blazi, M.; Jakopovic, M.; Tomas, D.; Stipancic, I.; Zarkovic, N. Altered iron metabolism, transferrin receptor 1 and ferritin in patients with colon cancer. *Cancer Lett.* **2006**, *238*, 188–196. [CrossRef]

9. The Human Protein Atlas. Available online: https://www.proteinatlas.org/ (accessed on 16 February 2021).

10. Harari, D.; Kuhn, N.; Abramovich, R.; Sasson, K.; Zozulya, A.L.; Smith, P.; Schlapschy, M.; Aharoni, R.; Koster, M.; Eilam, R.; et al. Enhanced *in vivo* efficacy of a type I interferon superagonist with extended plasma half-life in a mouse model of multiple sclerosis. *J. Biol. Chem.* **2014**, *289*, 29014–29029. [CrossRef]

11. Falvo, E.; Tremante, E.; Arcovito, A.; Papi, M.; Elad, N.; Boffi, A.; Morea, V.; Conti, G.; Toffoli, G.; Fracasso, G.; et al. Improved doxorubicin encapsulation and pharmacokinetics of ferritin-fusion protein nanocarriers bearing proline, serine, and alanine elements. *Biomacromolecules* **2016**, *17*, 514–522. [CrossRef]

12. Falvo, E.; Malagrino, F.; Arcovito, A.; Fazi, F.; Colotti, G.; Tremante, E.; Micco, P.D.; Braca, A.; Opri, R.; Giuffre, A.; et al. The presence of glutamate residues on the PAS sequence of the stimuli sensitive nano-ferritin improves in vivo biodistribution and mitoxantrone encapsulation homogeneity. *J. Control Release* **2018**, *275*, 177–185. [CrossRef]

13. Falvo, E.; Damiani, V.; Conti, G.; Conti, G.; Boschi, F.; Messana, K.; Giacomini, P.; Milella, M.; Laurenzi, V.; Morea, V.; et al. High activity and low toxicity of a novel CD71-targeting nanotherapeutic named The-0504 on preclinical models of several human aggressive tumors. *J. Exp. Clin. Cancer Res.* **2021**, *40*, 63. [CrossRef]

14. Yan, F.; Wu, H.; Liu, H.; Deng, Z.; Liu, H.; Duan, W.; Liu, X.; Zheng, H. Molecular imaging-guided photothermal/photodynamic therapy against tumor by iRGD-modified indocyanine green nanoparticles. *J. Control Release* **2016**, *224*, 217–228. [CrossRef] [PubMed]

15. Wang, K.; Zhang, X.; Liu, Y.; Liu, C.; Jiang, B.; Jiang, Y. Tumor penetrability and anti-angiogenesis using iRGD-mediated delivery of doxorubicin-polymer conjugates. *Biomaterials* **2014**, *35*, 8735–8747. [CrossRef]

16. Li, H.; Qiu, Z.; Li, F.; Wang, C. The relationship between MMP-2 and MMP-9 expression levels with breast cancer incidence and prognosis. *Oncol. Lett.* **2017**, *14*, 5865–5870. [CrossRef] [PubMed]

17. Caculitan, N.G.; Ma, Y.; Zhang, D.; Kozak, K.R.; Liu, Y.; Pillow, T.H.; Sadowsky, J.; Cheung, T.K.; Phung, Q.; Haley, B. Cathepsin B is dispensable for cellular processing of cathepsin B-cleavable antibody–drug conjugates. *Cancer Res.* **2017**, *77*, 7027–7037. [CrossRef] [PubMed]

18. Zhong, Y.J.; Shao, L.H.; Li, Y. Cathepsin B-cleavable doxorubicin prodrugs for targeted cancer therapy (Review). *Int. J. Oncol.* **2013**, *42*, 373–383. [CrossRef]

19. Mei, L.; Zhang, Q.; Yang, Y.; He, Q.; Gao, H. Angiopep-2 and activatable cell penetrating peptide dual modified nanoparticles for enhanced tumor targeting and penetrating. *Int. J. Pharm.* **2014**, *474*, 95–102. [CrossRef]

20. Liu, X.; Jiang, J.; Ji, Y.; Lu, J.; Chan, R.; Meng, H. Targeted drug delivery using iRGD peptide for solid cancer treatment. *Mol. Syst. Des. Eng.* **2017**, *2*, 370–379. [CrossRef]

21. He, Z.; Tessier-Lavigne, M. Neuropilin is a receptor for the axonal chemorepellent Semaphorin III. *Cell* **1997**, *90*, 739–751. [CrossRef]

22. Schneider, C.A.; Rasband, W.S.; Eliceiri, K.W. NIH Image to ImageJ: 25 years of image analysis. *Nat. Methods* **2012**, *9*, 671–675. [CrossRef]

23. Kilic, M.A.; Ozlu, E.; Calis, S. A novel protein-based anticancer drug encapsulating nanosphere: Apoferritin-doxorubicin complex. *J. Biomed. Nanotechnol.* **2012**, *8*, 508–514. [CrossRef]

24. Liu, L.; Zhang, C.; Li, Z.; Wang, C.; Bi, J.; Yin, S.; Wang, Q.; Yu, R.; Liu, Y.; Su, Z. Albumin binding domain fusing R/K-X-X-R/K sequence for enhancing tumor delivery of doxorubicin. *Mol. Pharm.* **2017**, *14*, 3739–3749. [CrossRef] [PubMed]

25. Schlapschy, M.; Binder, U.; Borger, C.; Theobald, I.; Wachinger, K.; Kisling, S.; Haller, D.; Skerra, A. PASylation: A biological alternative to PEGylation for extending the plasma half-life of pharmaceutically active proteins. *Protein Eng. Des. Sel.* **2013**, *26*, 489–501. [CrossRef] [PubMed]

26. Kettler, K.; Veltman, K.; Meent, D.V.D.; Wezel, A.V.; Hendriks, A.J. Cellular uptake of nanoparticles as determined by particle properties, experimental conditions, and cell type. *Environ. Toxicol. Chem.* **2014**, *33*, 481–492. [CrossRef]

27. Feron, O. Tumor-penetrating peptides: A shift from magic bullets to magic guns. *Sci. Transl. Med.* **2010**, *2*, 34ps26. [CrossRef] [PubMed]

pharmaceutics

Article

Microfluidic Synthesis and Purification of Magnetoliposomes for Potential Applications in the Gastrointestinal Delivery of Difficult-to-Transport Drugs

Carlos E. Torres [1,†], Javier Cifuentes [1,†], Saúl C. Gómez [1], Valentina Quezada [1], Kevin A. Giraldo [1], Paola Ruiz Puentes [1], Laura Rueda-Gensini [1], Julian A. Serna [1], Carolina Muñoz-Camargo [1], Luis H. Reyes [2,*], Johann F. Osma [3,*] and Juan C. Cruz [1,*]

[1] Department of Biomedical Engineering, School of Engineering, Universidad de Los Andes, Carrera 1 No. 18A-12, Bogotá 111711, Colombia; ce.torres10@uniandes.edu.co (C.E.T.); jf.cifuentes10@uniandes.edu.co (J.C.); sc.gomez11@uniandes.edu.co (S.C.G.); v.quezada@uniandes.edu.co (V.Q.); ka.giraldo@uniandes.edu.co (K.A.G.); p.ruiz@uniandes.edu.co (P.R.P.); l.ruedag@uniandes.edu.co (L.R.-G.); ja.serna10@uniandes.edu.co (J.A.S.); c.munoz2016@uniandes.edu.co (C.M.-C.)

[2] Department of Chemical and Food Engineering, School of Engineering, Universidad de Los Andes, Carrera 1 No. 18A-12, Bogotá 111711, Colombia

[3] Department of Electrical and Electronic Engineering, School of Engineering, Universidad de Los Andes, Carrera 1 No. 18A-12, Bogotá 111711, Colombia

[*] Correspondence: lh.reyes@uniandes.edu.co (L.H.R.); jf.osma43@uniandes.edu.co (J.F.O.); jc.cruz@uniandes.edu.co (J.C.C.); Tel.: +57-1-3394949 (ext. 1789) (J.C.C.)

[†] These authors contributed equally to this work.

Citation: Torres, C.E.; Cifuentes, J.; Gómez, S.C.; Quezada, V.; Giraldo, K.A.; Puentes, P.R.; Rueda-Gensini, L.; Serna, J.A.; Muñoz-Camargo, C.; Reyes, L.H.; et al. Microfluidic Synthesis and Purification of Magnetoliposomes for Potential Applications in the Gastrointestinal Delivery of Difficult-to-Transport Drugs. *Pharmaceutics* **2022**, *14*, 315. https://doi.org/10.3390/pharmaceutics14020315

Academic Editors: Francisco José Ostos, José Antonio Lebrón and Pilar López-Cornejo

Received: 29 November 2021
Accepted: 12 January 2022
Published: 28 January 2022

Publisher's Note: MDPI stays neutral with regard to jurisdictional claims in published maps and institutional affiliations.

Abstract: Magnetite nanoparticles (MNPs) have gained significant attention in several applications for drug delivery. However, there are some issues related to cell penetration, especially in the transport of cargoes that show limited membrane passing. A widely studied strategy to overcome this problem is the encapsulation of the MNPs into liposomes to form magnetoliposomes (MLPs), which are capable of fusing with membranes to achieve high delivery rates. This study presents a low-cost microfluidic approach for the synthesis and purification of MLPs and their biocompatibility and functional testing via hemolysis, platelet aggregation, cytocompatibility, internalization, and endosomal escape assays to determine their potential application in gastrointestinal delivery. The results show MLPs with average hydrodynamic diameters ranging from 137 ± 17 nm to 787 ± 45 nm with acceptable polydispersity index (PDI) values (below 0.5). In addition, we achieved encapsulation efficiencies between 20% and 90% by varying the total flow rates (TFRs), flow rate ratios (FRRs), and MNPs concentration. Moreover, remarkable biocompatibility was attained with the obtained MLPs in terms of hemocompatibility (hemolysis below 1%), platelet aggregation (less than 10% with respect to PBS $1\times$), and cytocompatibility (cell viability higher than 80% in AGS and Vero cells at concentrations below 0.1 mg/mL). Additionally, promising delivery results were obtained, as evidenced by high internalization, low endosomal entrapment (AGS cells: PCC of 0.28 and covered area of 60% at 0.5 h and PCC of 0.34 and covered area of 99% at 4 h), and negligible nuclear damage and DNA condensation. These results confirm that the developed microfluidic devices allow high-throughput production of MLPs for potential encapsulation and efficient delivery of nanostructured cell-penetrating agents. Nevertheless, further in vitro analysis must be carried out to evaluate the prevalent intracellular trafficking routes as well as to gain a detailed understanding of the existing interactions between nanovehicles and cells.

Keywords: magnetoliposomes; microfluidics; oral drug delivery; magnetite nanoparticles

1. Introduction

Oral drug administration is one of the most convenient routes of drug delivery due to patient preference, shelf life, sustained delivery, cost-effectiveness, and ease of large-scale manufacture [1,2]. Additionally, orally administered drugs can be directed through the gastrointestinal tract to allow localized treatment of different pathologies, such as cancer, infections, inflammations, and various digestive system diseases [1]. Nevertheless, the success of this delivery route depends on the physicochemical properties of such drugs and, particularly, their water solubility and cell-membrane permeability [3]. Different approaches have been proposed to control pharmacokinetics and improve release efficacy and safety [4]. Among these, encapsulation is one of the most attractive ones for preserving compounds with biological activity, especially when exposed to conditions that might be detrimental to their chemical stability [5,6]. In the pharmaceutical industry, the delivery of drugs has been significantly improved by encapsulation into polymeric capsules and liposomes [7,8]. These liposomal vehicles have been widely studied for pharmaceutical preparations with limited passing across biological barriers, such as the blood–brain barrier and the intestinal epithelium, due to attractive features such as the flexibility of changing their chemical composition, structure, and colloidal size [9–13].

Additionally, a significant challenge in drug delivery has been to achieve better internalization and high bioavailability [9,10]. This challenge has been addressed by an increasing number of delivery vehicles that include both viral and non-viral vectors [11]. Among these, magnetite nanoparticles (MNPs) functionalized with translocating proteins and peptides have been studied as potent vehicles for cell penetration and endosomal escape. Moreover, a possible enhancement of escape is expected if these vehicles are encapsulated into liposomes [12,14–16].

Liposomes with encapsulated MNPs, called magnetoliposomes (MLPs), have been extensively used as carriers in the pharmaceutical industry due to their ability to release various active molecules at a given site without the need for molecularly targeted agents [17]. This is in addition to the improvement in the biocompability, drug delivery rate for some compounds, and cellular uptake without a significant reduction in the activity of the functional compounds immobilized and delivered employing MNPs [18,19]. Moreover, these novel drug delivery vehicles might offer potential improvements in targeting, stabilization of antimicrobial agents, and gastroretention. This might help to reduce various possible side effects of oral administration, such as the uncontrolled destruction of both pathogenic and non-pathogenic microbiota, and, therefore, prevent the appearance of complications such as dysbiosis [20]. This correlation between the microbiota's metabolic activity and the improvement of the bioavailability is particularly relevant in determining the overall efficacy of these novel drug delivery vehicles. Therefore, it is important to determine if the carrier nanovehicle can increase the compound bioavailability and reduce the associated biotransformations, which in turn define the bioactivity expression in response to a particular microbiome [21–23].

Currently, MLPs have been explored as drug delivery carriers to treat conditions as diverse as cancer, Parkinson's, and Alzheimer's [24–26]. Over the past few years, several techniques have been proposed for preparing MLPs, in which nanoparticles can be encapsulated in the aqueous lumen, embedded in the lipid bilayer, or conjugated on the surface of the liposome [27–30]. By implementing these techniques, liposome solutions' parameters vary considerably, posing some challenges related to their particle size, dispersity, lamellarity, entrapment efficiency, and, most importantly, the difficulty in separating the non-encapsulated/unbound MNPs [31,32].

Currently, liposome preparation techniques using microfluidic systems have allowed greater control over physical properties, yielding massive and robust production of MLPs with uniform size distribution, high loading efficiencies, and reduced costs [33–35]. Nevertheless, there is still a challenge in the separation and sample purification methods due to the minor size differences between the MLPs and the non-encapsulated nanoparticles [32,36]. Due to several limitations of current separation techniques, microfluidic systems have been proposed as potential low-cost particle separation systems based on different active or passive methods to separate nanoscale objects such as DNA, viruses, proteins, exosomes, and nanoparticles [37]. Exploring the scope of microfluidics separation approaches, there has been a growing interest in using magnetic gradients to retain excess MNPs without compromising the integrity of the MLPs. The significant traction gained by this approach could be mainly attributed to its applicability, versatility, and ease of implementation in many areas of the biomedical field, including disease diagnostics, therapeutics, and cell sorting [38–40].

This study proposes the synthesis of MLPs using a microfluidic approach; FEM simulations implemented in COMSOL Multiphysics® to study the separation of MLPs from nanoconjugates aided by a magnetophoretic microfluidic system; the manufacture and experimental validation of two separation devices; and, finally, the in vitro testing of the synthesized MLPs to evaluate whether this delivery vehicle is biocompatible and improves the cell penetration of orally administered CefTRIAxone, a drug with exceedingly low intestinal absorption. The prototypes of the microfluidic devices for MLPs synthesis were manufactured by a low-cost method based on laser cutting techniques and led to MLPs with acceptable physical properties and encapsulation efficiency. Additionally, the separation devices showed different efficiencies depending on the implemented approach which varied considerably compared with those obtained during the experimental validation. Nevertheless, qualitatively, both methods led to similar results, confirming their suitability for the intended objective. Finally, the preliminary in vitro evaluation demonstrated that MLPs showed high biocompatibility, low endosomal entrapment, and high internalization rates, which are crucial factors in developing novel vehicles for delivering difficult-to-transport drugs.

2. Materials and Methods

2.1. Magnetite Nanoparticles Synthesis and Functionalization

Magnetite nanoparticles (MNPs) were synthesized by the chemical co-precipitation method. For this, $FeCl_2$ (0.34 g, J. T. Baker, Phillipsburg, NJ, USA) and $FeCl_3$ (0.93 g, Merck, Kenilworth, NJ, USA) were solubilized in 60 mL of type I water. In addition, 0.69 g of NaOH (PanReac AppliChem, Darmstadt, Germany) was added to 17 mL of type I water, and then both solutions were heated at 80 °C. Next, a NaOH solution was added dropwise to the iron chloride solution at a 5 mL/min rate under constant stirring. A black precipitate was observed, corresponding to the formation of MNPs. The obtained MNPs were then washed four times with NaCl solution (1.5% w/v) and twice with type I water aided by a neodymium magnet. Then, 100 mg of MNPs were silanized by adding 50 µL of glacial acetic acid (PanReac AppliChem, Barcelona, Spain) followed by 400 µL of (3-aminopropyl) triethoxysilane (APTES, 98%, Sigma-Aldrich, St. Louis, MO, USA). The MNPs solution was left to react under constant stirring (250 rpm) at 60 °C for 1 h and then washed as mentioned previously. Later, 100 mg of the silanized MNPs (MNP-APTES) was mixed with 2 mL of glutaraldehyde solution (2% v/v, Sigma-Aldrich, St. Louis, MO, USA) (solution previously stirred (220 rpm) at room temperature for 1 h to allow glutaraldehyde activation (MNP-APTES-GA)). Then, 5 mL of a NH_2-PEG-propionic acid (99%, Merck, Darmstadt, Germany) solution (2 mg/mL) was added dropwise to the MNP-APTES-GA conjugates under constant stirring to obtain MNP-APTES-PEG conjugates. The solution was left to react at 220 rpm and room temperature for 24 h and washed four times with NaCl solution (1.5% w/v) and twice with type I water. Similarly, (3-[(2-aminoethyl)dithio]propionic acid) (AEDP, ThermoFisher, Waltham, MA, USA) immobilization was carried out using 14 mg of N-[3-dimethylammino)-propyl]-N'-ethyl carbodiimide hydrochloride (EDC, 98%, Sigma-Aldrich, St. Louis, MO, USA) and 7 mg of N-hydroxy succinimide (NHS, 98%, Sigma-Aldrich, St. Louis, MO, USA) solution in 5 mL of type I water added to 100 mg of MNP-PEG in 50 mL of type I water to activate the terminal carboxyl groups. Nanoparticles were ultrasonicated (ultrasonic bath, Branson, Danbury, CT, USA) for 10 min, and 5 mL of an AEDP solution (5 mg/mL) was added dropwise under constant stirring. The solution was left to react at 220 rpm and room temperature for 24 h. The MNP-PEG-AEDP conjugates were washed with NaCl (1.5% w/v) and type I water. Finally, the antibiotic CefTRIAxone (CTA, Vitalis, 1 g I.M/I.V) was immobilized by following the same protocol for AEDP immobilization, using 5 mL of CefTRIAxone solution (2 mg/mL). Scheme 1 shows the complete methodology for the synthesis.

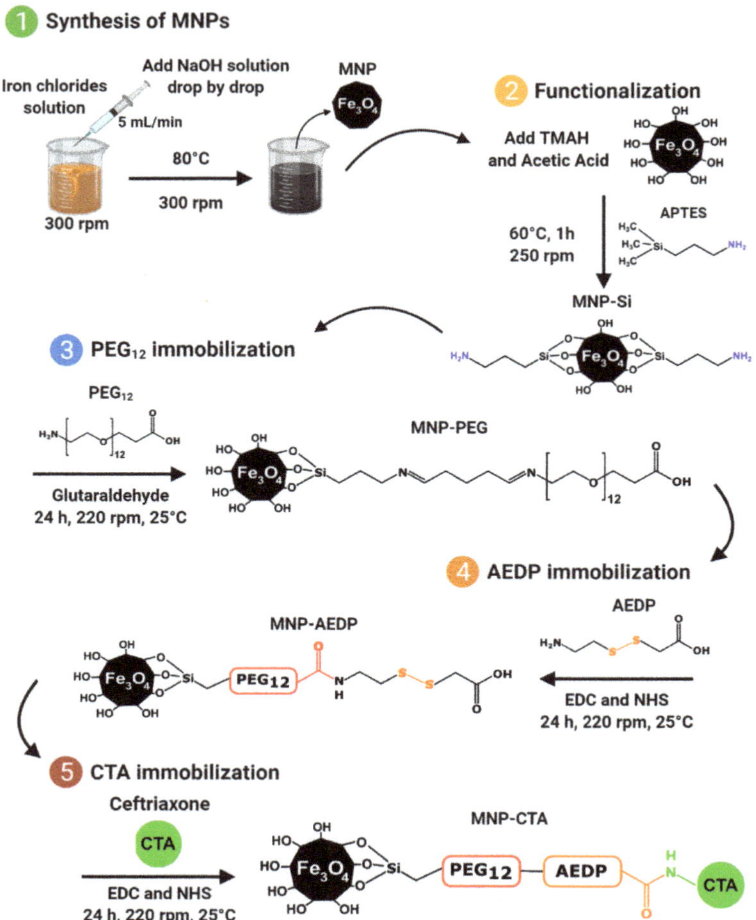

Scheme 1. Schematic of the developed workflow for the synthesis of MNP-PEG-AEDP-CTA nanoconjugate. (1) Magnetite nanoparticles (MNPs) are synthesized by co-precipitation. (2) MNPs are silanized with APTES and subsequently with (3) NH_2-PEG-propionic acid. This was followed by the conjugation of (4) AEDP and, finally, (5) the immobilization of the drug CTA.

The resulting MNP-PEG-AEDP-CTA conjugates were labeled with rhodamine B (95%, Sigma-Aldrich, St. Louis, MO, USA) for fluorescence-based assays. For this, 14 mg of EDC, 7 mg of NHS, and 5 mg of rhodamine B (RdB) were dissolved in 5 mL of type I water containing 2 mL of dimethylformamide (DMF, Supelco/Sigma-Aldrich, Bellefonte, PA, USA). Rhodamine B solution was left under constant stirring for 15 min to allow the activation of carboxylic groups. Next, the previously activated rhodamine B solution was added to 50 mL of MNP-PEG-AEDP-CTA aqueous solution (2 mg/mL) and left to react at 220 rpm, room temperature, and in complete darkness for 24 h. The resulting MNP-PEG-AEDP-CTA-RdB was washed several times with NaCl (1.5% w/v) and type I water to remove the excess reagents. MNP-PEG-AEDP-CTA-RdB nanoconjugates were resuspended in type I water and stored in complete darkness at 4 °C until further use for the MLPs preparation described below. A schematic of the synthesized nanoconjugate is shown in Figure 1A.

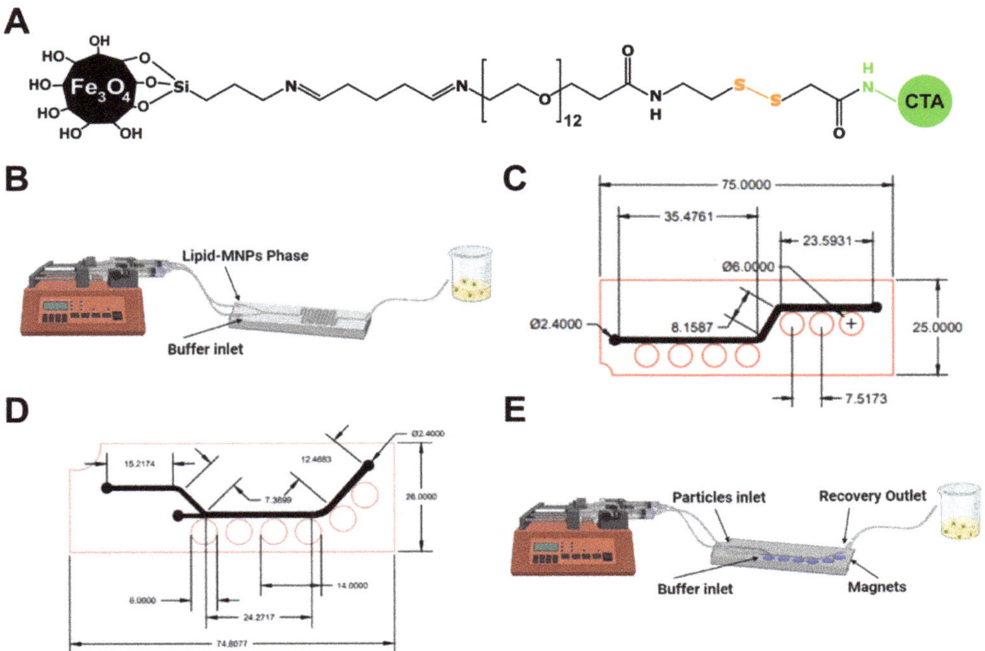

Figure 1. (**A**) Schematic of the synthesized nanoconjugate for the MLPs production. (**B**) Microfluidic experimental setup for the synthesis phase. Microfluidic separation designs proposed: (**C**) System 1 and (**D**) System 2. (**E**) Microfluidic experimental setup for the separation phase.

2.2. Magnetoliposomes Synthesis Using the Microfluidic Approach

2.2.1. Lipidic-MNPs Phase Preparation

First, 100 mg of soy lecithin (1-α-lecithin, soybean-cas 8002-43-5-calbiochem) (Merck, Kenilworth, NJ, USA) was dissolved in 10 mL of chloroform c2432 (>99.5%, Merck, Kenilworth, NJ, USA), and 1 mL and 2 mL of MNPs (1.7 mg/mL) were added. The sample was rotary evaporated for 1 h at 45 °C under a vacuum (rotary evaporator, Hei-VAP Value Digital Vertical, Heidolph, Schwabach, Germany). Then, 10 mL of ethanol (96% *v/v*) was added to the rotary evaporator flask. The sample was vigorously agitated and rotated for 30 min at atmospheric temperature and pressure.

2.2.2. Microfluidic System Manufacture and Experimental Setup

The manufacture and design of microfluidic devices for MLP synthesis was based on the study presented by Aranguren et al. [41]. A laser cutting machine (TROTEC® Speedy 100, 60 w laser cutter, TROTEC, Marchtrenk, Austria) was used for engraving and cutting the microfluidic channels proposed on a PMMA substrate. The two or three layers of the microfluidic devices were manually aligned and sealed using 96% (*v/v*) ethanol with a mechanical press placed on a hot plate (110 °C). For the synthesis, the microfluidic channels were purged with a 10 mL syringe filled with 96% (*v/v*) ethanol for 15 min. Then, a syringe filled with the lipidic-nanoconjugates phase and a syringe with NaCl (anhydrous, Redi-Dri™, free-flowing, ACS reagent, >99%) solution (0.05 M) were mounted on an infusion pump (MedCaptain MP30), as is shown in Figure 1B. The syringes were connected to the microsystems using two probes (Nelaton, Probes, Medex caliber 8) (Medex, Smiths Medical

Inc., Minneapolis, MN, USA). The synthesis was carried out using a total flow ratio (TFR) set at 2.5 mL/min and 5 mL/min with a varying flow rate ratio (FRR) from 1:1 to 5:1 (aqueous:solvent ratio).

2.3. Magnetoliposomes Characterization

Magnetoliposome size and polydispersity index were measured using the Zetasizer Nano ZS (Malvern, Panalytical, Egham, UK). Additionally, a morphology and size characterization using a TEM Tecnai F20 Super Twin TMP (FEI, Hillsboro, OR, USA) was performed to determine the synthesis effectiveness. For the TEM analysis, a sample drop was deposited on a copper grid with a carbon coating that was dried for 1 h. Next, the prepared sample was stained with 2% uranyl acetate by depositing one drop on the grid for 8 min, washed with deionized water, and left to dry for imaging at a total magnification of 71, 97, and 145 kX.

2.4. Magnetolipsomes Encapsulation Efficiency (EE%)

The encapsulation efficiency of synthesized MLPs was analyzed using a Spectrofluorometer (0239D-2219 FluoroMax plus C, Horiba, Miyanohigashi, Japan) to track changes in the fluorescence intensity before and after treatment with Triton X-100 (Sigma-Aldrich, St. Louis, MO, USA). For this characterization, 100 μL of MLPs prepared with MNP-PEG-AEDP-CTA-RdB was pipetted into a 96-well microplate for the first fluorescence-based analysis where the fluorescence intensity was measured. Then, 10 mL of Triton X-100 was added to the microplates to break the magnetoliposome membranes and allow the labeled nanoconjugates to escape. Finally, a second measure of the sample fluorescence intensity was carried out to analyze the changes compared with the initial intensity. The fluorescence spectrum of rhodamine B allowed tracking intensity by setting up excitation and emission filters at 546 nm and 568 nm, respectively. The encapsulation efficiency was calculated using Equation (1):

$$\text{EE (\%)} = 100 \times \frac{(Int\ (Final) - Int\ (Initial) - Int\ (Triton\ X\text{-}100))}{Int\ (Final)}, \tag{1}$$

where *Int (Final)* is the emission post-Triton *X-100*, *Int (Initial)* is the emission pre—*Triton X-100* treatment, and *Int (Triton X-100)* is the blank emission of *Triton X-100*.

2.5. Magnetoliposomes Purification

2.5.1. Lipidic-Nanoconjugates Phase Preparation

The design of the two components of the microfluidic devices proposed (System 1 and System 2) was conducted on two rectangular PMMA layers (width: 7.48 cm, height: 2.60 cm) via AutoCAD v23.0 (AutoDesk Inc., San Rafael, CA, USA). The microfluidic channel was superimposed over one of the two layers, followed by locating holes in each piece. The larger holes were to accommodate permanent neodymium magnets 6 mm in diameter and 8 mm deep, while the smaller ones were for the inlets and outlets of the microchannel. In this case, the diameter was 2.4 mm. (Figure 1C,D).

In these designs, the magnetophoretic separation was analyzed in silico with two different methods. The first one involved implementing the particle tracing module, where magnetoliposomes and nanoconjugates were considered separate components within the microfluidic device. The simulation was conducted by implementing the laminar flow, particle tracing, and the magnetic field without current modules of COMSOL Multiphysics® (COMSOL Inc., Stockholm, Sweden). The second method was based on a mixture model approach to simulate a dispersed phase, considered a ferrofluid (i.e., a suspension of magnetic nanoconjugates in water) due to the high concentration of nanoparticles in the domain. Several recent reports support this assumption (rather than considering individual particles) for FEM simulations, since it leads to results that are closer to those obtained experimentally [42–46]. In this case, the implemented COMSOL modules were the "mixture model with laminar flow", the "magnetic field without current", and "diluted species' transport".

2.5.2. Multiphysics Simulations of Magnetophoretic Separation via the Particle Tracing Module

The magnetic field was incorporated into the simulations through the "magnetic fields with no currents" model of the AC/DC module of COMSOL. In this case, the magnetic field intensity is calculated by solving the Maxwell equations for permanent magnets, resulting in the governing equations shown in Equations (2)–(4) [47].

$$H = -\nabla\, V_m,$$ (2)

$$\nabla \cdot B = 0,$$ (3)

where:

$$B = \mu_0\mu_r\vec{H} + B_r,$$ (4)

where $\vec{H}$ is the magnetic field distribution, V_m is the magnetic scalar potential, B is the magnetic flux density distribution, and μ_0 and μ_r the vacuum permeability and relative permeability. Finally, B_r represents the remanent flux density that, in this case, was set to 1 T. The second physic coupled to the model was the laminar flow governed by the Navier–Stokes conservation of momentum equation for incompressible fluids, Equation (5), which is accompanied by the conservation of mass by the continuity equation (Equation (6)).

$$\nabla[-PI + \mu\,(\nabla u + (\nabla u)^T)] + F = 0,$$ (5)

$$\rho\nabla \cdot (u) = 0,$$ (6)

where P is the pressure, μ is the fluid's dynamic viscosity, F accounts for the volumetric forces, and ρ is the fluid density. Finally, the particle tracing for fluid flow was coupled to the model, where Newton's second law governs the movement of the particles transported within the device. This is described by Equation (7):

$$\frac{d(m_p\upsilon)}{dt} = F_t,$$ (7)

where m_p is the mass of the particles, υ the velocity, and F_t the sum of all forces acting on the particle. In this case, the involved forces were the drag force, defined by Equations (8) and (9) by the Stokes law, and the magnetic force, dependent on the magnetic flux density distribution described by Equation (10).

$$F_D = \frac{1}{\tau_p}m_p(u - \upsilon),$$ (8)

$$\tau_p = \frac{\rho_p\, d_p^2}{18\mu},$$ (9)

where m_p is the particle mass, u is the velocity field, υ the particle velocity, ρ_p the particle density, d_p the particle diameter, and μ the dynamic viscosity.

$$F_M = \frac{V_m\Delta X}{\mu_0}(B\cdot\nabla)B$$ (10)

where ΔX is the magnetic susceptibility difference between the particle and the fluid. Finally, the FEM simulations to solve the set of equations for laminar flow and magnetic field were conducted via a stationary study. Additionally, for the particle tracing module for 1200 particles per component (i.e., nanoconjugates and MLPs), a bi-directionally coupled particle tracing was used with a MUMPS solver. The computational domain was meshed with 176,141 domain elements and 6523 boundary elements for System 1 and 47,152 domain elements and 1838 boundary elements for System 2. This module's boundary conditions were the drag force in all the microfluidic channel domains and the system's inlets as the main entrance for the particles into the system. The model parameters are summarized in Table 1:

Table 1. Parameters employed for the particle tracing module.

Parameter	Value	Units
B_r	1.00	T
μ	1.00×10^{-3}	Pa.s
μ_0	12.57×10^{-7}	H/m
μ_{rMNps}	2.50	DV
μ_{rMLs}	1.50	DV
ρ_{MNps}	5180	kg/m^3
ρ_{MLs}	3063	kg/m^3
d_{MNps}	1.00×10^{-7}	m
d_{MLs}	2.50×10^{-7}	m

2.5.3. Multiphysics Simulations of Magnetophoretic Separation via the Mixture Model

The magnetic field was established using a magnetic field, no currents physic, as described previously. The governing equations for this simulation are presented in Equations (1)–(3). The ferrofluid was simulated, aided by the mixture model, laminar flow physics. The interface solves a set of Navier–Stokes equations for the momentum of the mixture. The pressure distribution is calculated from a mixture-averaged continuity equation, and the velocity of the dispersed phase is described by a slip model [48]. The momentum conservation equation and the continuity equation are presented in Equations (11) and (12):

$$\rho \frac{du}{dt} + \rho(u \cdot \nabla)u = [-pl + \mu(\nabla u + (\nabla u)^T - \frac{2}{3}(\nabla \cdot u)] - \nabla \cdot [\rho C_d(1 - C_d)U_{slip}U_{slip}{}^T] + F, \tag{11}$$

$$(\rho_c - \rho_d) \{\nabla \cdot [\Phi_d (1 - C_d) U_{slip}] + \frac{m_{dc}}{\rho_d}\} + \rho_c (\nabla \cdot u) = 0, \tag{12}$$

where P is the pressure, μ is the dynamic viscosity of the fluid, ρ_c and ρ_d the continuous phase density and dispersed phase density, Φ_d the volume fraction of the dispersed phase, m_{dc} the turbulent dispersed phase diffusion, and F the body forces, which in this case are described by the Kelvin body force due to a spatially non-uniform magnetic field according to Equation (13) [46]:

$$F = (\vec{M} \cdot \nabla)\vec{B}, \tag{13}$$

where $\vec{B}$ is the magnetic flux density distribution and $\vec{M}$ the magnetization. Finally, the diluted species' transport was used to determine the effective concentration of the nanoparticles inside the channel. This solution inside the microchannel is described by the convective–diffusive Equation (14).

$$\frac{dC_p}{dt} + \nabla \cdot (-D_p C_p) + u \cdot \nabla C_p = 0, \tag{14}$$

where C_p is the concentration, u is the velocity field provided by the mixture model, and D_p is the effective diffusivity of the NPs as calculated by Equation (15).

$$D_p = \frac{K_B T}{3 \Pi \eta_{ff} d_p}, \tag{15}$$

Here, K_B is the Boltzmann constant, T is temperature, η_{ff} is the ferrofluid viscosity, and d_p is the diameter of the particles.

Finally, time-dependent simulations were carried out using a MUMPS solver with a phase volume fraction of 0.2 for each particle component entering the upper inlet. For System 1, complete mesh consists of 59,862 domain elements and 1845 boundary elements, while for System 2, the computational domain mesh comprised 66,735 domain elements and 1946 boundary elements. The final meshing is shown in Supplementary Figure S2, and the parameters used for this model are presented in Table 2.

Table 2. Parameters employed for the mixture model.

Parameter	Value	Units
B_r	1.00	T
μ	1.00×10^{-3}	Pa.s
μ_{rMNps}	2.50	DV
μ_{rMLs}	1.50	DV
μ_0	12.57×10^{-7}	H/m
ρ_C	1000	kg/m^3
ρ_{MNps}	5180	kg/m^3
ρ_{MLs}	3063	kg/m^3
d_{MNps}	1.00×10^{-7}	m
d_{MLs}	2.50×10^{-7}	m
D_{MNps}	4.83×10^{-12}	m^2/s
D_{MLs}	1.93×10^{-12}	m^2/s
ϕ_d	0.20	DV
x_{MNPs}	4.00×10^{-4}	m^3/kg
x_{MLs}	1.33×10^{-4}	m^3/kg

2.5.4. Microfluidic System Manufacture and Experimental Setup

The manufacture of the microfluidic device for the MLPs synthesis (Supplementary Figure S1) was based on the methodology presented above and reported previously by us in the studies of Aranguren et al. and Campaña et al. [41,49]. For the first design, seven neodymium magnets were in proximity to the microchannels, as is shown in Figure 1C. In parallel, one syringe of 10 mL was filled with the solution of LPs and MNP-PEG-AEDP-CTA nanoconjugates and connected to the system inlet. The solution was pumped into the device with syringe pumps (78-8110C Programmable Touch Screen, Cole-Parmer®, Vernon Hills, IL, USA, and B Braun Perfusor® compact, B. Braun, Melsungen, Germany) at total flow rates (TFRs) from 1 to 3 mL/min. The second design included six neodymium magnets, as is shown in Figure 1D. Two syringes of 10 mL were filled up and connected to the system's inlets. The first one was filled with the LPs and MNP-PEG-AEDP-CTA nanoconjugates, while the second one with a NaCl solution (0.05 M) (Figure 1E). The solutions were pumped into the device with the syringe pumps at total flow rates from 1 to 3 mL/min by maintaining a 1:1 FRR. The samples recovered from each design were analyzed using a spectrofluorometer (0239D-2219 FluoroMax plus C, Horiba, Miyanohigashi, Japan) to track changes in the fluorescence intensity compared with the control sample, which is the solution before injection into the system. As for the EE experiment, the fluorescence spectrum of rhodamine B allowed intensity tracking by setting up excitation and emission filters at 546 nm and 568 nm, respectively.

2.6. In Vitro Testing of MLPs

2.6.1. Hemocompatibility

To determine the hemocompatibility of the liposomes, MNP-PEG-AEDP-CTA nanoconjugates, and magnetoliposomes, a blood sample was extracted from a healthy human donor in a vacutainer tube containing EDTA. Erythrocytes were obtained by centrifugation at 1800 rpm for 5 min. The supernatant was discarded, and erythrocytes were washed five times with NaCl solution (0.9% w/v) and twice with PBS 1×. To form a stock solution, 1 mL of the washed erythrocytes was suspended in 9 mL of PBS 1× and carefully homogenized. The liposomes and magnetoliposomes were evaluated at 0.1, 0.05, and 0.025 mg/mL, while MNP-PEG-AEDP-CTA nanoconjugates were at concentrations ranging from 200 µg/mL to 12.5 µg/mL. Triton 100-X (10% v/v) and PBS 1× were used as positive and negative controls, respectively. To evaluate the hemolytic activity, 100 µL of the erythrocyte stock solution was seeded with 100 µL of the different treatments in a 96-well microplate. The microplate was then incubated under constant stirring at 37 °C for 1 h. The plate was centrifuged, and the supernatants were then transferred to another 96-well microplate. Finally, absorbance was read at 450 nm, and hemolysis percentage was calculated by following Equation (16):

$$\text{Hemolysis (\%)} = 100 \times \frac{(Abs\ (sample)\ -\ Abs\ (C-))}{(Abs\ (C+)\ -\ Abs\ (C-))} \tag{16}$$

2.6.2. Platelet Aggregation

The platelet aggregation capacities of the liposomes, MNP-PEG-AEDP-CTA, and magnetoliposomes were evaluated by exposing them to a blood sample extracted from a healthy human donor in a vacutainer tube containing sodium citrate. Platelet-rich plasma (PRP) was obtained by centrifuging the collected blood at 1000 rpm for 15 min. Erythrocytes were discarded, and the supernatant containing PRP was used to run the test. The liposomes and magnetoliposomes were evaluated at 0.1, 0.05, and 0.025 mg/mL and the MNP-PEG-AEDP-CTA at concentrations ranging from 200 µg/mL to 12.5 µg/mL. Thrombin and PBS 1× were used as positive and negative references, respectively. The aggregation capacity was evaluated by exposing 50 µL of PRP to 50 µL of the different treatments in a 96-well microplate. The microplate was incubated at 37 °C for 5 min, and then absorbance was read at 620 nm. Platelet aggregation percentage was calculated by following Equation (17):

$$\text{Platelet aggregation (\%)} = 100 \times \frac{Abs\ (sample)}{Abs\ (C+)}, \tag{17}$$

2.6.3. Cytotoxicity

The cytocompatibility of the liposomes, MNP-PEG-AEDP-CTA, and magnetoliposomes was determined as a measure of the impact on the metabolic activity in two different cell lines, namely, Vero (ATCC® CCL-81) and gastric cancer (AGS, ATCC® CRL-1739) cells, with the aid of a colorimetric assay based on 3-(4,5-dimethylthiazol-2-yl)-2,5-diphenyltetrazolium (MTT, Sigma-Aldrich, St. Louis, MO, USA). The liposomes and magnetoliposomes were evaluated at 0.1, 0.05, and 0.025 mg/mL and the MNP-PEG-AEDP-CTA at serial dilutions from 200 µg/mL to 12.5 µg/mL. Non-supplemented DMEM medium was used as the negative control. To evaluate cell viability in the different cell lines, 100 µL of a cell stock solution in DMEM medium supplemented with FBS (10%) was seeded in a 96-well microplate at a cell density of 10×10^4 cells/well. Microplates were incubated at 37 °C, 5% CO_2, and a humidified atmosphere for 24 h. After that, DMEM medium supplemented with FBS (10%) was extracted and replaced with a non-supplemented DMEM medium containing the different treatments. Viability was studied at 24 and 48 h after the exposure. To determine the viability percentage, 10 µL of MTT reagent (5 mg/mL) was added to each well, and the microplates were then incubated, under the same conditions described above, for 2 h. Finally, supernatants were discarded, and 100 µL of DMSO was added to each well to dissolve the formed formazan crystals. The absorbance was read at 595 nm with the aid of a microplate reader (Thermo Scientific Multiskan™ FC Microplate Photometer). Cell viability was calculated by following Equation (18):

$$\text{Cell viability (\%)} = 100 \times \frac{Abs\ (sample)}{Abs\ (C-)}, \tag{18}$$

2.6.4. Cell Internalization and Endosomal Escape Analysis

Cell internalization and endosomal escape abilities of MNP-PEG-AEDP-CTA-RdB nanoconjugates and magnetoliposomes were assessed by colocalization between the labeled nanoconjugates and Lysotracker Green® DND-26 (Thermo Fisher, Waltham, MA, USA) in Vero (ATCC® CCL-81) and gastric cancer cells (AGS, ATCC® CRL-1739). For this, cells were seeded on glass slides deposited into a 24-well microplate at a cell density of

5×10^4 cells/well. Cells were then incubated with DMEM medium supplemented with FBS (10% v/v) at 37 °C and 5% CO_2 for 24 h to allow cell adhesion. Once the incubation time was achieved, DMEM medium was extracted and replaced with supplemented DMEM medium containing the different treatments at 50 µg/mL and magnetoliposomes with an equivalent amount of MNP-PEG-AEDP-CTA-RdB of 25 µg/mL. Cells were incubated for 0.5 h and 4 h. Next, the medium was extracted, and cells were washed three times with PBS 1× to remove the excess of the treatments. After this, PBS 1× was drawn, and cells were exposed to a DMEM solution containing Hoechst 33,342 (Thermo Fisher, Waltham, MA, USA) (1:1000) and Lysotracker Green® DND-26 (1:10000) for 10 min before imaging via confocal microscopy. The images were acquired in an Olympus FV1000 confocal laser scanning microscope with a PlanApo 60× oil immersion objective. Imaging of nuclei, endosomes, and MNP-PEG-AEDP-CTA-RdB nanoconjugates was performed at the following excitation/emission wavelengths: 358 nm/461 nm, 488 nm/520 nm, and 546 nm/575 nm, respectively. Analysis was carried out by taking 10 images for each treatment with an average of 10 cells per image. The internalization and cytosol distribution were studied by calculating the surface area coverage. Image processing and analyses were performed on the software Fiji-ImageJ®. Statistical analyses and data processing were carried out on GraphPad Prism® V 6.01 software (GraphPad Software, La Jolla, CA, USA). Statistical comparisons were made using the unpaired t-test. Results of $p \leq 0.05$ (*) were considered significant.

2.7. Statistical Analyses

All data measurements are reported as mean ± standard deviation. Each experiment was carried out in triplicate. Data analysis was performed using the Graph Pad Prism V 6.01® software. Statistical comparisons were determined by running two-way ANOVA followed by post-treatment (Dunn's Multiple Comparison test). Results with p-value ≤ 0.05 (*) were considered significant. (*) corresponds to statistically significant difference with a p-value between 0.01 and 0.05; (**) represents $0.001 \leq p$-value < 0.01; (***) represents $0.0001 \leq p$-value ≤ 0.001; and (****) represents p-value < 0.0001. In addition, "ns" represents no statistically significant differences between the treatments.

3. Results and Discussion

3.1. Characterization of Magnetoliposomes Using the Microfluidic Approach

Figure 2A,B shows the size and PDI of the synthesized MLPs using the two-layer device. In this case, no apparent differences in size were identified for the evaluated TFRs and the concentration of the nanoconjugates used in the experiment for different FRRs. Nevertheless, there is a slight decrease in size with the increase of the FRR in almost all cases except for the TFR of 5 mL/min at a concentration of nanoconjugates of 0.17 mg/mL for the 1:1 to the 2:1 FRR, where it is comparable with the results for the synthesis of liposomes for different FRRs [31,41]. Additionally, the size of the MLPs synthesized is smaller than 400 nm except for the case where the TFR, concentration, and FRR are the lowest. The PDI of the synthesized MLPs presented values under 0.5 in almost all cases, which indicates that the MLPs samples obtained had acceptable polydispersity.

Figure 2C,D shows the size and PDI of the synthesized MLPs using the three-layer device. In this case, there is a significant difference in size for the evaluated TFRs and the concentration of the nanoconjugates used in the experiment for different FRRs. The 5 mL/min TFR led to lower size values than the ones obtained using the 2.5 mL/min TFR, and there is a slight increase in the size for the MLPs at a higher concentration of nanoconjugates. These observations are contrary to those reported by Joshi et al., according to which the TFR has no impact on the liposome sizes [33]. The dimensions of the channels and the incorporation of nanoconjugates into the lipid phase before entering the system might be relevant factors to explain the identified differences. The obtained MLPs sizes are larger than those obtained using the two-layer device, but in this case such sizes are under 400 nm. Additionally, the PDI values of the MLPs were below 0.5 in nearly all cases, which indicates that the MLPs samples obtained have an acceptable polydispersity. Compared with the two-layer device, there is a slight increase in the PDI values. This strongly suggests the two-layer system provides superior control over MLP sizes and their PDI values.

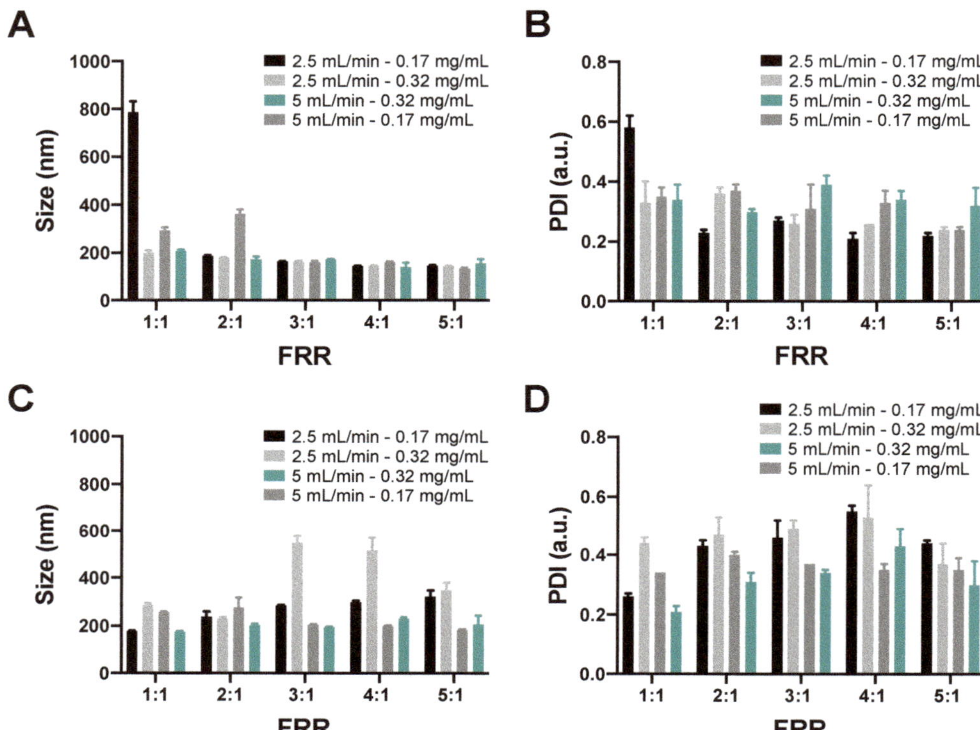

Figure 2. MLPs size and PDI for the two-layer device and the three-layer device. Two-layer device: (**A**) MLPs size using TFRs of 2.5 mL/min and 5 mL/min for nanoconjugates concentrations of 0.17 mg/mL and 0.32 mg/mL and FRRs from 1:1 to 5:1. (**B**) MLPs PDI using TFRs of 2.5 mL/min and 5 mL/min for nanoconjugates concentrations of 0.17 mg/mL and 0.32 mg/mL and FRRs from 1:1 to 5:1. Three-layer device: (**C**) MLPs size using TFRs of 2.5 mL/min and 5 mL/min for nanoconjugates concentrations of 0.17 mg/mL and 0.32 mg/mL and FRRs from 1:1 to 5:1. (**D**) MLPs PDI using TFRs of 2.5 mL/min and 5 mL/min for nanoconjugates concentrations of 0.17 mg/mL and 0.32 mg/mL and FRRs from 1:1 to 5:1.

Figure 3 shows the TEM characterization of the MLPs obtained with both the two-layer and three-layer devices. The images show that the MLPs formed correctly and in agreement with previous reports of MLPs synthesized via microfluidics [34,35]. The images also show that the size of the MLPs obtained for the three-layer device is slightly larger than that obtained with the two-layer device, which agrees well with the hydrodynamic diameters measured via DLS.

3.2. Magnetolipsomes Encapsulation Efficiency

Figure 4 shows the encapsulation efficiencies (EE%) for the MLPs synthesized with both devices at different TFR values, nanoconjugates concentrations in the lipid phase, and FRRs. The EE% obtained at 0.17 mg/mL nanoconjugates concentration with the two-layer device was higher for almost all evaluated FRRs, while no identifiable trend was observable for the three-layer device. This agrees well with the notion that a superior control of MLP assembly is achievable with the two-layer devices. Nevertheless, the results show efficiencies ranging from 20% to 90% for both devices for different FRRs, supporting the idea that the operating conditions strongly influence the performance of the devices. In addition, no correlation was found between FRR or TFR with the EE% values, which strongly suggests that the encapsulation process occurs randomly throughout the microfluidic device. However, the FRR is still an essential parameter in the size control of MLPs, which indicates that it is critical to define quality control strategies along with the nanoconjugates concentration [34]. Additionally, it is important to remark that the microfluidic synthesis of this type of drug delivery system provides a suitable route to enhance the therapeutic drug delivery efficiency compared with traditional methods. This provides further evidence for the relevance of the MLPs produced in this study, as they show consistent properties (e.g., size and morphology) without significant investments in infrastructure or instrumentation [34,50].

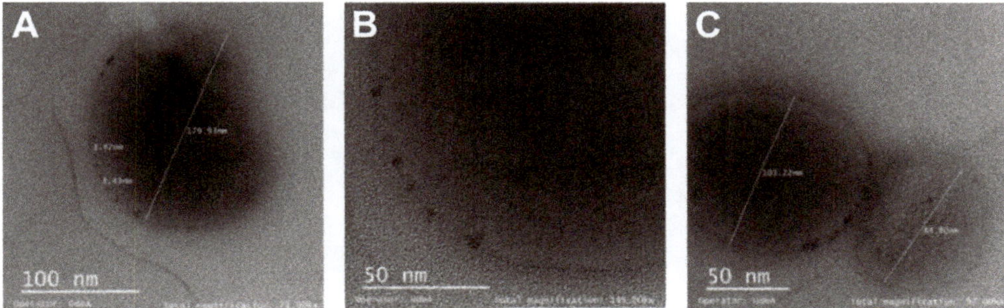

Figure 3. MLP characterization via TEM. (**A**) MLP synthesized with the three-layer device using TFR of 5 mL/min for nanoconjugates concentration of 0.32 mg/mL and a FRR set at 4:1 (**B**) Magnification by 145 k× of the MLP presented in A. (**C**) MLPs synthesized with the two-layer device using TFR of 5 mL/min for nanoconjugates concentration of 0.32 mg/mL and a FRR set at 4:1.

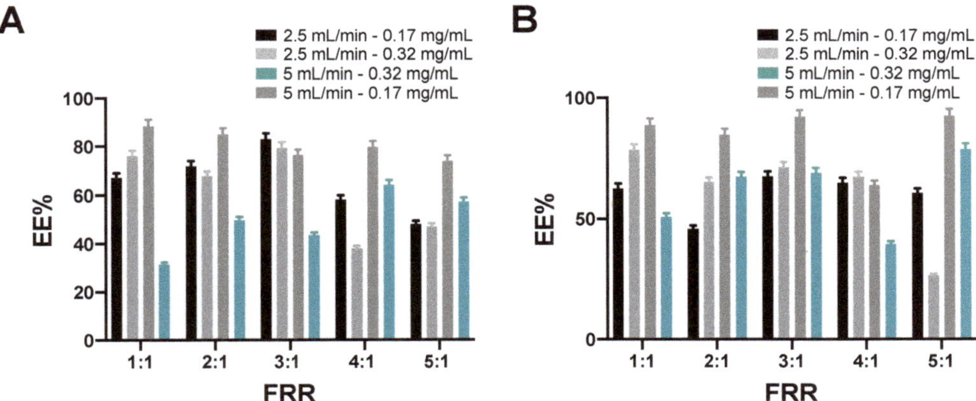

Figure 4. MLP encapsulation efficiency (EE%) for two-layer and three-layer devices. (**A**) Encapsulation efficiency for the MLPs synthesized using the two-layer device with TFRs of 2.5 mL/min and 5 mL/min at nanoconjugates concentrations of 0.17 mg/mL and 0.32 mg/mL and FRRs from 1:1 to 5:1. (**B**) Encapsulation efficiency for the MLPs synthesized using the three-layer device with TFRs of 2.5 mL/min and 5 mL/min at nanoconjugates concentrations of 0.17 mg/mL and 0.32 mg/mL and FRRs from 1:1 to 5:1.

3.3. Magnetoliposomes Purification

Supplementary Figure S3A shows the intensity of the magnetic field acting on the microfluidic separation channel for System 1. The magnetic field is higher in proximity to the permanent magnets, as was expected [51]. Figure 5A shows the experimental performance of System 1 and the identification of some of the particles' accumulation regions. Figure 5B shows the microfluidic system's separation performance where both nanoconjugates and MLPs are attracted to the channel wall near the higher magnetic flux density regions. However, calculations failed to show that the percentage of nanoconjugates trapped is higher compared to the MLPs. Figure 5C presents the obtained velocity fields for the dispersed phase within the channels for the mixture model approach. The results indicate that an increase in the dispersed phase's velocity matches each magnet's highest intensity locations. Finally, Figure 5D shows the concentration profile for the nanoconjugates within the channel as estimated by the transport of diluted species model during the first seconds of the study. This result indicates that the concentration tends to increase in regions where the magnetic flux density is higher, which directly results in streamlines targeting these regions along the entire channel.

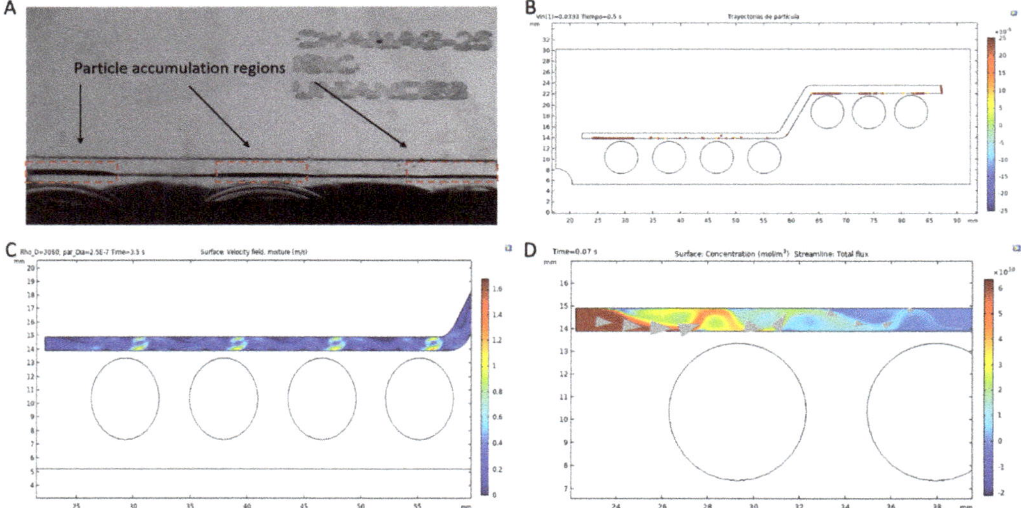

Figure 5. Qualitative results of the separation System 1. (**A**) Zoom of nanoconjugates accumulation regions inside the magnetophoretic microfluidic channel. (**B**) Particle trajectories in the magnetophoretic separation channel (nanoconjugates are shown in yellow and MLPs in red). (**C**) Velocity profile of the mixture. (**D**) Concentration profile in the microchannel.

Supplementary Figure S3B shows the intensity of the magnetic field acting on the microfluidic separation channel for System 2. Figure 6A shows the experimental performance of System 2 and the identification of some of the particles' accumulation regions. As for System 1, these regions are located in the channel sections near the magnets. Figure 6B shows the microfluidic system's separation performance, where the behavior presented is almost identical to that of System 1. In the case of the results for the mixture model, Figure 6C illustrates the velocity field results for the dispersed phase within the channels. In contrast, Figure 6D shows that the concentration profiles for the nanoconjugates and their streamlines tend to increase in regions of high magnetic flux density.

Figures 5 and 6 show qualitative results that illustrate the general performance of the proposed magnetophoretic separation devices and the trajectories of the nanoconjugates along the devices' microchannels. Nevertheless, a quantitative analysis is critical to estimate separation efficiencies and further verify them experimentally. For the case of the mixture model analysis, we selected several locations (Supplementary Figure S4A,B) along the computational domain to determine the concentration of each type of particle component (i.e., MLPs and nanoconjugates) and to calculate their concentration difference close to the zones where the magnetic field is the highest. The final separation efficiency percentage (SE%) was calculated as the average percentage difference in concentration for all of the selected locations. For the particle tracing model, a particle counter for each type of particle was set at the system's outlet.

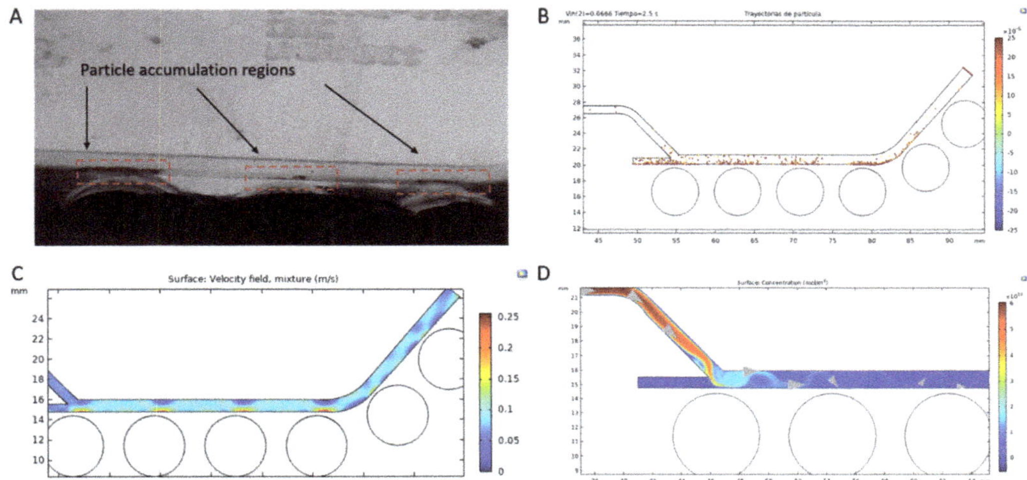

Figure 6. Qualitative results of the separation System 2. (**A**) Zoom of nanoconjugates accumulation regions inside the magnetophoretic microfluidic channel. (**B**) Particle trajectories in the magnetophoretic separation channel (nanoconjugates are shown in yellow and MLPs in red). (**C**) Velocity profile of the mixture. (**D**) Concentration profile inside the microchannel.

The separation efficiency is calculated as the percentage difference between each type of particle arriving at the outlet. Figure 7 shows the quantitative results for the separation efficiency calculated for the two simulation approaches implemented here and the corresponding comparison with the obtained experimental results for both systems. The results show that the best separation efficiency achieved was 31.55% for System 1 at a TFR of 1 mL/min, while for System 2 it was 51.22% at a TFR of 2 mL/min.

For System 1, the results show the dominance of magnetophoretic over hydrodynamic forces, as an increase in the TFR led to a decrease in the separation efficiency [52]. Despite the higher separation efficiencies obtained with System 2, such a correlation was not clear for this system. In addition, it is important to highlight that we found that separation with small TFRs might lead to a relatively large fraction of MLPs trapped along with nanoconjugates in the accumulation regions, thereby reducing the number of purified MLPs at the end of the process.

Compared with the mixture model, quantitative results for System 1 indicate that the particle tracing model led to results with a higher level of agreement with those observed experimentally. In contrast, the System 2 mixture model simulation approach showed better performance in predicting quantitative separation results than those obtained with the particle tracing model. Despite these results, it is important to highlight a significant difference in the choice of one approach over the other in terms of ease of implementation. In this regard, although the mixture model describes the suspended magnetic nanoconjugates as a ferrofluid, making the simulation more realistic, the computational cost of this modeling approach compared with particle tracing is much higher [42,43,52]. Our results recommend implementing both models for a more comprehensive understanding of the devices, as these two approaches complement each other and might provide much more robust insights for further manufacturing and experimental testing.

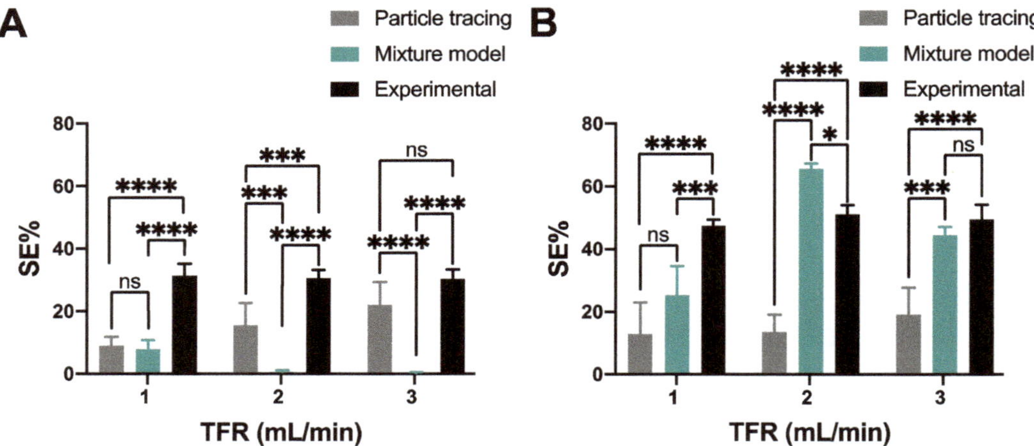

Figure 7. Microfluidic separation efficiency for Systems 1 and 2. (**A**) Comparison of both simulation approaches and the experimental results obtained for System 1 with a TFR ranging from 1 to 3 mL/min. (**B**) Comparison of both simulation approaches and the experimental results obtained for System 2 with a TFR ranging from 1 to 3 mL/min. Results with a *p*-value $\leq$ 0.05 (*) were considered significant. (*) corresponds to statistically significant difference with a *p*-value between 0.01 and 0.05; (***): $0.0001 \leq$ *p*-value $\leq$ 0.001; (****): *p*-value < 0.0001; "ns" represents no statistically significant differences between the treatments.

Because the obtained separation efficiencies with System 2 were higher than System 1, it was selected to produce the MLPs for further experimentation.

3.4. In Vitro Testing of MLPs

3.4.1. Biocompatibility

Figure 8 shows the cytocompatibility, hemocompatibility, and platelet aggregation results for the produced MLPs and the LPs. Figure 8A,B shows the viability of Vero cells after 24 and 48 h of exposure to the different treatments. MLPs and LPs show high biocompatibility at concentrations below 0.1 mg/mL. However, at concentrations above 0.1 mg/mL, the cell viability decreases to about 70%, showing a dose-dependent behavior. A similar tendency was observed for AGS cells (Figure 8C,D). In addition, AGS cells exhibited high tolerance to the treatments, reaching viability percentages above 80%, even at concentrations higher than 0.1 mg/mL. Figure 5A–D shows the cytocompatibility results for Vero and AGS cells exposed to nanoconjugates. Significant cytotoxicity levels were found at concentrations above 50 μg/mL in AGS cells, whereas in Vero cells, the viability remained above 80% even at higher concentrations. This result can be related to the significant sensitivity of AGS cells to CTA. Additionally, Figure 8E and Supplementary Figure S5E show the hemolysis percentage of the MLPs, LPs, and nanoconjugates compared with the positive and negative controls. The results show hemolysis percentages below 1% for MLPs and LPs and 1.16% for the nanoconjugates at the highest evaluated concentrations. Similar results were obtained previously for MLPs, where the hemolysis percentages were below 5% [12,18,53,54]. Figure 8F and Supplementary Figure S5F show the platelet aggregation percentages of the MLPs, LPs, and nanoconjugates compared with the positive control. The results indicate that the platelet aggregation of MNP, MLPs, and LPs remain below 55%

even at high concentrations. Compared with the negative control, the observed aggregation is acceptable even at the highest evaluated concentrations but is slightly higher than those reported previously [12,54].

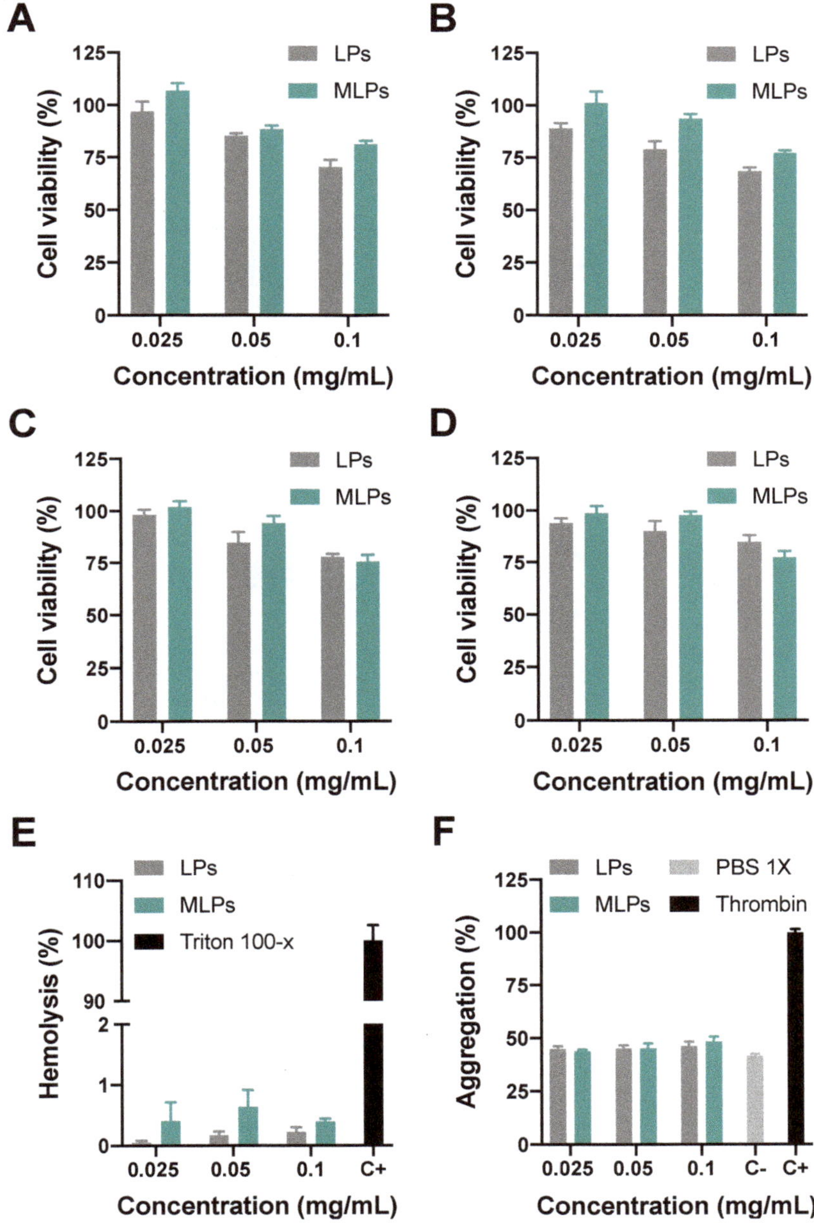

Figure 8. Biocompatibility assays for MLPs and LPs. Viability of Vero cells after 24 (**A**) and 48 h (**B**) of exposure. Viability of AGS cells after 24 (**C**) and 48 h (**D**) of exposure. (**E**) Hemolysis of MLPs and LPs with Triton X-100 as the positive control and PBS 1× as the negative control. (**F**) Platelet aggregation of MLPs and LPs with PBS 1× as the negative control and thrombin as the positive one.

3.4.2. Cell Internalization and Endosomal Escape Analysis

Figures 9 and 10 show the confocal images corresponding to the delivery of MNP-PEG-AEDP-CTA-RdB nanoconjugates (labeled MNPs in the figures) and MLPs on Vero and AGS cells at different times (i.e., 0.5 h and 4 h). Staining with Hoechst 33,342 allowed determination of the impact of the treatments on the cell viability by analyzing the nucleus morphology and distribution. Images clearly show nuclei with regular spherical-shaped morphology with no visible fragmentation or DNA condensation [55]. These results demonstrate non-apoptotic cells, confirming high biocompatibility in both cell lines even after 4 h of exposure. These results provide further evidence of the high cell viability levels obtained via MTT (Figure 8A–D and Supplementary Figure S5A–D).

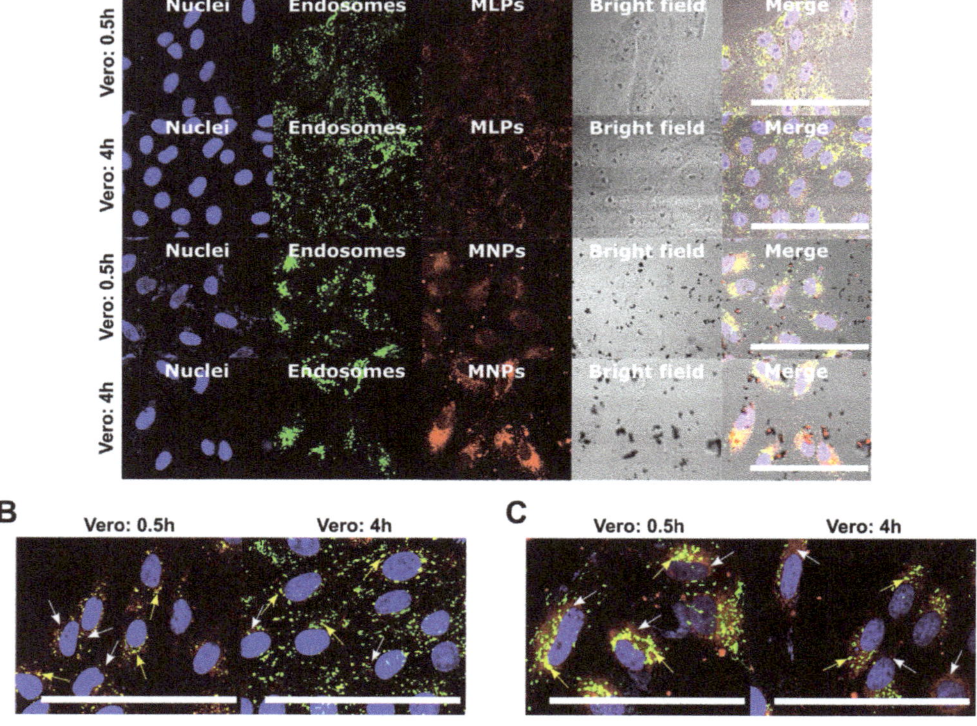

Figure 9. (**A**) Cell internalization and endosomal escape for magnetoliposomes (MLPs) and MNP-PEG-AEDP-CTA-RdB nanoconjugates (MNPs) in Vero cells with 40× magnification after 0.5 h and 4 h of exposure. The scale bar corresponds to 100 μm. Vero cells with 60× magnification and digital zoom to 120× after 0.5 h and 4 h of exposure to MLPs (**B**) and nanoconjugates (MNPs) (**C**). The scale bar for both (**B**,**C**) corresponds to 50 μm. The yellow arrows indicate colocalization between the green and the red channels, showing nanoparticles encapsulated into endosomes. The white arrows indicate non-colocalized zones, displaying nanoconjugates that escaped endosomes or reached the intracellular space by a different internalization mechanism.

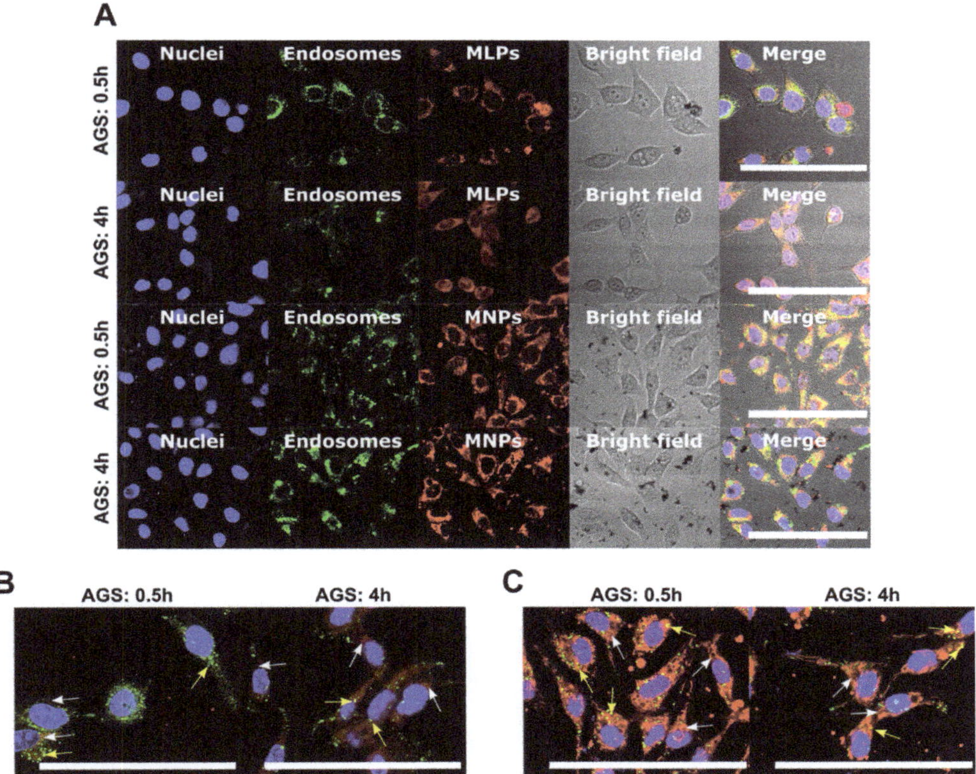

Figure 10. (**A**) Cell internalization and endosomal escape for magnetoliposomes and MNP-PEG-AEDP-CTA-RdB nanoconjugates (MNPs) in AGS cells with 40× magnification after 0.5 h and 4 h of exposure. The scale bar corresponds to 100 μm. AGS cells with 60× magnification and digital zoom to 120× after 0.5 h and 4 h of exposure to MLPs (**B**) and nanoconjugates (MNPs). (**C**). The scale bar for both (**B**,**C**) corresponds to 50 μm. The yellow arrows indicate colocalization between the green and the red channels, showing nanoparticles encapsulated into endosomes. The white arrows indicate non-colocalized zones, displaying nanoparticles that escaped endosomes or reached the intracellular space by a different internalization mechanism.

In addition, internalization and endosomal escape abilities were studied as a measure of the colocalization of Lysotracker Green® with the rhodamine B labeled nanoconjugates and their distribution intracellularly. Figures 9 and 10 show cells with a strong red fluorescent signal in the intracellular space, confirming the internalization of both MLPs and nanoconjugates. High cell penetration rates are most likely a consequence of employing PEG for the functionalization of MNPs to obtain the tested nanoconjugates and the use of LPs as powerful vehicles which favor membrane fusion and, consequently, the effective delivery of cargoes. LPs have also been reported to improve nanoparticle transport and plasma half-life [56]. The versatility of the developed vehicle allows the transport of CTA into the intracellular space even in the absence of LPs. This promising result presents the PEGylated magnetite-based nanovehicles as a fascinating tool for designing more potent oral delivery platforms to transport molecules of difficult intestinal absorption, such as CTA. This approach has been studied and validated previously by Kawish and colleagues [57].

Figure 11 shows the quantitative results for the analysis of endosomal escape and the distribution of the nanoparticles into the cells. Pearson's correlation coefficient (PCC) was used as a statistic tool for quantifying colocalization and covered area percentage to measure MLPs and nanoconjugates internalization and distribution. In AGS cells, a non-statistically significant difference was observed between the MLPs and the nanoconjugates, indicating that encapsulation into LPs failed to increase the endosomal escape of nanoconjugates. This was confirmed by an increase in the PCC after 4 h of exposure. However, the covered area of nanoconjugates slightly increased after 4 h, whereas the covered area of MLPs almost doubled for the same time. We hypothesize that these results might be a consequence of the interplay of different internalization routes. In this regard, as opposed to endocytic routes, it is very likely that nanoconjugates prevalently enter cells by a rapid and direct translocation mechanism [56]. In contrast, due to their negatively charged surface, LP and MLP internalization occurs mainly by endocytic routes. This is in line with recent reports that indicate that internalization rates for nanostructures with anionic coatings are lower than those of cationic and neutral coatings [56].

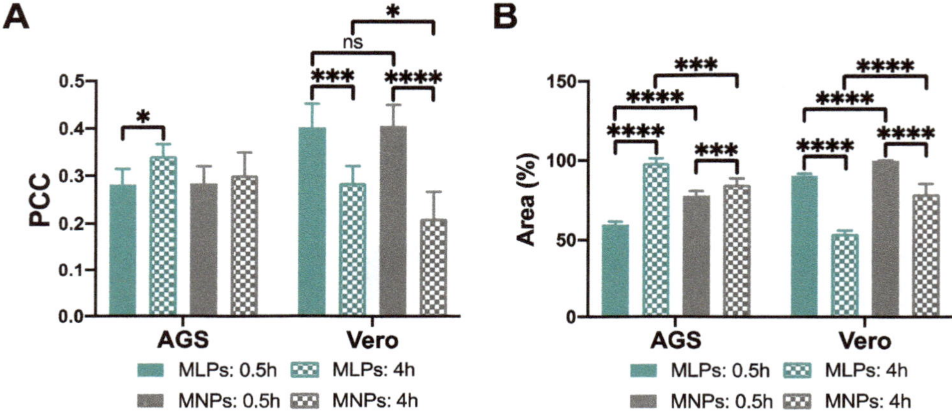

Figure 11. Pearson correlation coefficient (PCC) and percentage of area covered by the nanoconjugates in Vero and AGS cells. Higher PCC values indicate a higher amount of nanoconjugates trapped in endosomes. (**A**) PCC of MLPs and nanoconjugates in Vero and AGS cells for 0.5 h and 4 h. (**B**) Area covered by the MLPs and nanoconjugates in Vero and AGS cells for 0.5 h and 4 h. Results with a p-value ≤ 0.05 (*) were considered significant. (*) corresponds to statistically significant difference with a p-value between 0.01 and 0.05; (***): $0.0001 \leq p$-value ≤ 0.001; (****): p-value < 0.0001; "ns" represents no statistically significant differences between the treatments.

In Vero cells, the vehicles led to entirely different results, as evidenced by a statistically significant decrease in the PCC for both treatments after 4 h, confirming, therefore, endosomal escape. Somewhat surprisingly, for the same time, the covered area showed a statistically significant decrease. This suggests that, after internalization, the nanoconjugates escape endosomes and likely accumulate in different organelles. Future work will be dedicated to confirming this hypothesis.

The different penetration levels achieved for the two evaluated cell lines can be attributed to their significantly different cell membrane compositions, which might substantially alter the cell–nanoconjugate interactions. For example, the overexpression of claudin proteins in AGS cell membranes is likely to interfere with the internalization routes, rates, and achieved intracellular distributions [58]. Based on this, the rational design and development of novel vehicles for specific therapeutic applications must include a comprehensive analysis of membrane composition for the targeted cells. This is critical to engineer nanovehicles capable of taking advantage of and avoiding the possible interactions leading to cell penetration.

The obtained results clearly show the great potential of the developed nanovehicles as versatile carriers for difficult-to-transport drugs with high biocompatibility and the possibility to selectively internalize in cells with specific characteristics. Despite these promising results, future studies should include detailed studies on the impact of MLPs and the delivered compounds on microbiota bioactivity and bioavailability. This with the main objective of ensuring the homeostasis of patients' microbiomes, which is controlled by molecular interactions and plays a central role in modulating metabolism, immunity, and response to infections. In addition, this understanding is critical to determining the bioavailability of the functional compounds, as it largely depends on the metabolism exerted by the microbiota during the delivery process [20,22]. We expect that our MLPs will show no impact on the microbiome balance and, therefore, contribute to novel oral delivery routes for patients suffering from complex diseases, such as eczema, inflammatory disorders, hypertension, and chronic kidney disease [20,59].

4. Conclusions

Over the past few years, MLPs have been studied to improve the cell penetration efficiency of nanovehicles for drug delivery after administration. This study presents a low-cost microfluidic device to produce MLPs with sizes ranging from 136.87 ± 3.97 nm to 787.47 ± 45.65 nm and polydispersity values ranging from 0.21 ± 0.02 to 0.58 ± 0.04. The device operates by putting into intimate contact lecithin liposomes (LPs) and functionalized magnetite nanoparticles (MNPs) within a serpentine microchannel. As reported elsewhere, the size of the MLPs appears to be strongly influenced by the flow rate ratio (FRR) between the components infused into the system, which supports the importance of considering this parameter for designing and optimizing the device's performance. We prepared MNPs functionalized with a polymeric spacer (PEG) and a molecule containing a reducible disulfide bond (AEDP) to evaluate encapsulation. Additionally, we selected the immobilization of antibiotic CefTRIAxone (CTA) for proof-of-concept, considering its limited passing of the intestinal lumen after oral delivery. For all evaluated FRRs (i.e., from 1:1 to 5:1), the obtained nanoconjugates (i.e., MNP-PEG-AEDP-CTA) were encapsulated into the LPs (to form the MLPs) with efficiency above about 80% at a concentration of 0.17 mg/mL and while operating the device at a TFR of 5 mL/min.

However, this approach is challenging, as purifying the obtained MLPs is not a simple task, due to the proximity in properties of the involved components. Here, we decided to take advantage of the magnetic properties of the nanoconjugates to develop a robust and high-throughput separation scheme enabled by microfluidics and permanent magnets. Accordingly, we designed two magnetophoretic microfluidic separation devices where permanent magnets can be located along the main separation channels to retain excess nanoconjugates. To investigate the feasibility of this approach, we conducted two different multiphysics simulations approaches that provided complementary qualitative and quantitative information regarding the purification efficiency of the proposed devices. The first one was based on a particle tracing model, while the second relied on a multiphase mixture model. After conducting parametric sweeps for the FRRs, the results of the two approaches allowed us to confirm that the designed systems were well-suited to retain the nanoconjugates at hot spots of high magnetic field intensity. Although the qualitative results of both simulations show adequate behavior for the nanoconjugates within the systems, quantitative results varied considerably between approaches, due to the differences in inlet parameters and conditions used for each approximation. Conversely, it is highly suggested to combine these approaches for the rational design of microfluidic separation devices. The experimental testing showed separation efficiencies ranging from 30% to 31% for System 1 and 47% to 51% for System 2, showing a weak correlation between the TFR and separation efficiency in System 1. According to these results, System 2 operating at a TFR of 2 mL/min was employed for purifying MLPs for further in vitro testing.

Finally, we validated and demonstrated the great potential of the developed nanovehicles as a versatile carrier for difficult-to-transport drugs by showing high hemocompatibility, low platelet aggregation, and high cytocompatibility in two relevant cell lines (i.e., Vero and AGS). Despite different cell internalization and endosomal escape results in these two cell lines, the achieved coverage shows a promising potency, which is attractive for applications in gastrointestinal delivery. Furthermore, the results suggest unique nanoconjugate–cell membrane interactions and, consequently, interplay of different internalization mechanisms, which need to be considered for further surface engineering experiments.

Supplementary Materials: The following supporting information can be downloaded at: https://www.mdpi.com/article/10.3390/pharmaceutics14020315/s1, Figure S1: Manufactured separation Systems 1 and 2. (A) System 1 (B) System 2, Figure S2: Meshing used for multiphysics simulations. (A) System 1 for particle tracing; (B) System 2 for particle tracing; (C) System 1 for mixture model; (D) System 2 for mixture model, Figure S3: Magnetic flux density results for the separation Systems 1 and 2. (A) System 1 and (B) System 2, Figure S4: Evaluation points for mixture model simulations separation efficiency. (A) System 1 and (B) System 2, Figure S5: Biocompatibility assays for nanoconjugates (MNPs). Viability of Vero cells after 24 (A) and 48 h (B) of exposure. Viability of AGS cells after 24 (C) and 48 h (D) of exposure. (E) Hemolysis of MNPs with Triton X-100 as the positive control and PBS $1\times$ as the negative control. (F) Platelet aggregation of nanoconjugates with PBS $1\times$ as the negative control and thrombin as the positive one.

Author Contributions: Conceptualization, J.C.C. and J.F.O.; methodology, C.E.T., J.C., P.R.P., J.C.C. and J.F.O.; formal analysis, C.E.T., J.C. and J.C.C.; investigation, C.E.T., J.C., S.C.G., V.Q., K.A.G., P.R.P., J.A.S. and L.R.-G.; resources, C.M.-C., L.H.R., J.F.O. and J.C.C.; writing—original draft preparation, C.E.T. and J.C.; writing—review and editing, C.E.T., J.C., C.M.-C., L.H.R., J.F.O. and J.C.C.; visualization, J.C.; supervision, C.M.-C., L.H.R., J.C.C. and J.F.O.; project administration, J.C.C. and L.H.R.; funding acquisition, C.M.-C., J.C.C. and L.H.R. All authors have read and agreed to the published version of the manuscript.

Funding: This research was funded by the Colombian Ministry of Science, Technology, and Innovation (Minciencias), Grant IDs 782-2019 and 845-2018. Additional funding was provided by the 2019 Fundación Santafé de Bogotá-Uniandes grant "Production of recombinant antimicrobial peptides to modify materials of biomedical interest" and 2018 Fundación Santafé de Bogotá-Uniandes grant "Development of multifunctional magnetoliposomes as vehicles for the delivery of combined therapies of low dosage and high bioavailability for the treatment of Parkinson's disease".

Institutional Review Board Statement: Human blood samples were collected under the permission granted by the ethics committee at Universidad de los Andes (minute number 928–2018).

Informed Consent Statement: Informed consent was obtained from all subjects involved in the study.

Acknowledgments: We would like to thank the Department of Biomedical Engineering, the Department of Food and Chemical Engineering, and the Department of Electrical and Electronics Engineering at Universidad de los Andes for the financial and technical support. Additionally, we thank Juan Camilo Orozco for performing the confocal experiments and Johana Arboleda for TEM imaging.

Conflicts of Interest: The authors declare no conflict of interest.

References

1. Alqahtani, M.S.; Kazi, M.; Alsenaidy, M.A.; Ahmad, M.Z. Advances in Oral Drug Delivery. *Front. Pharmacol.* **2021**, *12*, 62. [CrossRef] [PubMed]
2. Homayun, B.; Lin, X.; Choi, H.-J. Challenges and Recent Progress in Oral Drug Delivery Systems for Biopharmaceuticals. *Pharmaceutics* **2019**, *11*, 129. [CrossRef]
3. Youhanna, S.; Lauschke, V.M. The Past, Present and Future of Intestinal in Vitro Cell Systems for Drug Absorption Studies. *J. Pharm. Sci.* **2021**, *110*, 50–65. [CrossRef] [PubMed]
4. Vyas, S.P.; Gupta, S. Optimizing Efficacy of Amphotericin B through Nanomodification. *Int. J. Nanomed.* **2006**, *1*, 417. [CrossRef] [PubMed]
5. Liu, W.; Wang, J.; McClements, D.J.; Zou, L. Encapsulation of β-Carotene-Loaded Oil Droplets in Caseinate/Alginate Microparticles: Enhancement of Carotenoid Stability and Bioaccessibility. *J. Funct. Foods* **2018**, *40*, 527–535. [CrossRef]

6.   Feng, J.; Wu, Y.; Zhang, L.; Li, Y.; Liu, S.; Wang, H.; Li, C. Enhanced Chemical Stability, Intestinal Absorption, and Intracellular Antioxidant Activity of Cyanidin-3-O-Glucoside by Composite Nanogel Encapsulation. *J. Agric. Food Chem.* **2019**, *67*, 10432–10447. [CrossRef]

7.   Kouhi, M.; Prabhakaran, M.P.; Ramakrishna, S. Edible Polymers: An Insight into Its Application in Food, Biomedicine and Cosmetics. *Trends Food Sci. Technol.* **2020**, *103*, 248–263. [CrossRef]

8.   Cheng, Y.; Ji, Y. RGD-Modified Polymer and Liposome Nanovehicles: Recent Research Progress for Drug Delivery in Cancer Therapeutics. *Eur. J. Pharm. Sci.* **2019**, *128*, 8–17. [CrossRef]

9.   Czuba, E.; Diop, M.; Mura, C.; Schaschkow, A.; Langlois, A.; Bietiger, W.; Neidl, R.; Virciglio, A.; Auberval, N.; Julien-David, D.; et al. Oral Insulin Delivery, the Challenge to Increase Insulin Bioavailability: Influence of Surface Charge in Nanoparticle System. *Int. J. Pharm.* **2018**, *542*, 47–55. [CrossRef] [PubMed]

10.   Moura, R.P.; Pacheco, C.; Pêgo, A.P.; des Rieux, A.; Sarmento, B. Lipid Nanocapsules to Enhance Drug Bioavailability to the Central Nervous System. *J. Control. Release* **2020**, *322*, 390–400. [CrossRef]

11.   Ho, B.X.; Loh, S.J.H.; Chan, W.K.; Soh, B.S. In Vivo Genome Editing as a Therapeutic Approach. *Int. J. Mol. Sci.* **2018**, *19*, 2721. [CrossRef]

12.   Ramírez-Acosta, C.M.; Cifuentes, J.; Castellanos, M.C.; Moreno, R.J.; Muñoz-Camargo, C.; Cruz, J.C.; Reyes, L.H. Ph-Responsive, Cell-Penetrating, Core/Shell Magnetite/Silver Nanoparticles for the Delivery of Plasmids: Preparation, Characterization, and Preliminary in Vitro Evaluation. *Pharmaceutics* **2020**, *12*, 561. [CrossRef]

13.   Hidai, C.; Kitano, H. Nonviral Gene Therapy for Cancer: A Review. *Diseases* **2018**, *6*, 57. [CrossRef] [PubMed]

14.   Perez, J.; Cifuentes, J.; Cuellar, M.; Suarez-Arnedo, A.; Cruz, J.C.; Muñoz-Camargo, C. Cell-Penetrating and Antibacterial BUF-II Nanobioconjugates: Enhanced Potency via Immobilization on Polyetheramine-Modified Magnetite Nanoparticles. *Int. J. Nanomed.* **2019**, *14*, 8483. [CrossRef] [PubMed]

15.   Cuellar, M.; Cifuentes, J.; Perez, J.; Suarez-Arnedo, A.; Serna, J.A.; Groot, H.; Muñoz-Camargo, C.; Cruz, J.C. Novel BUF2-Magnetite Nanobioconjugates with Cell-Penetrating Abilities. *Int. J. Nanomed.* **2018**, *13*, 8087. [CrossRef]

16.   Lopez-Barbosa, N.; Suárez-Arnedo, A.; Cifuentes, J.; Gonzalez Barrios, A.F.; Silvera Batista, C.A.; Osma, J.F.; Muñoz-Camargo, C.; Cruz, J.C. Magnetite–OmpA Nanobioconjugates as Cell-Penetrating Vehicles with Endosomal Escape Abilities. *ACS Biomater. Sci. Eng.* **2019**, *6*, 415–424. [CrossRef]

17.   Choi, W.I.; Sahu, A.; Wurm, F.R.; Jo, S.-M. Magnetoliposomes with Size Controllable Insertion of Magnetic Nanoparticles for Efficient Targeting of Cancer Cells. *RSC Adv.* **2019**, *9*, 15053–15060. [CrossRef]

18.   Garcia-Pinel, B.; Jabalera, Y.; Ortiz, R.; Cabeza, L.; Jimenez-Lopez, C.; Melguizo, C.; Prados, J. Biomimetic Magnetoliposomes as Oxaliplatin Nanocarriers: In Vitro Study for Potential Application in Colon Cancer. *Pharmaceutics* **2020**, *12*, 589. [CrossRef] [PubMed]

19.   Cardoso, B.D.; Rodrigues, A.R.O.; Bañobre-López, M.; Almeida, B.G.; Amorim, C.O.; Amaral, V.S.; Coutinho, P.J.; Castanheira, E. Magnetoliposomes Based on Shape Anisotropic Calcium/Magnesium Ferrite Nanoparticles as Nanocarriers for Doxorubicin. *Pharmaceutics* **2021**, *13*, 1248. [CrossRef]

20.   Karavolos, M.; Holban, A. Nanosized Drug Delivery Systems in Gastrointestinal Targeting: Interactions with Microbiota. *Pharmaceuticals* **2016**, *9*, 62. [CrossRef]

21.   Vamanu, E.; Gatea, F. Correlations between Microbiota Bioactivity and Bioavailability of Functional Compounds: A Mini-Review. *Biomedicines* **2020**, *8*, 39. [CrossRef] [PubMed]

22.   Dabulici, C.M.; Sârbu, I.; Vamanu, E. The Bioactive Potential of Functional Products and Bioavailability of Phenolic Compounds. *Foods* **2020**, *9*, 953. [CrossRef]

23.   Vamanu, E.; Rai, S.N. The Link between Obesity, Microbiota Dysbiosis, and Neurodegenerative Pathogenesis. *Diseases* **2021**, *9*, 45. [CrossRef]

24.   Pereira, D.S.; Cardoso, B.D.; Rodrigues, A.R.O.; Amorim, C.O.; Amaral, V.S.; Almeida, B.G.; Queiroz, M.-J.R.; Martinho, O.; Baltazar, F.; Calhelha, R.C.; et al. Magnetoliposomes Containing Calcium Ferrite Nanoparticles for Applications in Breast Cancer Therapy. *Pharmaceutics* **2019**, *11*, 477. [CrossRef]

25.   Wu, Y.; Lu, Z.; Li, Y.; Yang, J.; Zhang, X. Surface Modification of Iron Oxide-Based Magnetic Nanoparticles for Cerebral Theranostics: Application and Prospection. *Nanomaterials* **2020**, *10*, 1441. [CrossRef] [PubMed]

26.   De Jesus, P.D.; Pellosi, D.S.; Tedesco, A.C. Magnetic Nanoparticles: Applications in Biomedical Processes as Synergic Drug-Delivery Systems. In *Materials for Biomedical Engineering*; Elsevier: Amsterdam, The Netherlands, 2019; pp. 371–396.

27.   Rodrigues, A.R.O.; Almeida, B.G.; Araújo, J.P.; Queiroz, M.-J.R.; Coutinho, P.J.; Castanheira, E.M. Magnetoliposomes for Dual Cancer Therapy. In *Inorganic Frameworks as Smart Nanomedicines*; Elsevier: Amsterdam, The Netherlands, 2018; pp. 489–527.

28.   Rodrigues, A.R.O.; Gomes, I.; Almeida, B.G.; Araújo, J.P.; Castanheira, E.M.; Coutinho, P.J. Magnetoliposomes Based on Nickel/Silica Core/Shell Nanoparticles: Synthesis and Characterization. *Mater. Chem. Phys.* **2014**, *148*, 978–987. [CrossRef]

29.   Bonnaud, C.; Monnier, C.A.; Demurtas, D.; Jud, C.; Vanhecke, D.; Montet, X.; Hovius, R.; Lattuada, M.; Rothen-Rutishauser, B.; Petri-Fink, A. Insertion of Nanoparticle Clusters into Vesicle Bilayers. *ACS Nano* **2014**, *8*, 3451–3460. [CrossRef] [PubMed]

30.   Floris, A.; Ardu, A.; Musinu, A.; Piccaluga, G.; Fadda, A.M.; Sinico, C.; Cannas, C. SPION@ Liposomes Hybrid Nanoarchitectures with High Density SPION Association. *Soft Matter* **2011**, *7*, 6239–6247. [CrossRef]

31. Carugo, D.; Bottaro, E.; Owen, J.; Stride, E.; Nastruzzi, C. Liposome Production by Microfluidics: Potential and Limiting Factors. *Sci. Rep.* **2016**, *6*, 25876.

32. Mota-Cobián, A.; Velasco, C.; Mateo, J.; España, S. Optimization of Purification Techniques for Lumen-Loaded Magnetoliposomes. *Nanotechnology* **2020**, *31*, 145102. [CrossRef] [PubMed]

33. Joshi, S.; Hussain, M.T.; Roces, C.B.; Anderluzzi, G.; Kastner, E.; Salmaso, S.; Kirby, D.J.; Perrie, Y. Microfluidics Based Manufacture of Liposomes Simultaneously Entrapping Hydrophilic and Lipophilic Drugs. *Int. J. Pharm.* **2016**, *514*, 160–168. [CrossRef] [PubMed]

34. Al-Ahmady, Z.S.; Donno, R.; Gennari, A.; Prestat, E.; Marotta, R.; Mironov, A.; Newman, L.; Lawrence, M.J.; Tirelli, N.; Ashford, M.; et al. Enhanced Intraliposomal Metallic Nanoparticle Payload Capacity Using Microfluidic-Assisted Self-Assembly. *Langmuir* **2019**, *35*, 13318–13331. [CrossRef]

35. Conde, A.J.; Batalla, M.; Cerda, B.; Mykhaylyk, O.; Plank, C.; Podhajcer, O.; Cabaleiro, J.M.; Madrid, R.E.; Policastro, L. Continuous Flow Generation of Magnetoliposomes in a Low-Cost Portable Microfluidic Platform. *Lab. Chip* **2014**, *14*, 4506–4512. [CrossRef]

36. Takagi, J.; Yamada, M.; Yasuda, M.; Seki, M. Continuous Particle Separation in a Microchannel Having Asymmetrically Arranged Multiple Branches. *Lab. Chip* **2005**, *5*, 778–784. [CrossRef] [PubMed]

37. Salafi, T.; Zeming, K.K.; Zhang, Y. Advancements in Microfluidics for Nanoparticle Separation. *Lab. Chip* **2017**, *17*, 11–33. [CrossRef]

38. Kye, H.G.; Park, B.S.; Lee, J.M.; Song, M.G.; Song, H.G.; Ahrberg, C.D.; Chung, B.G. Dual-Neodymium Magnet-Based Microfluidic Separation Device. *Sci. Rep.* **2019**, *9*, 9502. [CrossRef]

39. Khashan, S.A.; Dagher, S.; Alazzam, A.; Mathew, B.; Hilal-Alnaqbi, A. Microdevice for Continuous Flow Magnetic Separation for Bioengineering Applications. *J. Micromechan. Microeng.* **2017**, *27*, 55016. [CrossRef]

40. Guo, J.; Lin, L.; Zhao, K.; Song, Y.; Huang, M.; Zhu, Z.; Zhou, L.; Yang, C. Auto-Affitech: An Automated Ligand Binding Affinity Evaluation Platform Using Digital Microfluidics with a Bidirectional Magnetic Separation Method. *Lab. Chip* **2020**, *20*, 1577–1585. [CrossRef]

41. Aranguren, A.; Torres, C.E.; Muñoz-Camargo, C.; Osma, J.F.; Cruz, J.C. Synthesis of Nanoscale Liposomes via Low-Cost Microfluidic Systems. *Micromachines* **2020**, *11*, 1050. [CrossRef]

42. Garg, H.; Kharola, A.S.; Karar, V. Numerical Analysis of Different Magnet Shapes on Heat Transfer Application Using Ferrofluid. In Proceedings of the 2015 COMSOL Conference, Pune, India, October 29–30 2015; Volume 51, p. 160030.

43. Bruno, N.M.; Ciocanel, C.; Kipple, A. Modeling Flow of Magnetorheological Fluid through a Micro-Channel. In Proceedings of the 2009 COMSOL Conference, Boston, MA, USA, October 8–10 2009; pp. 1–7.

44. Yang, M.; O'Handley, R.; Fang, Z. Modeling of Ferrofluid Passive Cooling System. In Proceedings of the 2010 COMSOL Conference, Boston, MA, USA, October 7–9 2010; pp. 1–6.

45. Torres-Díaz, I.; Rinaldi, C. Recent Progress in Ferrofluids Research: Novel Applications of Magnetically Controllable and Tunable Fluids. *Soft Matter* **2014**, *10*, 8584–8602. [CrossRef]

46. Kumar, C.; Hejazian, M.; From, C.; Saha, S.C.; Sauret, E.; Gu, Y.; Nguyen, N.-T. Modeling of Mass Transfer Enhancement in a Magnetofluidic Micromixer. *Phys. Fluids* **2019**, *31*, 63603. [CrossRef]

47. García-Jimeno, S.; Escribano, E.; Queralt, J.; Estelrich, J. Magnetoliposomes Prepared by Reverse-Phase Followed by Sequential Extrusion: Characterization and Possibilities in the Treatment of Inflammation. *Int. J. Pharm.* **2011**, *405*, 181–187. [CrossRef]

48. Multiphysics, C. Comsol Multiphysics 5.3 CFD Module User's Guide. 2014, pp. 274–289. Available online: https://doc.comsol.com/5.3/doc/com.comsol.help.cfd/CFDModuleUsersGuide.pdf (accessed on 28 November 2021).

49. Campaña, A.L.; Sotelo, D.C.; Oliva, H.A.; Aranguren, A.; Ornelas-Soto, N.; Cruz, J.C.; Osma, J.F. Fabrication and Characterization of a Low-Cost Microfluidic System for the Manufacture of Alginate–Lacasse Microcapsules. *Polymers* **2020**, *12*, 1158. [CrossRef] [PubMed]

50. Allen, T.M.; Cullis, P.R. Liposomal Drug Delivery Systems: From Concept to Clinical Applications. *Adv. Drug Deliv. Rev.* **2013**, *65*, 36–48. [CrossRef] [PubMed]

51. Wu, J.; Yan, Q.; Xuan, S.; Gong, X. Size-Selective Separation of Magnetic Nanospheres in a Microfluidic Channel. *Microfluid. Nanofluid.* **2017**, *21*, 47. [CrossRef]

52. Guo, J.; Wang, Y.; Xue, Z.; Xia, H.; Yang, N.; Zhang, R. Numerical Analysis of Capture and Isolation of Magnetic Nanoparticles in Microfluidic System. *Mod. Phys. Lett. B* **2018**, *32*, 1840075. [CrossRef]

53. Lorente, C.; Cabeza, L.; Clares, B.; Ortiz, R.; Halbaut, L.; Delgado, Á.V.; Perazzoli, G.; Prados, J.; Arias, J.L.; Melguizo, C. Formulation and in Vitro Evaluation of Magnetoliposomes as a Potential Nanotool in Colorectal Cancer Therapy. *Colloids Surf. B Biointerfaces* **2018**, *171*, 553–565. [CrossRef]

54. Ye, H.; Tong, J.; Wu, J.; Xu, X.; Wu, S.; Tan, B.; Shi, M.; Wang, J.; Zhao, W.; Jiang, H.; et al. Preclinical Evaluation of Recombinant Human IFNα2b-Containing Magnetoliposomes for Treating Hepatocellular Carcinoma. *Int. J. Nanomed.* **2014**, *9*, 4533.

55. Crowley, L.C.; Marfell, B.J.; Waterhouse, N.J. Analyzing Cell Death by Nuclear Staining with Hoechst 33342. *Cold Spring Harb. Protoc.* **2016**, *2016*, pdb-prot087205. [CrossRef]

56. Rueda-Gensini, L.; Cifuentes, J.; Castellanos, M.C.; Puentes, P.R.; Serna, J.A.; Muñoz-Camargo, C.; Cruz, J.C. Tailoring Iron Oxide Nanoparticles for Efficient Cellular Internalization and Endosomal Escape. *Nanomaterials* **2020**, *10*, 1816. [CrossRef]

57. Kawish, M.; Elhissi, A.; Jabri, T.; Muhammad Iqbal, K.; Zahid, H.; Shah, M.R. Enhancement in Oral Absorption of Ceftriaxone by Highly Functionalized Magnetic Iron Oxide Nanoparticles. *Pharmaceutics* **2020**, *12*, 492. [CrossRef] [PubMed]

58. Zavala-Zendejas, V.E.; Torres-Martinez, A.C.; Salas-Morales, B.; Fortoul, T.I.; Montaño, L.F.; Rendon-Huerta, E.P. Claudin-6, 7, or 9 Overexpression in the Human Gastric Adenocarcinoma Cell Line AGS Increases Its Invasiveness, Migration, and Proliferation Rate. *Cancer Investig.* **2011**, *29*, 1–11. [CrossRef] [PubMed]
59. Oyama, J.; Node, K. Gut Microbiota and Hypertension. *Hypertens. Res.* **2019**, *42*, 741–743. [CrossRef] [PubMed]

Article

Complementary Nucleobase Interactions Drive Co-Assembly of Drugs and Nanocarriers for Selective Cancer Chemotherapy

Fasih Bintang Ilhami [1], Enyew Alemayehu Bayle [1] and Chih-Chia Cheng [1,2,*]

[1] Graduate Institute of Applied Science and Technology, National Taiwan University of Science and Technology, Taipei 10607, Taiwan; fasihilhami17@gmail.com (F.B.I.); enyewalemayehu@gmail.com (E.A.B.)
[2] Advanced Membrane Materials Research Center, National Taiwan University of Science and Technology, Taipei 10607, Taiwan
* Correspondence: cccheng@mail.ntust.edu.tw

Abstract: A new concept in cooperative adenine–uracil (A–U) hydrogen bonding interactions between anticancer drugs and nanocarrier complexes was successfully demonstrated by invoking the co-assembly of water soluble, uracil end-capped polyethylene glycol polymer (BU-PEG) upon association with the hydrophobic drug adenine-modified rhodamine (A-R6G). This concept holds promise as a smart and versatile drug delivery system for the achievement of targeted, more efficient cancer chemotherapy. Due to A–U base pairing between BU-PEG and A-R6G, BU-PEG has high tendency to interact with A-R6G, which leads to the formation of self-assembled A-R6G/BU-PEG nanogels in aqueous solution. The resulting nanogels exhibit a number of unique physical properties, including extremely high A-R6G-loading capacity, well-controlled, pH-triggered A-R6G release behavior, and excellent structural stability in biological media. Importantly, a series of in vitro cellular experiments clearly demonstrated that A-R6G/BU-PEG nanogels improved the selective uptake of A-R6G by cancer cells via endocytosis and promoted the intracellular release of A-R6G to subsequently induce apoptotic cell death, while control rhodamine/BU-PEG nanogels did not exert selective toxicity in cancer or normal cell lines. Overall, these results indicate that cooperative A–U base pairing within nanogels is a critical factor that improves selective drug uptake and effectively promotes apoptotic programmed cell death in cancer cells.

Keywords: adenine–uracil base pair; complementary hydrogen bonded drug carrier system; controlled drug delivery; supramolecular nanogels; selective cytotoxicity

Citation: Ilhami, F.B.; Bayle, E.A.; Cheng, C.-C. Complementary Nucleobase Interactions Drive Co-Assembly of Drugs and Nanocarriers for Selective Cancer Chemotherapy. *Pharmaceutics* **2021**, *13*, 1929. https://doi.org/10.3390/pharmaceutics13111929

Academic Editors: Francisco José Ostos, José Antonio Lebrón and Pilar López-Cornejo

Received: 19 October 2021
Accepted: 12 November 2021
Published: 15 November 2021

Publisher's Note: MDPI stays neutral with regard to jurisdictional claims in published maps and institutional affiliations.

1. Introduction

The specific sequences present in biopolymers such as DNA, RNA, and proteins are responsible for the survival of complex, adaptable living organisms [1]. Analogous synthetic polymers with well-controlled, designed sequences have been predicted to function as components in a wide range of applications [2–4]. Complementary, noncovalent multiple hydrogen bonding interactions—such as adenine-thymine (A–T), guanine-cytosine, and adenine–uracil (A–U) base pairing between nucleic acids—provide versatile tools to control and tune the structure and function of polymers containing complementary nucleobases [5–8]. In recent decades, numerous research groups have reported that "bio-constituted" hydrogen bonding interactions could facilitate the self-assembly of various structures with nanometer-scale features. Moreover, well-controlled, dynamic physical properties and varied stimuli-responsive properties in response to environmental changes could be achieved by varying the amounts and strength of the nucleobase pairs [9–11]. For instance, supramolecular amphiphilic block copolymers with complementary adenine and thymine hydrogen bonding interactions that spontaneously self-assemble into nano-sized micelles in aqueous solution were reported by Kuang et al., and a well-controlled drug release profile could be obtained by tuning the content of A–T in the polymer structure [12]. Wang et al. designed a unique supramolecular phospholipid with moderate hydrogen bonding interaction recognition between adenosine and uridine; this phospholipid could self-assemble into spherical liposomes with highly pH-responsive ability, and had high potential for controlled drug release [13]. Fan and co-workers blended an adenine-containing amphiphilic block copolymer and uracil-functionalized crosslinking agent to self-assemble new micelles that exhibited rapid, pH-controlled drug release to significantly reduce cell viability [14]. Based on the examples above, synthetic supramolecular polymers containing nucleobase pairs are a promising concept in various areas of research, and the exploitation of non-covalent interactions to confer polymeric structures with dynamic characteristics may expand their applications.

The development of polymeric nanocarriers in order to improve chemotherapeutic efficacy and intracellular delivery for cancer therapy has attracted overwhelming enthusiasm in modern pharmaceutical technology. In terms of biocompatible nanocarriers, polyethylene glycol (PEG) is regarded as one of the most important components of nanocarriers for drug delivery, owing to its high level of water solubility, considerable chain mobility, and low toxicity, among other suitable physicochemical properties [15–17]. Moreover, the introduction of PEG segments into nanocarrier structures decreases their tendency to form aggregates, and thus results in enhanced structural stability and avoids the clearance of the nanocarriers by the reticuloendothelial system (RES) [18–20]. Koo et al. noted that shell cross-linked polymer micelles based on a combination of PEG and polyamino acids exhibited high stability in water, improved biocompatibility towards normal and cancer cells, and enhanced the intracellular release triggered by glutathione (GSH) [21]. Yokoyama et al. developed a new PEG-*b*-poly(α,β-aspartic acid) block copolymer; the resulting drug-loaded micelles were extremely stable in aqueous solution, and their size and distribution could be easily tuned [22]. Nevertheless, previous reports mentioned that not all hydrophilic PEG segments provide advantageous behavior in an aqueous environment [23–25]. A range of hydrophilic PEG segments have varied stabilization effects, including invoking side-effects via immunological responses, inhibiting the intracellular uptake of the nanocarrier, and non-biodegradability. Thus PEG segments with a molecular weight below 20 kDa are preferable for use in drug carriers [23–27]. Inspired by the specific features of nucleobases, we confidently speculated that the incorporation of nucleobase molecules into the terminal end-groups of PEG may confer unique self-assembly behavior and physical properties in aqueous solution, leading to potential candidates for drug delivery applications.

A recent series of studies in our laboratory demonstrated that the introduction of complementary A–U base pairs within polymer structures confers the ability to spontaneously self-assemble into stable, physically crosslinked networks [28], resulting in excellent film-forming capability, tailorable mechanical performance and intriguing self-healing capacity upon tuning of the content of the A–U complexes in the matrix [29]. To further extend the concept of complementary nucleobase interactions to drug carrier systems, we herein design and synthesize a new difunctional, uracil-terminated PEG macromer (BU-PEG) that associates with adenine-functionalized rhodamine (A-R6G) in an aqueous environment via complementary A–U interactions, and thus results in the formation of self-assembled spherical nanogels with high structural stability (Scheme 1) [30]. In addition, the resulting A-R6G/BU-PEG nanogels possess a number of unique physical properties, including extremely high A-R6G-loading capacity, wide-range tailorable A-R6G-loading content, distinct green fluorescence behavior, and well-controlled pH-triggered A-R6G release. Importantly, a series of in vitro experiments clearly confirmed that the A-R6G-loaded BU-PEG nanogels highly selectively targeted cancer cells, and were selectively internalized by them, and could thus promote high levels of rapid apoptotic death through an endocytotic pathway, but they were not internalized by and did not harm normal cells. As far as we are aware, this is the first study to develop a complementary drug_carrier system based on hydrogen bonding interactions between stable A–U base pairs, with the goals of improving both the safety and effectiveness of cancer chemotherapy. Thus, this new concept for the fabrication of drug_carrier systems via complementary A–U base pairing offers in-depth insight into the benefits of manipulating the drug delivery behavior of nanocarriers with potential in various biological and biomedical applications.

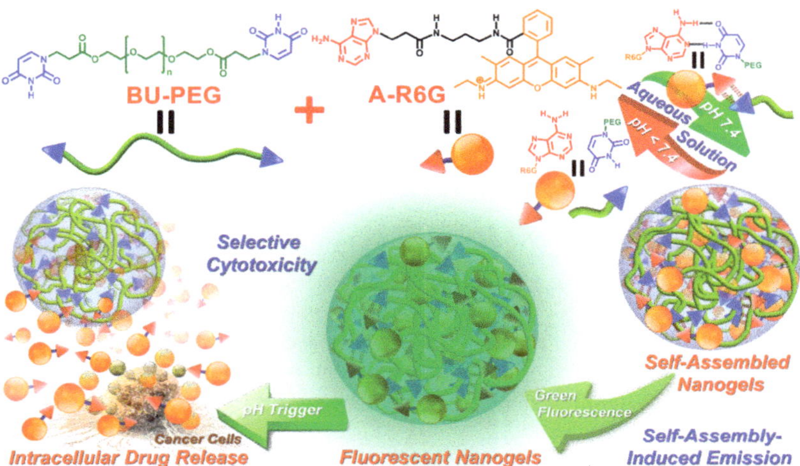

Scheme 1. Schematic illustration of the co-assembly process, fluorescent properties, and cancer cell-selective cytotoxic behavior of hydrogen-bonded A-R6G/BU-PEG complexes. The upper-right green and red arrows represent the association and dissociation of the complementary A–U base pairs under various pH conditions.

2. Materials and Methods

2.1. Chemicals and Materials

Uracil ($\geq$99% purity) and adenine (>99.5% purity) were purchased from Acros Organics (Geel, Belgium). Rhodamine 6G (R6G), polyethylene glycol (PEG, average molecular mass: 1900–2200 g/mol), dimethylformamide (DMF), potassium *tert*-butoxide (*t*-BuOK), triethylamine (TEA), deuterated chloroform (CDCl$_3$), and deuterium oxide (D$_2$O) were obtained from Sigma-Aldrich Chemical (Milwaukee, WI, USA) at the highest purity available. The HPLC-grade organic solvents were used as received from TEDIA (Fairfield, OH, USA). Phosphate-buffered saline (PBS), Dulbecco's modified Eagle's medium (DMEM), fetal bovine serum (FBS), penicillin-streptomycin, trypsin-EDTA, trypan blue, 4′,6-diamidino-2-phenylindole (DAPI), the Dead Cell Apoptosis Kit with Brilliant Violet-421™ Annexin V (BV421-Annexin V), and Ghost Dye™ Red 780 (GDR780) were purchased from Thermo Fisher Scientific (Waltham, MA, USA). All chemicals and reagents were employed as received. Adenine-functionalized rhodamine derivative (A-R6G) and PEG diacrylate (PEGDA, number average molecular weight (M$_n$) = ~2000) were synthesized and characterized according to procedures that have been described previously [31–33].

2.2. Synthesis of BU-PEG

PEGDA (4 g, 2 mmol) and uracil (0.5 g, 4.5 mmol) were dissolved in 300 mL of DMF, and agitated for 48 h at 60 °C with a small quantity of *t*-BuOK as a catalyst (0.04 g, 0.003 mmol). After removing DMF by vacuum distillation, the crude product was dissolved in chloroform (200 mL) and the insoluble impurities were removed by filtration through a Büchner funnel. Subsequently, the chloroform was evaporated by rotary evaporation, then the obtained product was washed three times with diethyl ether and dried in an oven at 30 °C for 1 day. The yield was 83% (3.8 g).

2.3. Characterization

2.3.1. Proton and Carbon Nuclear Magnetic Resonance (^{1}H NMR and ^{13}C NMR)

The ^{1}H NMR and ^{13}C NMR spectra were obtained on a Bruker AVIII NMR spectrometer (Billerica, MA, USA) at 500 MHz; the 25 mg samples were dissolved in 1 mL of CDCl$_3$ and D$_2$O, respectively.

2.3.2. Gel Permeation Chromatography (GPC)

Molecular weight information was obtained using a Waters Alliance 2690 HPLC Separation Module (Waters Corporation, Milford, MA, USA) with tetrahydrofuran (THF) as the mobile phase at a flow rate of 1.0 mL/min and 40 °C. The M_n and the polydispersity index (PDI) were determined by comparison against a series of narrow distribution polystyrene standards.

2.3.3. Critical Micelle Concentration (CMC)

Pyrene was utilized as a fluorescent probe to measure the CMC of the PEG and BU-PEG polymers. Varied concentrations of samples (0.00001 to 0.4 mg/mL) were prepared in water in advance. Next, 10 μL of pyrene solution was dropped into the tubes, then the mixtures were sonicated and incubated at 4 °C overnight to allow the polymers and aqueous phase to stabilize fully. All samples were examined using a fluorescence spectrometer (Jasco FP-8300 Spectrophotometer Hitachi, Tokyo, Japan) at an excitation wavelength of 335 nm. The emission intensities at 373 nm and 392 nm were recorded and plotted against the sample concentration in order to determine the CMC values.

2.3.4. Particle Size and Surface Charge

Hydrodynamic particle size, PDI, and zeta (ζ) potential values were measured with a dynamic light scattering particle analyzer (DLS, Nano Brook 90Plus PALS, Brookhaven, Holtsville, NY, USA) connected to a 632-nm He-Ne laser beam (scattering angle: 90°). DLS measurements for each sample were repeated ten times, and averaged.

2.3.5. Photoluminescence (PL) and Ultraviolet-Visible (UV-Vis)

The PL and UV-Vis spectra of samples in aqueous solution were acquired at 25 °C using Hitachi F4500 luminescence and Jasco FP-8300 spectrophotometers (Hitachi, Tokyo, Japan), respectively.

2.3.6. Atomic Force Microscopy (AFM) and Scanning Electron Microscopy (SEM)

The surface morphologies of the samples were assessed by AFM (NX10; AFM Park Systems, Suwon, Korea) and SEM (JSM-6500F system JEOL, Tokyo, Japan). Specimens were prepared by spin-coating diluted aqueous solutions onto silicon wafers at 1250 rpm for 15 s, and subsequently vacuum-dried at ambient temperature overnight.

2.4. Preparation of A-R6G/BU-PEG and R6G/BU-PEG Complexes

Different amounts of A-R6G or R6G (0.1 mg, 0.5 mg, and 1 mg) were added to BU-PEG (1 mg) in DMF (2 mL), stirred for 24 h, and then dialyzed against PBS (pH 7.4, 10 mM) or distilled water (DW) for 24 h (1000 Da molecular weight cut-off (MWCO)); the PBS (or DW) was replenished every 4 h. Finally, the absorption spectra of the A-R6G/BU-PEG solutions were obtained via UV/Vis spectrophotometry at λ = 525 nm to establish the concentration–absorbance standard curves of R6G and A-R6G. The absorption spectra of the A-R6G/BU-PEG and R6G/BU-PEG complexes (1 mg/mL) were compared to the standard curves for R6G and A-R6G. The following equation was used to calculate the drug loading content (DLC) and drug loading efficiency (DLE):

$$DLC\% = \frac{\text{Weight of drug loaded in polymeric nanogels}}{\text{Weight of drug loaded polymeric nanogels}} \times 100$$

$$DLE\% = \frac{\text{Weight of drug loaded in polymeric nanogels}}{\text{weight of drug input}} \times 100$$

2.5. Evaluation of the Stability of Pristine BU-PEG and R6G/A-R6G-Loaded BU-PEG Nanogels

The stability of blank and R6G/A-R6G-loaded BU-PEG nanogels in DMEM were investigated in the presence of FBS, which functions as a nanoparticle-destabilizing agent [34]. Pristine BU-PEG and A-R6G-loaded or R6G-loaded BU-PEG nanogels were mixed with DMEM containing 2:1 v/v serum, and the particle size and distributions were measured over 24 h by DLS.

2.6. Study of the Drug Release Behavior of Drug-Loaded Nanogels

The A-R6G-loaded or R6G-loaded BU-PEG nanogels (5 mL) were placed in dialysis tubing (MWCO = 1000 Da) and immersed in 50 mL of PBS with various pH values (7.4, 6.5, or 6.0) at 25 °C; the PBS was stirred (100 rpm) using a magnetic stirrer. At predefined intervals, 5 mL of external buffer solution was sampled and replaced with 5 mL of new PBS solution with the same pH. The amount of released R6G (or A-R6G) was measured by UV-Vis spectrometry at λ = 525 nm against a standard calibration curve for free R6G in PBS, and plotted vs. time.

2.7. Hemolysis Assay

Sheep red blood cells (SRBCs, Cosmo Bio, Tokyo, Japan) were used to assess the hemolytic activity of A-R6G-loaded and R6G-loaded BU-PEG nanogels. Briefly, 1 mL of the SRBCs and 0.5 mL of PBS were added into a microcentrifuge tube, centrifuged at 12,000 rpm for 15 min, and the supernatant was removed as plasma. Next, 1.5 mL of PBS was added, vortexed, and centrifuged; this wash step was repeated three times until the supernatant was clear. Subsequently, the resulting SRBC solutions (500 µL) were added to various concentrations of drug-loaded nanogels (10, 20, 40, 100, and 150 µg/mL). Triton X-100 solution (1%) was used as a positive control and PBS as a negative control. All samples were placed in a 5% CO_2 incubator at 37 °C for 4 h, then centrifuged, and then 100 µL of the supernatants were transferred into a 96-well plate and the absorbance values were measured at a wavelength of 540 nm on an ELISA reader. The hemolysis index was calculated as follows:

$$\text{Hemolysis}\% = \frac{A_{\text{sample}} - A_{\text{negative}}}{A_{\text{positive}} - A_{\text{negative}}} \times 100\%$$

where a represents the optical density (OD) values of the test sample, positive control (1% Triton X-100), or negative control (PBS).

2.8. Cell Lines and Culture Conditions

HeLa cells (human cervical cancer cell lines), MG-63 (human bone cancer cell lines), and NIH/3T3 cells (mouse embryonic fibroblast cell lines) were obtained from the ATCC (American Type Culture Collection, Manassas, VA, USA) and routinely cultured in DMEM supplemented with 10% FBS containing 1% penicillin-streptomycin in T-75 culture flasks at 37 °C in a 5% CO_2 incubator.

2.9. In Vitro Cell Cytotoxicity Assay

The cytotoxicity of the A-R6G-loaded and R6G-loaded BU-PEG nanogels were investigated against normal cell lines (NIH/3T3 cells) and cancer cell lines (HeLa and MG-63 cells) using the 3-(4,5-dimethylthiazol-2-yl)-2,5-diphenyltetrazolium bromide (MTT) method. Briefly, 1×10^5 cells/well were seeded into 96-well plates in 100 µL of complete DMEM medium, allowed to adhere for 24 h, then the culture media were replaced with new media containing 0.1–100 µg/mL of blank BU-PEG, free R6G or A-R6G, or A-R6G-loaded or R6G-loaded BU-PEG nanogels. The plates were incubated for 24 h, the media were removed, and 100 µL of MTT assay solution (5 mg/mL) was added per well, incubated for 4 h, then 100 µL of dimethyl sulfoxide was added to dissolve the formazan crystals in each well. The optical densities were measured using an ELx800 microplate reader (BioTek, Winooski, VT, USA) at 570 nm.

2.10. Cellular Uptake of R6G/A-R6G-Loaded BU-PEG Nanogels

Intracellular uptake of A-R6G-loaded and R6G-loaded BU-PEG nanogels was evaluated in NIH/3T3 and HeLa cells by confocal laser scanning microscopy (CLSM) and flow cytometry.

CLSM: NIH/3T3 and HeLa cells were seeded into 35-mm microscopy dishes at a density of 1×10^5, cultured for 24 h, then the medium was replaced with fresh serum-free medium containing A-R6G-loaded or R6G-loaded BU-PEG nanogels and incubated for various times (3, 12, or 24 h). Subsequently, the NIH/3T3 and HeLa cells were gently rinsed three times with cold PBS and fixed in 4% paraformaldehyde for 30 min at 25 °C. Thereafter, the nuclei of the cells were stained using 4′,6-diamidino-2-phenylindole (DAPI) for 30 min. The fluorescence of the cells was visualized via confocal microscopy (iRiS™ Digital Cell Imaging System, Logos Biosystems, Anyang-si, Korea).

Flow cytometry: NIH/3T3 and HeLa cells were seeded into 6-well plates at a density of 2×10^6 cells/well, incubated for 24 h at 37 °C, then A-R6G-loaded or R6G-loaded BU-PEG nanogels dissolved in fresh free-serum medium were added to the wells and cultured for 1, 3, 6, 12, or 24 h. The cells were detached using trypsin, re-suspended with cold PBS (0.5 mL) and analyzed on a BD FACSAria III flow cytometer (BD Biosciences, San Jose, CA, USA). Data events were collected and determined by FlowJo software.

2.11. Analysis of Apoptosis Induced by R6G/A-R6G-Loaded BU-PEG Nanogels

The BV421-Annexin and Ghost Red Dye-780 Detection Kit was used to double stain the NIH/3T3 and HeLa cells. Briefly, the cells were seeded in 6-well plates at 1×10^6 cells/well, allowed to adhere for 24 h, then incubated with A-R6G-loaded or R6G-loaded BU-PEG nanogels for different periods of time (1, 3, 6, or 12 h). Untreated NIH/3T3 and HeLa cells were prepared as controls. Next, the cells were re-suspended in binding buffer and stained using BV421-Annexin (15 min), followed by Ghost Red Dye-780 (15 min) in the dark, in accordance with the manufacturer's protocol, then analyzed by flow cytometry (BD FACSAria III).

2.12. Statistical Analysis

All results are provided as the means and standard deviations of at least three independent experiments.

3. Results and Discussion

Herein, our research aims and objectives focused on the development of cooperative multiple hydrogen bonds in nucleobase-functionalized groups between a nanocarrier and drug molecules to enhance the safety and efficiency of cancer chemotherapy, as illustrated in Scheme 1. A new a water-soluble uracil-end-capped BU-PEG polymer was developed and was prepared via a one-step Michael addition reaction of PEGDA [32,33] to uracil using a *t*-BuOK catalyst, resulting in a high-yielding product (83%). The resulting BU-PEG polymer showed the desired chemical structures and molecular weight, as determined by ^{1}H NMR, ^{13}C NMR, and GPC (see Supplementary Materials, Figure S1). In addition, BU-PEG is a white, semi-crystalline powder and can easily dissolve in water at 25 °C, even at concentrations as high as 50 mg/mL. Due to the formation of A–U base-pairings between BU-PEG and adenine-functionalized molecules, a strong cooperative hydrogen bonding partner drug, A-R6G, was synthesized according to our previous report [31]. After the introduction of an adenine moiety into the R6G structure, the drug exhibited extremely poor solubility in water, buffer and biological media, demonstrated unique green-fluorescent behavior, and exerted strong cytotoxic effects in various normal and cancer cell lines [31,35–37]. Due to the significant differences in the water solubility between the BU-PEG polymer and A-R6G drug, these intriguing findings motivated our interest in exploring the co-assembly behavior of BU-PEG/A-R6G complexes in water.

Before discussing the hydrogen-bonded complexes of BU-PEG with A-R6G in water, we first determined the basic physical characteristics of BU-PEG in water to help our understanding and control the self-assembly structures and dynamics of BU-PEG in water and to guide the construction of a drug_carrier system based on the combination of BU-PEG and A-R6G. To investigate the effect of the hydrogen-bonded uracil groups on the amphiphilic features of the hydrophilic PEG backbone in aqueous environments, UV-Vis measurements using hydrophobic pyrene as a fluorescent probe were employed to determine the CMC values of pristine PEG (M_n = ~2000) and BU-PEG in water [38]. As shown in Figure 1a, BU-PEG showed a low CMC value of 2.6×10^{-2} mg/mL, whereas pristine PEG did not exhibit any CMC characteristics across a broad concentration range from 10^{-5} to 1 mg/mL, suggesting that the introduction of a uracil moiety into the PEG end-groups significantly impacted the amphiphilicity of the PEG backbone in water. This observation was possibly due to the presence of the self-complementary, interpolymer uracil hydrogen bonding interactions between the chains, which thus prompted the formation of relatively longer linear polymer chains connected by hydrogen-bonded uracil dimers that subsequently altered the overall amphiphilicity of BU-PEG. Next, we studied the self-organized structure of BU-PEG in water via complementary DLS, AFM and SEM measurements. When the concentration of the BU-PEG solution was 0.5 mg/mL (above its CMC value), BU-PEG had a mean hydrodynamic diameter of 86 nm, a PDI value of 0.27 and a ζ-potential value of -41.12 ± 0.36 mV (Figure 1b, Table S1), suggesting that BU-PEG could self-construct uniform, mono-distributed nanogels in water; this observation was attributed to the presence of the self-complementary uracil hydrogen bonding interactions within the polymer structure. In validation of the DLS results, AFM and SEM images confirmed that BU-PEG formed spherical nanogels with a diameter ranging from 35 to 70 nm (Figure 1c,d). Thus, these results revealed that the uracil units within the polymer structure served as a key governing force in hydrogen bonding molecular recognition to induce the self-assembly of the polymer in an aqueous environment.

After exploring the self-assembled structures and the characteristics of BU-PEG in water, we examined the molecular recognition of the A–U base-pairing between BU-PEG and A-R6G in D_2O using ^{1}H NMR spectroscopy. When a binary mixture of A-R6G and BU-PEG was prepared at a 1:10 blending weight ratio in D_2O, the hydrophobic A-R6G almost completely dissolved in D_2O and the mixed solution had a light orange color, while pristine A-R6G entirely precipitated in D_2O, implying that A-R6G has a high tendency to interact with BU-PEG via strong complementary A–U interactions, thus leading to a significant increase in the solubility of A-R6G in D_2O. Further ^{1}H NMR analysis led to clear observation of the characteristic A-R6G peaks for the A-R6G/BU-PEG blend in D_2O, whereas no peaks were detected for pristine A-R6G due to its poor water solubility (Figure S2). These results clearly and unequivocally demonstrate that the adenine units of A-R6G underwent complementary hydrogen bonding interactions with the uracil end-groups of BU-PEG in the aqueous solution, which thus promoted the formation of the A-R6G/BU-PEG complex and drastically improved the water-solubility of A-R6G. In addition, these results also implied that the existence of the complementary A–U interactions within the A-R6G/BU-PEG complexes may significantly improve the encapsulation capacity and stability of A-R6G after blending/mixing with BU-PEG. Thus, we subsequently performed an A-R6G encapsulation experiment to evaluate the drug-loading performance of BU-PEG nanogels via a dialysis method (see further details in the Experimental Section) in order to validate whether the complementary A–U interactions within the drug–carrier system successfully conferred high drug entrapment efficiency and improved drug-entrapment stability. As expected, the A-R6G-loaded BU-PEG complexes with a weight ratio of 1:1 had the maximal DLC of $69.04 \pm 2.89\%$, whereas the R6G-loaded BU-PEG complexes with the same mixing ratio only achieved an R6G-loading content of $22.67 \pm 3.45\%$ (Table S1). Additionally, the resulting A-R6G-loaded BU-PEG complexes exhibited a wide-range tunable DLC, and the desired A-R6G-loading content could be achieved by controlling the A-R6G and BU-PEG blending ratio. In contrast to the A-R6G/BU-PEG system, the R6G and BU-PEG complexes prepared using different ratios of both materials exhibited similar DLC values after purification by dialysis (Table S1), indicating that R6G-loaded BU-PEG cannot stably encapsulate R6G due to the lack of complementary A–U interactions within the complexes, thus explaining the relatively low DLC and non-tailorable drug-loading content. In other words, the complementary hydrogen bonding interactions between uracil and the adenine moieties within the complexes increase the affinity and specificity of the nanocarriers for encapsulated hydrophobic drugs, and thus conferred extremely high and tunable DLC.

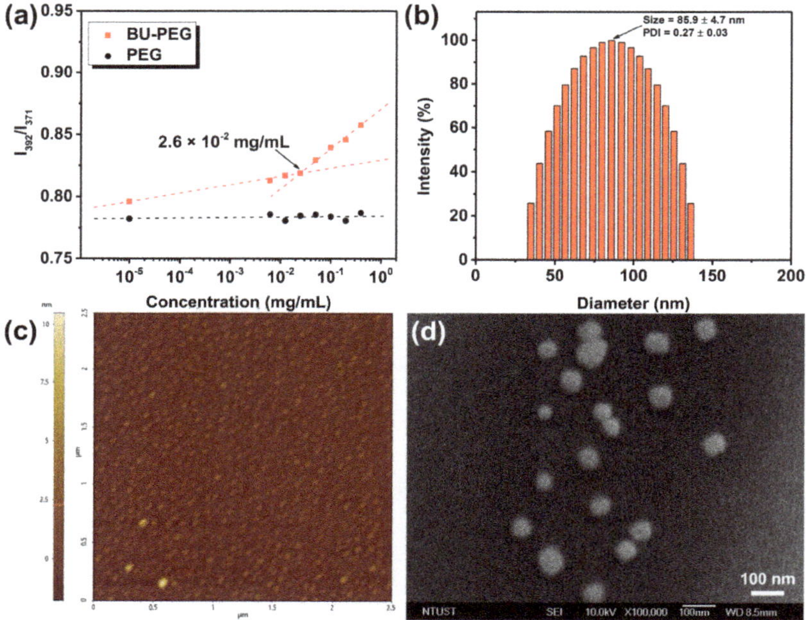

Figure 1. (**a**) Determination of the CMC values of PEG and BU-PEG in water in the presence of pyrene. (**b**) Hydrodynamic particle size of the BU-PEG polymer in water as determined by DLS at 25 °C. Surface morphologies of spin-coated BU-PEG polymers obtained by (**c**) AFM and (**d**) SEM at 25 °C.

The DLS analysis further demonstrated that A-R6G-loaded BU-PEG (containing 69% A-R6G) and R6G-loaded BU-PEG (containing 23% R6G) displayed mean hydrodynamic diameters of 181 ± 5.97 nm (PDI = 0.461) and 160 ± 5.53 nm (PDI = 0.237), respectively (Figure S3a, Table S1), suggesting that the complexes increased in size to offer a relatively large capacity to accommodate a large number of drug molecules compared with the pristine BU-PEG nanogels. In addition, the ζ-potential value of the A-R6G-loaded BU-PEG increased progressively from −38.29 ± 6.44 mV to 51.21 ± 3.66 mV with a gradual increase of DLC. All of the A-R6G-loaded complexes exhibited significantly larger ζ-potential values compared to the R6G-loaded BU-PEG and the pristine BU-PEG (Table S1), indicating the ζ-potential value of the A-R6G-loaded BU-PEG complexes gradually increased to improve the A-R6G encapsulation efficiency and stability of the complexes. In order to verify the results obtained by DLS, we examined the morphological structure of A-R6G-loaded and R6G-loaded BU-PEG complexes using AFM and SEM. As illustrated in Figure 2a and Figure S3b–d, both systems had nearly spherical shapes with sizes varying between 120–150 nm and 90–130 nm, respectively, consistent with the DLS results (Figure S3a). Overall, these findings further revealed that dynamic complexes of BU-PEG with hydrophobic A-R6G suspended in water were successfully constructed due to the complementary A–U interactions, and resulted in formation of a spherical-like A-R6G-encapsulated nanogel with tunable DLC capacity.

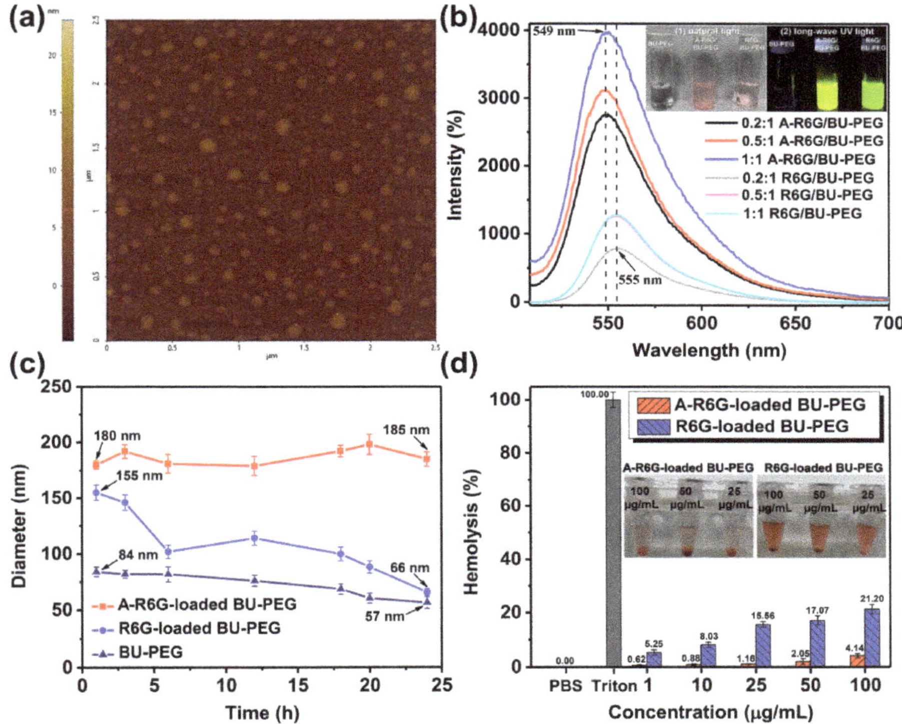

Figure 2. (**a**) AMF images of A-R6G-loaded BU-PEG nanogels measured at 25 °C. (**b**) PL spectra of A-R6G-loaded and R6G-loaded BU-PEG nanogels in water at 25 °C. Inset: photographs of A-R6G-loaded and R6G-loaded BU-PEG nanogels in water exposed to (1) natural lighting and (2) broadband UV lighting conditions. (**c**) Time-dependent DLS analysis of the structural stability of A-R6G-loaded and R6G-loaded BU-PEG nanogels in media containing 10% FBS at pH 7.4 for 24 h. (**d**) In vitro hemolytic assay of different concentrations of A-R6G-loaded and R6G-loaded BU-PEG nanogels on SRBCs. Inset: photographs showing SRBCs incubated with varying concentrations (1–100 µg/mL) of A-R6G-loaded or R6G-loaded BU-PEG nanogels at 37 °C in 5% CO_2 for 4 h.

Due to the intrinsic fluorescence emission of R6G and A-R6G, we further investigated the effect of A–U base pairing on the fluorescence properties of the A-R6G-loaded and R6G-loaded BU-PEG nanogels in water with different DLC values using PL spectroscopy. As shown in the right-upper inset of Figure 2b, when A-R6G and the BU-PEG were blended at a 1:1 weight ratio, the resulting aqueous solution (containing 51% A-R6G) exhibited a strong, bright-green fluorescence under excitation with a broadband UV, while the R6G-loaded BU-PEG solution (weight ratio of 1:1; containing 19% R6G) exhibited relatively weak green fluorescence. These results indicated the complementary A–U interactions within the nanogels dramatically enhanced the fluorescence emission behavior of A-R6G in aqueous solution. A quantitative analysis of the fluorescence enhancement of A-R6G-loaded BU-PEG nanogels in water was conducted by PL measurements with excitation at 480 nm. As shown in Figure 2b, the PL spectra of R6G-loaded BU-PEG nanogels containing 13% and 19% R6G (weight ratios of 1:1 and 0.5:1, respectively) in water exhibited maximum fluorescence peaks at 555 nm with the same intensity of around 1300, possibly due to the presence of the similar DLC of the nanogels. Interestingly, the maximum PL fluorescence peaks and intensities of the A-R6G-loaded BU-PEG nanogels with various DLC values were remarkably different to those of the R6G-loaded BU-PEG system: the A-R6G-loaded nanogels exhibited a significant blue-shift in their maximum fluorescence peaks from 555 nm to 549 nm, and a gradual increase in the maximum fluorescence intensity from 2740 to 3980 as the DLC increased, suggesting that the physical encapsulation of the A-R6G in the nanogels through complementary A–U interactions efficiently prevented the aggregation of the polycyclic aromatic rings of A-R6G within the nanogels [39,40], resulting in a significant blue-shift of the maximum fluorescence peak and a progressive enhancement in fluorescence intensity with the DLC, even with A-R6G-loading contents as high as 51%. Thus, these findings further demonstrate that the A–U interactions within this complementary drug-nanocarrier system manipulate the drug-encapsulation and fluorescence behavior of the nanogels in aqueous environments.

An ideal nanoparticulate drug-loaded delivery system must maintain high drug-entrapment stability in the normal cellular environment to ensure safe, effective delivery of drugs. Therefore, we studied the structural stability of A-R6G-loaded and R6G-loaded BU-PEG nanogels in culture media (DMEM supplemented with 10% serum FBS) and the SRBC hemolysis assay. FBS and SRBCs were employed as nanoparticle-destructuring agents to induce the rapid disassembly of the self-assembled nanogels [34]. As shown in Figure 2c, pristine BU-PEG exhibited a gradual decrease in the mean hydrodynamic diameter from 84 nm to 57 nm after 24 h incubation with FBS-containing DMEM, implying that the self-complementary hydrogen bonding uracil moieties could not preserve the structural integrity of the BU-PEG nanogels in FBS/PBA-mixed medium. After the encapsulation process, A-R6G-loaded BU-PEG nanogels remained at an almost constant mean hydrodynamic diameter after 24 h monitoring, whereas a progressive decrease in the particle size of the R6G-loaded BU-PEG nanogels from 155 nm to 66 nm was clearly observed, indicating the stable complementary A–U interactions between the A-R6G and the BU-PEG complex substantially improved drug-retention stability and prevented initial drug leakage from the nanogels. Similar trends in the improvement of structural stability were obtained in the hemolysis assay. As presented in Figure 2d and the inset photographs, the SRBC hemolytic assay indicated that a broad range of concentrations (1–100 μg/mL) of A-R6G-loaded BU-PEG nanogels did not exert significant hemolytic activity. Even at high concentrations up to 100 μg/mL, the A-R6G-loaded nanogels showed a low hemolytic activity of 4.1%, indicating the excellent compatibility of the nanogels with blood, which is potentially favorable for in vivo applications [41,42]. In contrast to the A-R6G-loaded BU-PEG system, the percentage hemolysis gradually increased as the concentration of R6G-loaded BU-PEG nanogels increased. The percentage hemolysis was up to 21.2% at 100 μg/mL of R6G-loaded BU-PEG, indicating that BU-PEG nanogels cannot maintain their structural stability due to the lack of complementary interactions between BU-PEG and R6G, thus resulting in significant hemolysis. The combination of long-term drug-entrapment stability and low-level hemolysis for A-R6G-loaded BU-PEG is an extremely attractive set of features that are rarely observed in traditional drug carrier systems. Therefore, A-R6G-loaded BU-PEG nanogels may represent a potential drug-delivery system that can provide a safe, reliable and efficient delivery of A-R6G within cellular environments.

The above findings prompted us to further evaluate the drug release behavior of A-R6G (or R6G) from BU-PEG nanogels in PBS solutions at different pH values (pH 7.4, 6.5, 6.0) at 25 °C using a dialysis method. At pH 7.4, the cumulative release of A-R6G from BU-PEG nanogels was only 30% after 48 h, whereas R6G release of up to 56% was observed from the BU-PEG nanogels, suggesting that the encapsulated A-R6G within the nanogels exhibited high structural stability under normal physiological conditions due to the complementary A–U interactions, thus leading to slow drug release and low cumulative drug release (Figure 3a,b). Interestingly, when the environmental pH was decreased to 6.5 or 6.0, the A-R6G or R6G was released much more quickly than at a normal physiological pH of 7.4. Both systems exhibited a significant initial burst, with over 50% of the drug released in the first 12 h at either pH 6.5 or 6.0, followed by cumulative release of 94% and 88% after 48 h, respectively, implying that weakly acidic environments can trigger the structural disassembly of both drug-loaded nanogels, and thus induce a rapid drug release rate and high cumulative drug release. Overall, due to the transient A–U interactions within their structure, A-R6G-loaded BU-PEG nanogels exhibit high A-R6G-loading capacity and excellent structural stability that directly inhibits premature drug release and a confers a low drug release rate under normal physiological pH 7.4. However, the A-R6G-loaded BU-PEG nanogels disassembled and release the drug rapidly in mildly acidic conditions, and thereby possessed well-controlled pH-triggered drug release properties. Thus, this self-assembled drug–carrier system based on complementary A–U interactions may represent an attractive and potential route for safer, more efficient in vitro and in vivo delivery and release of medication.

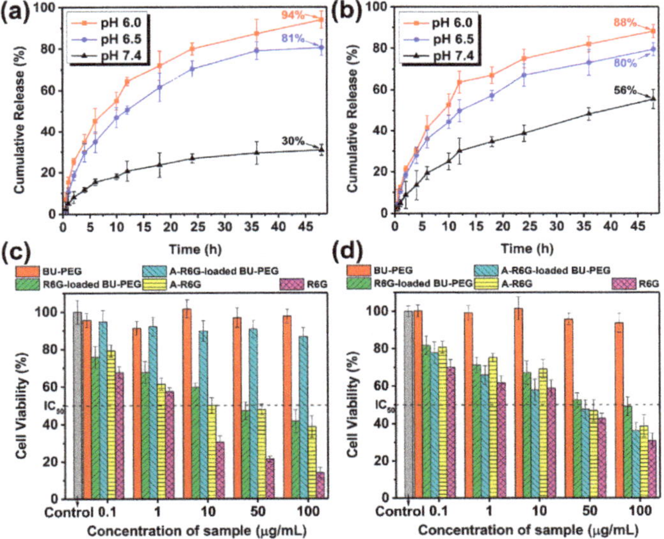

Figure 3. In vitro drug release profiles of (**a**) A-R6G-loaded and (**b**) R6G-loaded BU-PEG nanogels. In vitro cytotoxicity of A-R6G-loaded and R6G-loaded BU-PEG nanogels against (**c**) NIH/3T3 cells and (**d**) HeLa cells after 24 h incubation.

Potential drug delivery nanogels must be highly biocompatible and exert low cytotoxicity against normal and cancer cells, and they must also only release the encapsulated drug under specific conditions in the cellular environment. Thus, an MTT-based chromogenic assay was used to quantify the cytotoxic activity of pristine BU-PEG, A-R6G-loaded-, and R6G-loaded BU-PEG nanogels toward normal NIH/3T3 cells and cancerous HeLa and MG-63 cells. As indicated in Figure 3c,d and Figure S4, a range of concentrations of BU-PEG nanogels exerted negligible cytotoxic effects in normal and cancer cells after 24 h, indicating the BU-PEG was highly biocompatible. In contrast, pristine the A-R6G and R6G showed highly potent cytotoxic activities against normal and cancer cells, with half-maximal inhibitory concentrations (IC$_{50}$) ranging from 1 to 43 µg/mL. Extraordinarily, after culture with A-R6G-loaded BU-PEG nanogels at concentrations up to 100 µg/mL for 24 h, the viability of NIH/3T3 cells remained above 85%, while the R6G-loaded BU-PEG nanogels strikingly reduced the viability of NIH/3T3 cells, with an IC$_{50}$ value of 40.5 µg/mL (Figure 3c). These observations clearly suggested the complementary A–U interactions within the A-R6G-loaded BU-PEG nanogels significantly improved the structural stability of the nanogels and minimized premature A-R6G leakage under physiological conditions in NIH/3T3 cells. Conversely, the lack of complementary interactions within the R6G-loaded nanogels led to a significant reduction in NIH/3T3 cell viability. However, both the A-R6G-loaded and R6G-loaded BU-PEG nanogels exerted significant cytotoxic activities against HeLa and MG-63 cancer cells, with remarkable IC_{50} values of 39.7 µg/mL and 79.3 µg/mL in HeLa cells and 2.8 µg/mL and 9.3 µg/mL in MG-63 cells, respectively (Figure 3d and Figure S4). These results suggest that the pH-induced structural disassembly of both drug-loaded nanogels in the acidic extracellular environment of cancer cells led to rapid drug release, subsequently resulting in selective cytotoxic effects [43,44]. Thus, the weakly acidic extracellular cancer cell environment may facilitate the release of A-R6G within the interior of the cells and thus facilitate selective, potent cytotoxicity toward cancer cells, while reducing the adverse impacts of A-R6G-loaded nanogels in normal cells. Therefore, A-R6G-loaded BU-PEG nanogels could potentially significantly enhance the chemotherapeutic safety and effectiveness of anticancer drugs. However, the exact mechanisms of action of the selective cytotoxicity of A-R6G-loaded BU-PEG nanogels still remain uncertain. We are conducting research to more precisely define the structural features and in vivo cytotoxic effects of this drug–carrier system in order to confirm the selective and targeted cytotoxic effects of A-R6G towards cancer cells for chemotherapy applications.

In order to obtain further insight into the mechanisms of cellular uptake and intracellular drug release by A-R6G-loaded BU-PEG towards normal and cancer cells, NIH/3T3 and HeLa cells cultured with the drug-loaded nanogels for 3, 12, or 24 h, and then analyzed by CLSM to observe the cellular morphology and internalization of the nanogels [45]. Blue-fluorescent DAPI was used to stain the nuclei; A-R6G and R6G exhibit strong green fluorescence emission. As indicated in Figure 4a,b, the CLSM images clearly indicated no significant green fluorescent could be observed after 24 h incubation of NIH/3T3 cells with A-R6G-loaded BU-PEG nanogels. In contrast, remarkable green fluorescence was randomly distributed throughout the cytoplasm of the HeLa cells after 3 h incubation with A-R6G-loaded nanogels, and the green fluorescent signal gradually shifted into the nucleus after 24 h of incubation. In contrast to the A-R6G-loaded BU-PEG nanogels, a gradual increase in the green fluorescent signal and intensity was observed within the nuclei of NIH/3T3 or HeLa cells incubated with R6G-loaded BU-PEG nanogels for 3 to 24 h (Figure S5a,b). These findings are in good agreement with the MTT assay, and suggest the A-R6G-loaded BU-PEG nanogels were selectively internalized by the cancer cells and did not internalize in the normal cellular environment [46]. In contrast, the R6G-loaded BU-PEG nanogels underwent intensive, non-specific uptake by normal and cancer cells, possibly due to specific interactions between the A–U base-paring moieties of the nanogels and the surface of cancer cells. Overall, the above-mentioned findings indicate this supramolecular drug–carrier system containing complementary A–U pairs could promote the selective uptake of drugs by cancer cells and effectively induce cancer cell death while minimizing the cytotoxicity in normal cells.

In order to verify the CLSM images, we performed quantitative and qualitative flow cytometry analysis to further investigate the selective internalization of A-R6G-loaded BU-PEG nanogels by HeLa cells. As shown in Figure 5a,b, after incubation with A-R6G-loaded BU-PEG nanogels for 24 h, NIH/3T3 cells exhibited almost no change in A-R6G fluorescence intensity, whereas the A-R6G fluorescence intensity of HeLa cells gradually increased with the duration of incubation, indicating the complementary A–U base-pairs within the nanogels were endowed with a strong affinity for HeLa cells and thereby promoted selective, rapid internalization of the nanogels and subsequently led to the death of the cancer cells. In contrast, flow cytometry of NIH/3T3 and HeLa cells incubated with R6G-loaded BU-PEG nanogels for various periods of time revealed a gradual increase in R6G fluorescence intensity between 1 and 24 h, suggesting that the R6G/BU-PEG system was not able to selectively promote the internalization of nanogels by cancer cells. In addition, these results also revealed that R6G-loaded BU-PEG nanogels exhibited much higher and faster cellular uptake in HeLa cells than in NIH/3T3 cells (the lower left and right regions of Figure 5), possibly due to the differences in the surface charge and affinity between the cells. Moreover, we further assessed how the complementary hydrogen bonding interactions between the drug and carrier affected cellular uptake ability by plotting the average fluorescence intensity of the flow cytometry data versus incubation time. As shown in Figure S6, the internalization rate of the A-R6G-loaded BU-PEG nanogels was approximately 74 times higher in HeLa cells than NIH/3T3 cells after 24 h, whereas the internalization rate of the R6G-loaded BU-PEG nanogels was only 5.7 times higher in HeLa cells than NIH/3T3 cells after 24 h, which is in good agreement with the CLSM results (Figure 4 and Figure S5). These results further demonstrate that the A-R6G-loaded BU-PEG nanogels were selectively internalized into the HeLa cells but only minimally taken up by NIH/3T3 cells, whereas the R6G-loaded BU-PEG nanogels were non-specifically internalized by cells. Collectively, these findings clearly prove the complementary A–U interactions within the nanogels not only controlled their drug delivery and release properties, but also promoted the selective uptake of drugs into cancer cells and accelerated cell death; they may thus potentially enhance the overall efficiency of chemotherapy.

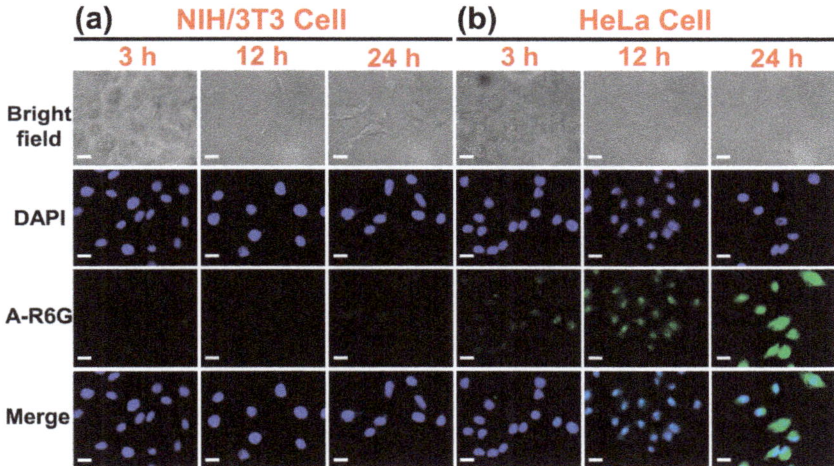

Figure 4. CLSM images of (**a**) NIH/3T3 cells and (**b**) HeLa cells cultured with A-R6G-loaded BU-PEG nanogels at normal physiological conditions (pH 7.4 and 37 °C) for 3, 12, or 24 h. The scale bars in all CLSM images are 20 μm.

To further identify the cytotoxic pathways and assess the mechanisms of cell death for A-R6G-loaded and R6G-loaded BU-PEG nanogels, dual fluorescent staining and flow cytometry were used to quantify viable, dead, and total cells after exposure of NIH/3T3 and HeLa cells to A-R6G-loaded and R6G-loaded BU-PEG nanogels for various periods of time. BV421 Annexin-V was used to detect phosphatidylserine expression on early apoptotic cells, while GDR-780 was used to label intracellular DNA, which is released after the integrity of the plasma membrane has been compromised in late apoptotic cells [47,48]. As shown in Figure 6d,h, after the incubation of HeLa cells with A-R6G-loaded or R6G-loaded BU-PEG nanogels for 12 h, respectively, over 90% of the NIH/3T3 cells incubated with A-R6G-loaded nanogels survived, whereas the R6G-loaded nanogels increased the proportions of early and late apoptotic cells to 27.7% and 29.4%, respectively, indicating that the A-R6G-loaded BU-PEG nanogels had extremely stable drug entrapment stability in a normal cellular environment, which significantly reduced the leakage of hydrophobic A-R6G during the drug delivery process. In contrast to the general trend in NIH/3T3 cells, the overall proportion of apoptotic HeLa cells gradually increased between 1 and 12 h, suggesting that the A-R6G-loaded nanogels taken up by the HeLa cells moved progressively toward the nucleus, and that the A-R6G was gradually released from the nanogels (Figure 6i–l). After

12 h of incubation with A-R6G-loaded nanogels, the proportions of early and late apoptotic HeLa cells dramatically increased to 38.3% and 1.53%, respectively, while the proportion of viable cells still remained high, at 56% (Figure 6l). By comparison, the apoptotic trends in the HeLa cells cultured with R6G-loaded BU-PEG nanogels were similar to the results for the NIH/3T3 cells presented in Figure 6h,p. These observations clearly confirm that the mildly acidic cancer microenvironment promotes the rapid intracellular release of A-R6G from BU-PEG by inducing dissociation of the complementary A–U interactions within the nanogels, and that the A-R6G subsequently promotes programmed cell death. Thus, this newly developed system based on complementary hydrogen bonding A–U interactions between A-R6G and BU-PEG may potentially remarkably enhance the effects of chemotherapy in cancer cells while substantially reducing adverse effects in healthy cells.

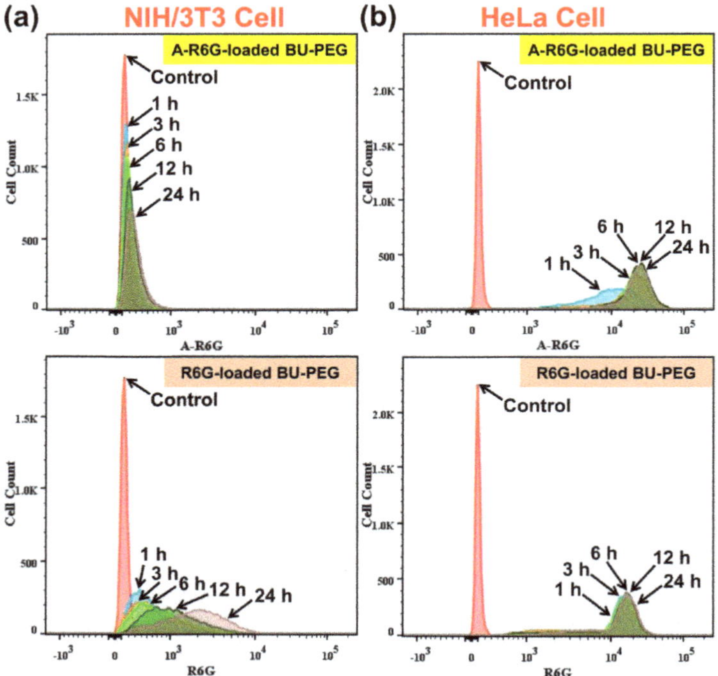

Figure 5. Flow cytometry histograms of (**a**) NIH/3T3 cells and (**b**) HeLa cells cultured with A-R6G-loaded or R6G-loaded BU-PEG nanogels at normal physiological conditions (pH 7.4 and 37 °C) for 1, 3, 6, 12, or 24 h.

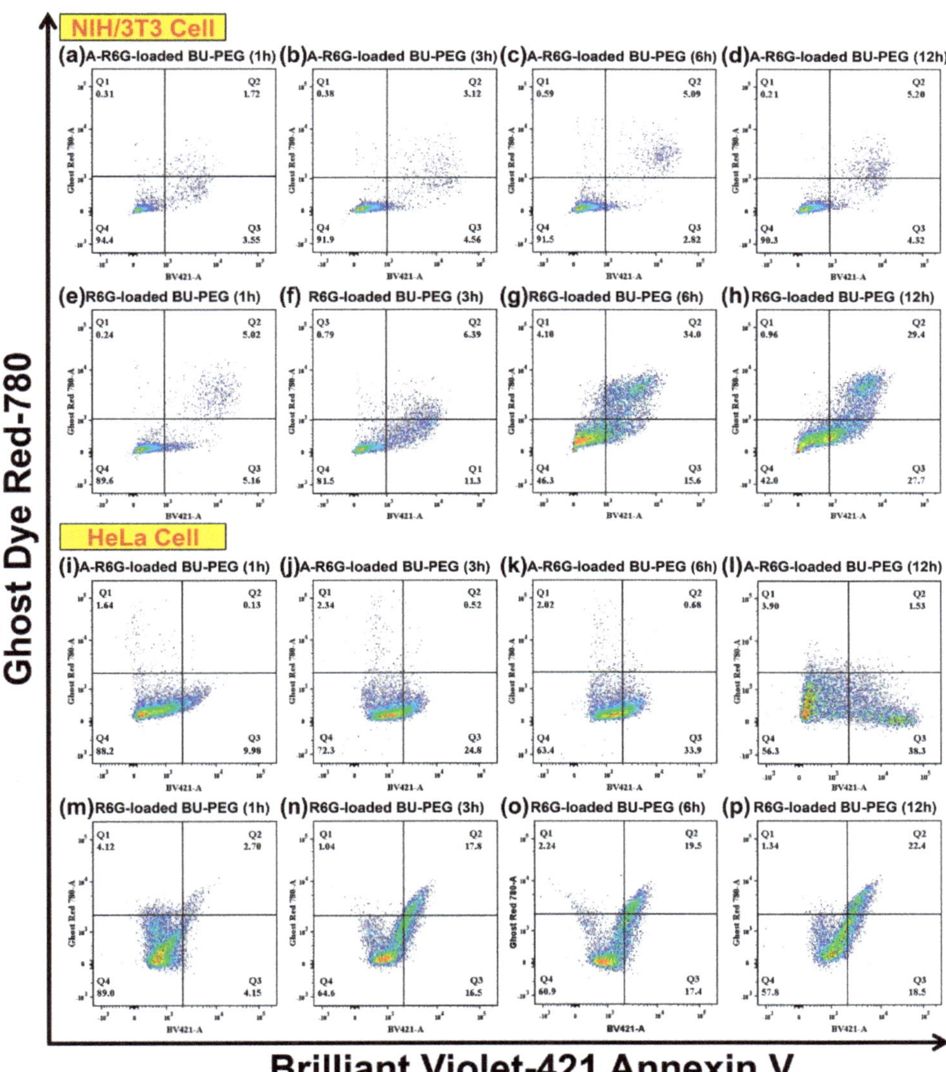

Figure 6. Flow cytometric dot plot quadrant charts of NIH/3T3 and HeLa cells cultured with A-R6G-loaded and R6G-loaded BU-PEG nanogels at normal physiological conditions (pH 7.4 and 37 °C) for 1, 3, 6, or 12 h before staining with BV421 Annexin V and GDR780. Figures (**a**)–(**p**) represent the results of flow cytometry for different time points during the co-culture period. The graph quadrants from the lower left to the upper left (turning anti-clockwise) represent viable cells (BV421 Annexin V⁻, GDR780⁻), early apoptotic cells (BV421 Annexin V⁺, GDR780⁻), late apoptotic cells (BV421 Annexin V⁺, GDR780⁺), and necrotic cells (BV421 Annexin V⁻, GDR780⁺). The numbers inside each quadrant refer to the proportions of cells.

4. Conclusions

We successfully developed a complementary drug delivery system based on cooperative hydrogen bonding A–U interactions between a drug and nanocarrier to achieve selective uptake by cancer cells, improve chemotherapeutic efficacy, and reduce adverse effects in normal cells. A R6G-based anticancer agent containing a hydrogen-bonded adenine unit (A-R6G) was obtained via a simple three-step chemical reaction. A-R6G displays poor aqueous solubility and unique green fluorescent behavior, and exerts potent cytotoxic effects against a variety of cell lines. A complementary hydrogen bonding partner, water-soluble uracil-end-capped BU-PEG polymer, was prepared at high yield (83%) via a one-step Michael addition reaction. BU-PEG can spontaneously self-assemble into well-defined nanoparticles with a variety of unique physical properties in water, such as interesting amphiphilic and morphological characteristics. The formation of A–U base pairs between BU-PEG and A-R6G leads to the formation of well-dissolved A-R6G-loaded BU-PEG nanogels in water. Interestingly, the resulting A-R6G-loaded BU-PEG nanogels exhibited a number of unique physical properties, including extremely high A-R6G-loading capacity (69.4%), a widely tunable A-R6G-loading content, singular green-fluorescence characteristics, and well-controlled pH-responsive drug-release behavior. Moreover, A-R6G-loaded BU-PEG nanogels showed excellent structural stability in cell culture media and low hemolytic activity towards SRBCs. The combination of long-term drug-entrapment stability and low-level hemolytic activity offered by A-R6G-loaded BU-PEG is extremely attractive, but rare in traditional drug–carrier systems. In vitro cytotoxicity studies clearly confirmed that A-R6G-loaded BU-PEG nanogels exhibit potent cytotoxic activity against cancerous HeLa and MG-63 cells, and only minimal cytotoxic effects in normal NIH/3T3 cells, whereas R6G-loaded BU-PEG nanogels did not show selective cytotoxicity against these cell lines. Thus, the complementary A–U interactions critically improve the selective uptake of A-R6G into cancer cells and selectively induce cancer cell death. Importantly, analysis of intracellular cellular uptake using CLSM and flow cytometry clearly demonstrated that A-R6G-loaded BU-PEG nanogels enabled the selective uptake of A-R6G by HeLa cells via endocytosis and promoted controlled intracellular release of A-R6G in the weakly acidic microenvironment of the cancer cells, which subsequently induced programmed cell death in the cancer cells. The opposite trends were observed in NIH/3T3 cells, i.e., poor internalization and extremely low toxicity. Thus, this study clearly demonstrates that introduction of complementary A–U interactions within this drug–carrier system provides an effective approach to selectively delivering anticancer drugs into cancer cells, subsequently improving the safety and effectiveness of chemotherapy, without the need to incorporate targeting moieties onto the surface of drug carriers.

Supplementary Materials: The following are available online at https://www.mdpi.com/article/10.3390/pharmaceutics13111929/s1, Scheme S1: Process for the synthesis of BU-PEG, Figure S1: (**a**) ^{1}H and ^{13}C NMR spectra of BU-PEG in CDCl$_3$ obtained at 25 °C. (**b**) GPC curves for PEG and BU-PEG in THF at 40 °C, Figure S2: ^{1}H NMR spectra of BU-PEG, A-R6G, and A-R6G/BU-PEG complexes in D$_2$O obtained at 25 °C, Figure S3: (**a**) DLS profiles for A-R6G-loaded and R6G-loaded BU-PEG nanogels in water at 25 °C. (**b**) AFM image of R6G-loaded BU-PEG. SEM images of (**c**) A-R6G-loaded BU-PEG and (**d**) R6G-loaded BU-PEG nanogels, Figure S4: In vitro MTT assay cytotoxicity of BU-PEG, A-R6G, R6G, and A-R6G-loaded and R6G-loaded BU-PEG nanogels towards MG-63 cells, Figure S5: CLSM images of (**a**) NIH/3T3 cells and (**b**) HeLa cells cultured with R6G-loaded BU-PEG nanogels at normal physiological conditions (pH 7.4 and 37 °C) for 3 h, 12 h, or 24 h. The scale bars in all CLSM images are 20 μm, Figure S6: Fluorescence intensity of NIH/3T3 and HeLa cells after incubation with A-R6G-loaded and R6G-loaded BU-PEG nanogels for different periods of time (1, 3, 6, 12, or 24 h), Table S1: Hydrodynamic particle size, zeta potential, drug-loading content (DLC) and drug-loading efficiency (DLE) of A-R6G-loaded and R6G-loaded BU-PEG nanogels.

Author Contributions: Conceptualization, C.-C.C.; investigation, F.B.I. and E.A.B.; writing—original draft preparation, F.B.I.; writing—review and editing, C.-C.C.; supervision, C.-C.C. All authors have read and agreed to the published version of the manuscript.

Funding: Ministry of Science and Technology, Taiwan (contract no. MOST 107-2221-E-011-041-MY3 and MOST 110-2221-E-011-003-MY3).

Institutional Review Board Statement: Not applicable.

Informed Consent Statement: Not applicable.

Data Availability Statement: Not applicable.

Acknowledgments: The authors would like to acknowledge funding from the Ministry of Science and Technology in Taiwan for this study (contract no. MOST 107-2221-E-011-041-MY3 and MOST 110-2221-E-011-003-MY3). This study was also partially supported by the Yushan Scholar Program by the Ministry of Education in Taiwan.

Conflicts of Interest: The authors declare no conflict of interest.

References

1. Blanchard, A.T.; Salaita, K. Emerging uses of DNA mechanical devices. *Science* **2019**, *365*, 1080–1081. [CrossRef]
2. Ouchi, M.; Badi, N.; Lutz, J.-F.; Sawamoto, M. Single-chain technology using discrete synthetic macromolecules. *Nat. Chem.* **2011**, *3*, 917–924. [CrossRef] [PubMed]
3. Badi, N.; Lutz, J.-F. Sequence control in polymer synthesis. *Chem. Soc. Rev.* **2009**, *38*, 3383–3390. [CrossRef] [PubMed]
4. Merrifield, R.B. Solid phase peptide synthesis. I. The synthesis of a tetrapeptide. *J. Am. Chem. Soc.* **1963**, *85*, 2149–2154. [CrossRef]
5. Müller, A.; Talbot, F.; Leutwyler, S. Hydrogen bond vibrations of 2-aminopyridine·2-pyridone, a Watson−Crick analogue of adenine·uracil. *J. Am. Chem. Soc.* **2002**, *124*, 14486–14494. [CrossRef]
6. McHale, R.; O'Reilly, R.K. Nucleobase containing synthetic polymers: Advancing biomimicry via controlled synthesis and self-assembly. *Macromolecules* **2012**, *45*, 7665–7675. [CrossRef]
7. Sivakova, S.; Rowan, S.J. Nucleobases as supramolecular motifs. *Chem. Soc. Rev.* **2005**, *34*, 9–21. [CrossRef]
8. Binder, W.H.; Zirbs, R. Supramolecular polymers and networkswith hydrogen bonds in the main-and side-chain. *Adv. Polym. Sci.* **2007**, *207*, 1–78.
9. Li, J.; Wang, Z.; Hua, Z.; Tang, C. Supramolecular nucleobase-functionalized polymers: Synthesis and potential biological applications. *J. Mater. Chem. B* **2020**, *8*, 1576–1588. [CrossRef]
10. Sivakova, S.; Bohnsack, D.A.; Mackay, M.E.; Suwanmala, P.; Rowan, S.J. Utilization of a combination of weak hydrogen-bonding interactions and phase segregation to yield highly thermosensitive supramolecular polymers. *J. Am. Chem. Soc.* **2005**, *127*, 18202–18211. [CrossRef]
11. Ishikawa, N.; Furutani, M.; Arimitsu, K. Adhesive materials utilizing a thymine–adenine interaction and thymine photodimerization. *ACS Macro Lett.* **2015**, *4*, 741–744. [CrossRef]
12. Kuang, H.; Wu, S.; Xie, Z.; Meng, F.; Jing, X.; Huang, Y. Biodegradable amphiphilic copolymer containing nucleobase: Synthesis, self-assembly in aqueous solutions, and potential use in controlled drug delivery. *Biomacromolecules* **2012**, *13*, 3004–3012. [CrossRef]
13. Wang, D.; Tu, C.; Su, Y.; Zhang, C.; Greiser, U.; Zhu, X.; Yan, D.; Wang, W. Supramolecularly engineered phospholipids constructed by nucleobase molecular recognition: Upgraded generation of phospholipids for drug delivery. *Chem. Sci.* **2015**, *6*, 3775–3787. [CrossRef] [PubMed]
14. Fan, J.; Zeng, F.; Wu, S.; Wang, X. Polymer micelle with pH-triggered hydrophobic-hydrophilic transition and de-cross-linking process in the core and its application for targeted anticancer drug delivery. *Biomacromolecules* **2012**, *13*, 4126–4137. [CrossRef] [PubMed]
15. Greenwald, R.B.; Choe, Y.H.; McGuire, J.; Conover, C.D. Effective drug delivery by PEGylated drug conjugates. *Adv. Drug Deliv. Rev.* **2003**, *55*, 217–250. [CrossRef]
16. Gabizon, A.; Shmeeda, H.; Grenader, T. Pharmacological basis of pegylated liposomal doxorubicin: Impact on cancer therapy. *Eur. J. Pharm. Sci.* **2012**, *45*, 388–398. [CrossRef]
17. Lee, J.S.; Feijen, J. Polymersomes for drug delivery: Design, formation and characterization. *J. Control. Release* **2012**, *161*, 473–483. [CrossRef]
18. Abuchowski, A.; van Es, T.; Palczuk, N.C.; Davis, F.F. Alteration of immunological properties of bovine serum albumin by covalent attachment of polyethylene glycol. *J. Biol. Chem.* **1977**, *252*, 3578–3581. [CrossRef]
19. Abuchowski, A.; McCoy, J.R.; Palczuk, N.C.; Davis, F.F. Effect of covalent attachment of polyethylene glycol on immunogenicity and circulating life of bovine liver catalase. *J. Biol. Chem.* **1977**, *252*, 3582–3586. [CrossRef]
20. Maeda, H.; Wu, J.; Sawa, T.; Matsumura, Y.; Hori, K. Tumor vascular permeability and the EPR effect in macromolecular therapeutics: A review. *J. Control. Release* **2000**, *65*, 271–284. [CrossRef]
21. Koo, A.N.; Lee, H.J.; Kim, S.E.; Chang, J.H.; Park, C.; Kim, C.; Park, J.H.; Lee, S.C. Disulfide-cross-linked PEG-poly(amino acid)s copolymer micelles for glutathione-mediated intracellular drug delivery. *Chem. Commun.* **2008**, *44*, 6570–6572. [CrossRef]
22. Yokoyama, M.; Miyauchi, M.; Yamada, N.; Okano, T.; Sakurai, Y.; Kataoka, K.; Inoue, S. Characterization and anticancer activity of the micelle-forming polymeric anticancer drug adriamycin-conjugated poly(ethylene glycol)-poly(aspartic acid) block copolymer. *Cancer Res.* **1990**, *50*, 1693–1700. [PubMed]
23. Torchilin, V.P. Multifunctional nanocarriers. *Adv. Drug Deliv. Rev.* **2006**, *58*, 1532–1555. [CrossRef] [PubMed]
24. Veronese, F.M.; Pasut, G. PEGylation, successful approach to drug delivery. *Drug Discov. Today* **2005**, *10*, 1451–1458. [CrossRef]
25. Parveen, S.; Sahoo, S.K. Nanomedicine. *Clin. Pharmacokinet.* **2006**, *45*, 965–988. [CrossRef] [PubMed]
26. Zeng, X.; Liu, G.; Tao, W.; Ma, Y.; Zhang, X.; He, F.; Pan, J.; Mei, L.; Pan, G. A Drug-self-gated mesoporous antitumor nanoplatform based on pH-sensitive dynamic covalent bond. *Adv. Funct. Mater.* **2017**, *27*, 1605985. [CrossRef]

27. Torchilin, V.P.; Trubetskoy, V.S. Which polymers can make nanoparticulate drug carriers long-circulating? *Adv. Drug Deliv. Rev.* **1995**, *16*, 141–155. [CrossRef]

28. Cheng, C.-C.; Yen, Y.C.; Ye, Y.S.; Chang, F.C. Biocomplementary interaction behavior in DNA-like and RNA-like polymers. *J. Polym. Sci. A Polym. Chem.* **2009**, *47*, 6388–6395. [CrossRef]

29. Cheng, C.-C.; Chang, F.C.; Dai, S.A.; Lin, Y.L.; Lee, D.J. Bio-complementary supramolecular polymers with effective self-healing functionality. *RSC Adv.* **2015**, *5*, 90466–90472. [CrossRef]

30. Cheng, C.C.; Gebeyehu, B.T.; Huang, S.Y.; Alemayehu, Y.A.; Sun, Y.T.; Lai, Y.C.; Chang, Y.H.; Lai, J.Y.; Lee, D.J. Entrapment of an adenine derivative by a photo-irradiated uracil-functionalized micelle confers controlled self-assembly behavior. *J. Colloid Interface Sci.* **2019**, *552*, 166–178. [CrossRef]

31. Ilhami, F.B.; Chung, A.; Alemayehu, Y.A.; Lee, A.W.; Chen, J.K.; Lai, J.Y.; Cheng, C.C. Self-assembled nanoparticles formed via complementary nucleobase pair interactions between drugs and nanocarriers for highly efficient tumor-selective chemotherapy. *Mater. Chem. Front.* **2021**, *5*, 5442–5451. [CrossRef]

32. Naga, N.; Sato, M.; Mori, K.; Nageh, H.; Nakano, T. Synthesis of network polymers by means of addition reactions of multifunctional-amine and poly(ethylene glycol) diglycidyl ether or diacrylate compounds. *Polymers* **2020**, *12*, 2047. [CrossRef] [PubMed]

33. Teng, D.; Wu, Z.; Zhang, X.; Wang, Y.; Zheng, C.; Wang, Z.; Li, C. Synthesis and characterization of in situ cross-linked hydrogel based on self-assembly of thiol-modified chitosan with PEG diacrylate using Michael type addition. *Polymer* **2010**, *51*, 639–646. [CrossRef]

34. Yang, C.; Tan, J.P.K.; Cheng, W.; Attia, A.B.E.; Ting, C.T.Y.; Nelson, A.; Hedrick, J.L.; Yang, Y.Y. Supramolecular nanostructures designed for high cargo loading capacity and kinetic stability. *Nano Today* **2010**, *5*, 515–523. [CrossRef]

35. Magut, P.K.S.; Das, S.; Fernand, V.E.; Losso, J.; McDonough, K.; Naylor, B.M.; Aggarwal, S.; Warner, I.M. Tunable cytotoxicity of rhodamine 6G via anion variations. *J. Am. Chem. Soc.* **2013**, *135*, 15873–15879. [CrossRef]

36. Bhattarai, N.; Chen, M.; Pérez, R.L.; Ravula, S.; Chhotaray, P.; Hamdan, S.; McDonough, K.; Tiwari, S.; Warner, I.M. Enhanced chemotherapeutic toxicity of cyclodextrin templated size-tunable rhodamine 6G nanoGUMBOS. *J. Mater. Chem. B* **2018**, *6*, 5451–5459. [CrossRef] [PubMed]

37. Bhattarai, N.; Mathis, J.M.; Chen, M.; Pérez, R.L.; Siraj, N.; Magut, P.K.S.; McDonough, K.; Sahasrabudhe, G.; Warner, I.M. Endocytic selective toxicity of rhodamine 6G nanoGUMBOS in breast cancer cells. *Mol. Pharm.* **2018**, *15*, 3837–3845. [CrossRef]

38. Wilhelm, M.; Zhao, C.L.; Wang, Y.; Xu, R.; Winnik, M.A.; Mura, J.L.; Riess, G.; Croucher, M.D. Poly (styrene-ethylene oxide) block copolymer micelle formation in water: A fluorescence probe study. *Macromolecules* **1991**, *24*, 1033–1040. [CrossRef]

39. Yang, Y.; Li, D.; Li, C.; Liu, Y.F.; Jiang, K. Hydrogen bond strengthening induces fluorescence quenching of PRODAN derivative by turning on twisted intramolecular charge transfer. *Spectrochim. Acta A Mol. Biomol. Spectrosc.* **2017**, *187*, 68–74. [CrossRef]

40. Waluk, J. Hydrogen-bonding-induced phenomena in bifunctional heteroazaaromatics. *Acc. Chem. Res.* **2003**, *36*, 832–838. [CrossRef]

41. Yu, Z.; Ye, S.; Li, G.; Zhang, M. Construction of an environmentally friendly octenylsuccinic anhydride modified pH-sensitive chitosan nanoparticle drug delivery system to alleviate inflammation and oxidative stress. *Carbohydr. Polym.* **2020**, *236*, 115972. [CrossRef] [PubMed]

42. Singh, N.; Sahoo, S.K.; Kumar, R. Hemolysis tendency of anticancer nanoparticles changes with type of blood group antigen: An insight into blood nanoparticle interactions. *Mater. Sci. Eng. C Mater. Biol. Appl.* **2020**, *109*, 110645. [CrossRef] [PubMed]

43. Gerweck, L.E.; Seetharaman, S. Cellular pH Gradient in Tumor versus Normal Tissue: Potential Exploitation for the Treatment of Cancer. *Cancer Res.* **1996**, *56*, 1194–1198. [PubMed]

44. Cheng, R.; Meng, F.; Deng, C.; Klok, H.-A.; Zhong, Z. Dual and multi-stimuli responsive polymeric nanoparticles for programmed site-specific drug delivery. *Biomaterials* **2013**, *34*, 3647–3657. [CrossRef]

45. Xue, J.; Guan, Z.; Lin, J.; Cai, C.; Zhang, W.; Jiang, X. Cellular internalization of rod-like nanoparticles with various surface patterns: Novel entry pathway and controllable uptake capacity. *Small* **2017**, *13*, 1604214. [CrossRef]

46. Bondar, O.V.; Saifullina, D.V.; Shakhmaeva, I.I.; Mavlyutova, I.I.; Abdullin, T.I. Monitoring of the zeta potential of human cells upon reduction in their viability and interaction with polymers. *Acta Nat.* **2012**, *4*, 78–81. [CrossRef]

47. Van Engeland, M.N.; Nieland, L.J.; Ramaekers, F.C.; Schutte, B.; Reutelingsperger, C.P. Annexin V-affinity assay: A review on an apoptosis detection system based on phosphatidylserine exposure. *Cytometry* **1998**, *31*, 1–9. [CrossRef]

48. Vermes, I.; Haanen, C.; Steffens-Nakken, H.; Reutelingsperger, C. A novel assay for apoptosis. Flow cytometric detection of phosphatidylserine expression on early apoptotic cells using fluorescein labelled Annexin V. *J. Immunol. Methods* **1995**, *184*, 39–51. [CrossRef]

pharmaceutics

MDPI

Article

Interaction of Supramolecular Congo Red and Congo Red-Doxorubicin Complexes with Proteins for Drug Carrier Design

Anna Jagusiak [1,*], Katarzyna Chłopaś [1], Grzegorz Zemanek [1], Izabela Kościk [1] and Irena Roterman [2]

[1] Jagiellonian University Medical College, Faculty of Medicine, Chair of Medical Biochemistry, Kopernika 7, 31-034 Krakow, Poland; kchlopas@su.krakow.pl (K.C.); grzegorz.zemanek@uj.edu.pl (G.Z.); izabela.koscik@uj.edu.pl (I.K.)

[2] Jagiellonian University Medical College, Faculty of Medicine, Department of Bioinformatics and Tele-Medicine, Lazarza 16, 31-530 Krakow, Poland; myroterm@cyf-kr.edu.pl

* Correspondence: anna.jagusiak@uj.edu.pl

Citation: Jagusiak, A.; Chłopaś, K.; Zemanek, G.; Kościk, I.; Roterman, I. Interaction of Supramolecular Congo Red and Congo Red-Doxorubicin Complexes with Proteins for Drug Carrier Design. *Pharmaceutics* **2021**, *13*, 2027. https://doi.org/10.3390/pharmaceutics13122027

Academic Editors: Francisco José Ostos, José Antonio Lebrón and Pilar López-Cornejo

Received: 19 October 2021
Accepted: 22 November 2021
Published: 28 November 2021

Publisher's Note: MDPI stays neutral with regard to jurisdictional claims in published maps and institutional affiliations.

Abstract: Targeted immunotherapy has expanded to simultaneous delivery of drugs, including chemotherapeutics. The aim of the presented research is to design a new drug carrier system. Systems based on the use of proteins as natural components of the body offer the chance to boost safety and efficacy of targeted drug delivery and excess drug removal. Congo red (CR) type supramolecular, self-assembled ribbon-like structures (SRLS) were previously shown to interact with some proteins, including albumin and antibodies complexed with antigen. CR can intercalate some chemotherapeutics including doxorubicin (Dox). The goal of this work was to describe the CR-Dox complexes, to analyze their interaction with some proteins, and to explain the mechanism of this interaction. In the present experiments, a model system composed of heated immunoglobulin light chain Lλ capable of CR binding was used. Heat aggregated immunoglobulins (HAI) and albumin were chosen as another model system. The results of experiments employing methods such as gel filtration chromatography and dynamic light scattering confirmed the formation of the CR-Dox complex of large size and properties different from the free CR structures. Electrophoresis and chromatography experiments have shown the binding of free CR to heated Lλ while CR-Dox mixed structures were not capable of forming such complexes. HAI was able to bind both free CR and CR-Dox complexes. Albumin also bound both CR and its complex with Dox. Additionally, we observed that albumin-bound CR-Dox complexes were transferred from albumin to HAI upon addition of HAI. DLS analyses showed that interaction of CR with Dox distinctly increased the hydrodynamic diameter of CR-Dox compared with a free CR supramolecular structure. To our knowledge, individual small proteins such as Lλ may bind upon heating a few molecules of Congo red tape penetrating protein body due to the relatively low cohesion of the dye micelle. If, however, the compactness is high (in the case of, e.g., CR-Dox) large ribbon-like, micellar structures appear. They do not divide easily into smaller portions and cannot attach to proteins where there is no room for binding large ligands. Such binding is, however, possible by albumin which is biologically adapted to form complexes with different large ligands and by tightly packed immune complexes and heat aggregated immunoglobulin-specific protein complex structures of even higher affinity for Congo red than albumin. The CR clouds formed around them also bind the CR-Dox complexes. The presented research is essential in the search for optimum solutions for SRLS application in immuno-targeting therapeutic strategies, especially with the use of chemotherapeutics.

Keywords: supramolecular self-assembled ribbon-like structures (SRLS); Congo red (CR); doxorubicin (Dox); bovine serum albumin (BSA); immunoglobulin light chain λ (Lλ); heat aggregated immunoglobulins (HAI); dynamic light scattering (DLS); elution volume (V_e)

193

1. Introduction

Drugs designed to reach molecular targets, among which monoclonal antibodies and kinase inhibitors are most frequently used, are the basis of modern therapy. Targeted immunotherapy is also expanded to simultaneous delivery of drugs, including chemotherapeutics. Immuno-targeting, defined as the use of immunological specificity directed to target connected with therapy, is still the subject of many investigations. Design and development of efficient carriers of anti-inflammatory and anticancer drugs are now extensively studied in order to increase the effectiveness and safety of the targeted therapies [1–5].

Self-assembled structures presented in this work are the group of compounds (polyaromatic molecules of an elongated shape with appropriately located polar groups) showing a tendency to self-associate via non-covalent interactions thus creating greater supramolecular systems. This phenomenon is also observed during the formation of microtubules or biological membrane structures. Some of these systems form elongated structures referred to as self-assembled ribbon-like structures (SRLS). These kinds of structure have the potential to be a part of systems delivering chemotherapeutics to cancerous tissue by immuno-targeting.

This is possible because of their ability to selectively interact with immune complexes. SRLS are examples of a novel type of protein ligand, as they bind to proteins via different interactions than the classic type [6]. SRLS systems bind to proteins at sites of local structural instability caused by unfolding conditions or function-derived structural changes in the protein molecule. The binding of SRLS to antigen-antibody complexes, with simultaneous lack of binding of free antibodies, can serve as an example. The described interaction is a foundation for using those compounds in immuno-targeting [7]. At the same time, SRLS systems can intercalate other molecules, including drugs, forming co-micellar systems [8].

Previous research has shown that SRLS can be applied in vivo as potential drug carriers. Such systems were easily bound to immune complexes formed in the body and then were gradually eliminated. Immune complexes are highly complex systems and their structural analysis is difficult. Conformational changes observed in Lλ under sub-denaturing conditions, which mimic those in antigen-complexed antibodies, contribute to CR binding. This is the reason why immunoglobulin light chain (Lλ) heated to 45 °C was used for the research on interaction between SRLS and antigen-bound antibodies as a model system [9]. Heat aggregated immunoglobulins such as immunoglobulin G (HAI) were investigated as another immune complex model system. One more protein that can be used in targeted therapies, albumin, was also found to bind supramolecular ligands. Some therapeutic agents, especially of cationic nature (like the widely used chemotherapeutic doxorubicin) cannot bind to albumin directly. Thus, albumin can be used in targeted therapy other than in combination with negatively charged carriers. In particular, when SRLS bind drugs, co-micellar structures are formed that can interact with albumin which allows for effective drug delivery and also protects the body against uncontrolled drug action [6,10–16].

The standard example of SRLS is Congo red (CR), which forms assemblages of elongated shape, with a high level of plasticity [6]. CR is a molecule with polyaromatic rings of elongated planar symmetric structures, substituted by amino and sulfonic groups [17,18]. The presence of sulfonic groups makes the structure polyanionic at neutral pH. Symmetrical charge arrangement and hydrophobic interaction between groups in the central part of the molecules, which is noncovalent, make the ribbon-like structure stable but also guarantees its high plasticity [6]. In the presented work, only CR was used as a model because its properties as SRLS are well known. In the future, it can be replaced by other, more biocompatible compounds with similar properties (e.g., Evans blue) [19–21].

The special property of supramolecular CR is that it is able to interact with proteins especially those containing β structure fragments. The mechanism of interaction of supramolecular ribbon-like CR with protein is different from classic protein–ligand interaction in the protein active site. Such complexes can be formed on the condition that protein β-sheet part is at least partially destabilized, which allows for function-derived conformational rearrangement. Thus, susceptibility of CR structures to deformations allowing the best fitting to the protein binding site is the important condition for optimal CR-protein binding. Since there is no defined, specific amino acid sequence that binds CR and the supramolecular ligand is capable of changes due to its plasticity, a variety of proteins are able to form such complexes. Proteins binding to supramolecular CR are native proteins (normal and pathological) as well as proteins treated with denaturing factors [6,22,23].

Examples of native CR binding proteins include molecules in which supramolecular CR binding site arises from function-derived intramolecular strain. Antibodies belong to such group of proteins. In antibodies, antigen binding induces some structural changes [24,25]. Native, antigen-unbound immunoglobulins G do not form complexes with CR but gain this ability upon antigen binding. In the presence of CR, the effect of antibody–antigen interaction enhancement was observed [26]. The property of selective CR binding shown by antigen-bound antibodies gives opportunities for its application in targeted drug delivery [6]. The model system for such interaction was developed as an isolated immunoglobulin light chain (Lλ) heated to 45 °C or immunoglobulin G heated to 63 °C (HAI). Upon heating, the protein is destabilized. The N-terminus is locally unfolded, which opens up the V domain. The same phenomenon takes place when antibodies are bound to the antigen. At the Fab ends associated with the antigen, the beta structure rich polypeptide is revealed and those structures are penetrated by and bound to CR. CR-Dox co-micelles can bind at the same region. However, it should be underlined that it is just a model system, and the real mechanism of CR-Dox co-micelle binding to immunological complexes might be different than that described as binding to the cavity emptied by the N-terminal chain fragment [9]. To our knowledge, there is no other research on the binding of CR complexed with drugs to heated Lλ or HAI.

Another example of CR-binding native protein is serum albumin. This universal carrier of many hydrophobic compounds (especially anionic) can bind CR as single molecules as well as a supramolecular ligand thanks to its structure with the binding cavity [6,11–16]. Albumin is a protein adapted to the transport of a variety of anionic compounds that can bind CR without preliminary structure change. The molecule of serum albumin binds up to 16 CR molecules and at most nine molecules of Evans blue, which has a similar structure to CR but shows weaker self-association properties [10]. Until now there have been no investigations of albumin ability to bind CR-drug complexes. There have also been no studies on drug transfer from a carrier such as albumin to the immune complex (presented here as the HAI model).

SRLS shows the ability to form complexes with planar molecules including chemotherapeutics, such as doxorubicin [8]. Doxorubicin (Dox) is one of the most effective drugs used in the therapy for many types of cancer [27]. At the same time, it is a highly cardiotoxic drug, and its use can result in the inhibition of hematopoiesis and gastrointestinal disorders [28,29]. Despite this, it is still widely used because of its high effectiveness and wide spectrum of anticancer effects. That is why further investigations of doxorubicin carriers are very important as they can improve the efficiency of its delivery and reduce its toxic effects. Earlier experiments have shown that doxorubicin binds to supramolecular ribbon-like CR structure (Figure 1) [30–33].

Previously published research on the structure of free CR and CR-Dox complex showed large differences between these systems. Positively charged Dox binds to negatively charged CR and the complex migrates faster than free CR towards the anode during electrophoresis [34]. The absorption spectra of CR and its complex with Dox also differ (hypochromic effect). CR-Dox complexes have increased size compared to free CR, which was confirmed using DLS and molecular modeling methods [8]. The fuzzy oil drop model, applied in the previous study, aimed to detect binding sites for supramolecular ligands in albumin (particularly between its pseudo-symmetrical fragments) as well as in V domains, indicated that complexation of dye molecules led to the formation of a stable supramolecular structure, anchored between antibodies that participate in the immune complex [35].

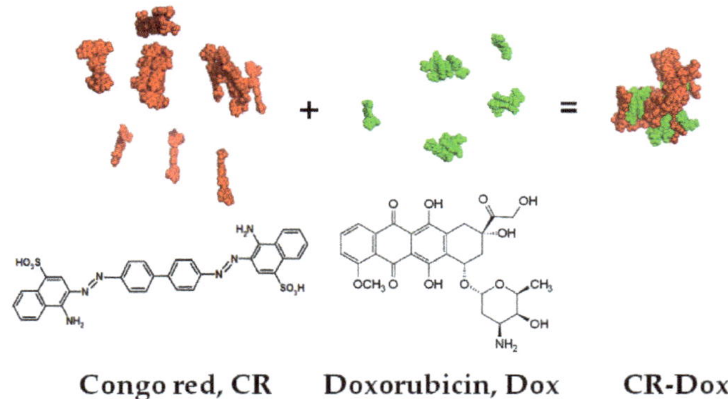

Figure 1. Structure of Congo red and Doxorubicin and the supramolecular complex CR-Dox.

The main goal of the present analysis was to compare the capability of drug binding mediated by CR to proteins: albumin, heated light chain, which is a model system of antibody bound to antigen, and antibodies bound to surface-immobilized antigens. Using DLS and molecular sieve methods, sizes of the free supramolecular system (CR) and its complex with a drug (CR-Dox, molar ratio 2:1) were compared at various concentrations of components and ionic strength of buffers. Subsequently, the interaction of free CR and its drug-bound complexes (CR-Dox) with proteins was investigated. The properties of SRLS, particularly their ability to bind drugs, antibodies in immune complexes and albumin justify research into the application of the described systems in targeted therapy (immuno-targeting). This research with the use of carriers specifically interacting with some proteins is important for targeted transport of drugs to the desired parts of the body, where inflammation or neoplastic process continues.

2. Materials and Methods

2.1. Materials

Congo red (CR, 96% purity, Aldrich Chemical Company, Inc., Milwaukee WI 53233, USA), doxorubicin hydrochloride (Dox, 98% purity, Sigma-Aldrich, Co., 3050 Spruce Street, St. Louis, MO 63103, USA), bovine serum albumin (BSA, 96% purity, Sigma-Aldrich, Co., 3050 Spruce Street, St. Louis, MO 63103, USA), and immunoglobulin light chain λ dimer (Lλ) were obtained from the urine of a patient with multiple myeloma. After salting out and dialysis it was purified on Sephacryl S300 column (Pharmacia). Immunoglobulin G was obtained from Baxter Healthcare Corporation, Hyland Division Glendale, CA 91203, USA. All other reagents used were of analytical grade and were purchased from commercial sources.

2.2. Methods of CR-Dox Preparation

CR-Dox complexes of the 2:1 molar ratio were used because in the previously optimized CR:Dox ratio doxorubicin is completely bound to CR. CR-Dox complexes were created by adding 2 volumes of preheated (2 min. at 100 °C) Congo red (1.43 mM CR in 0.05 M Tris/HCl buffer pH 7.4, 0.154 M NaCl) to 1 volume of 1.43 mM Dox dissolved in the same buffer. The mixture was incubated for 15 min at room temperature. The cohesion of the CR molecules forming the ribbon-like structure is not high, but in the presence of alkaline doxorubicin complexes are formed and its cohesion increases significantly. The complexes were passed through a Sephadex G-200 column to remove unbound components.

For DLS analysis of the effect of concentration on the hydrodynamic diameters of the analyzed probes, different concentrations were used. The final concentrations of higher concentration probes were: CR (1.43 mM), Dox (0.715 mM), and CR:Dox (molar ratio = 2:1, CR = 1.43 mM, Dox = 0.715 mM). All probes were dissolved in 0.05 M Tris/HCl. pH 7.4 buffer with 0.154 M NaCl.

The final concentration of lower concentration probes were: CR (0.31 mM), Dox (0.15 mM) and CR:Dox (molar ratio = 2:1, CR = 0.31 mM, Dox = 0.15 mM). All probes were also dissolved in 0.05 M Tris/HCl. pH 7.4 buffer with 0.154 M NaCl.

For DLS analysis of the effect of buffer ionic strength on the hydrodynamic radius, different concentrations of NaCl in the buffer were used (0.05 M Tris/HCl, pH 7.4 buffer with 0.154 M NaCl or 0.3 M NaCl). The final concentrations of probes in this experiment were lower: CR (0.31 mM), Dox (0.15 mM) and CR:Dox (molar ratio = 2:1, CR = 0.31 mM, Dox = 0.15 mM).

2.3. Methods of CR and CR-Dox Binding with Protein

Complexes of CR and CR-Dox with Proteins

CR and CR-Dox complexes with three different types of acceptor proteins were analyzed: 1. partly unfolded immunoglobulin chain Lλ; 2. partly unfolded immunoglobulin G (HAI); 3. plasma albumin.

1.   The immunoglobulin light chain λ (Lλ)

Partly unfolded Lλ (dimer) was used as a model protein that binds CR or CR-Dox complexes according to the same mechanism as the one observed in antigen-complexed antibodies, where natural structural destabilization is caused by intramolecular constraints evoked by simultaneous interaction with two antigenic determinants [25]. Lλ was isolated from the urine of a patient with multiple myeloma. Complexes were formed by mixing partially unfolded Lλ (obtained by 20 min. incubation at 45 °C (sub-denaturing conditions) with a ten-fold molar excess of CR or CR-Dox complexes (2:1 molar ratio). Under such conditions the N-terminal polypeptide loop of the Lλ undergoes local structural destabilization creating the binding site for 4 CR molecules (per monomer) [9].

2.   Heat aggregated immunoglobulin G (HAI)

Human immunoglobulins G at a concentration of 10 mg/mL (0.05 M PBS buffer) were heated for 20 min at 63 °C. The aggregate dissolves upon the addition of 100-fold molar excess of CR. To remove free or weakly bound dye molecules the protein-dye complex was filtered through Sephadex G-200.

3.   Albumin

Albumin is a model system for studying the CR and CR-Dox interaction with a typical carrier protein. Bovine serum albumin (BSA) in 0.05 M Tris/HCl buffer, 0.145 M NaCl, pH 7.4 was mixed and incubated (15 min.) with CR or CR-Dox at 10-fold molar excess of CR to BSA, and 2:1 ratio of CR to Dox.

2.4. Characterization of Free CR-Dox Complexes or CR-Dox Bound to Albumin, Lλ Light Chain, or HAI

2.4.1. Dynamic Light Scattering (DLS)

Hydrodynamic radii of CR, Dox, and CR-Dox complexes were measured by using the dynamic light scattering (DLS) method (detector Zetasizer Nano ZSP, Malvern, United Kingdom) with laser incident beam at λ = 633 nm and a fixed scattering angle of 173°. For the measurements, dispersants with the following parameters of viscosity and refractive index were used: (1) Tris/HCl 0.05 M with NaCl 0.3 M; viscosity 0.9208 cP; Refractive Index = 1.334; and (2) Tris HCl 0.05 M with NaCl 0.154 M; viscosity 0.9068 cP; Refractive Index = 1.332. Each measuring probe was incubated inside the instrument (3 min/25 °C). A measurement comprised 5–9 repetitions each of which was an average of 15 records measured for 9 s. Outliers were rejected from analysis and the results were averaged.

2.4.2. Gel-Filtration Chromatography (BioGel P-10 and BioGel P-300)

Elution volume (V_e) analysis for fractionation within a different size range was performed on 100 mm Bio®Spin columns (BioRad, Hercules, CA, USA). Columns were filled with 4 mL of BioGel P-10 in 0.01 M PBS buffer, pH 7.4 for supramolecular compound solutions (CR, CR-Dox). For complexes of CR and CR-Dox with proteins (BSA and Lλ), the BioGel P-300 in 0.01 M PBS buffer, pH 7.4 was used. The flow rate reflects the size of the complex (or the supramolecular entity). 80 μL of the sample was loaded onto the column. Then, it was rinsed with 0.01 M PBS buffer, pH 7.4.

The presence of the protein in the eluate was detected by dot-staining with bromophenol blue, CR concentration was determined spectrophotometrically and Dox fluorometrically.

2.4.3. Gel Electrophoresis

Electrophoresis was carried on 1% agarose plates (in 0.06 M sodium barbital buffer, pH 8.6) at 160 V for about 40 min. The position of CR-complexes was recorded, plates were then fixed with picric acid and the excess of CR was removed by reduction with sodium dithionate followed by staining for protein with bromophenol blue.

2.4.4. TEM

Transmission Electron Microscopy (TEM; JEOL JSM-7500F) was used to evaluate the structure of the HAI-CR complexes. Samples for TEM were prepared by heating the aqueous solution of immunoglobulin G at 63 °C for 20 min. After gradual cooling, the CR solution was added (the CR was labeled with $AgNO_3$). The resulting complexes of HAI-CR were purified from the excess of unbound CR using a thin layer chromatography on Sephadex G200. The procedure allows for obtaining well-dispersed HAI loaded with supramolecular CR "clouds". For TEM analysis, 1 μL of suspension was applied to the surface of the copper grid (300 mesh) and dried in vacuum.

2.5. Methods of Analysis of Competition between BSA and HAI for Binding of the CR-RhoB Complexes (Congo Red-Rhodamine B) or CR-Dox (Congo Red-Doxorubicin) Complexes

Additionally, BSA, BSA-CR complex (10:1), and BSA-CR-Dox (or BSA-CR-RhoB) complexes (CR-Dox or CR-RhoB ratios were 1:1, 2:1, or 5:1) were prepared. To remove excess CR unbound with the triple complex, gel-filtration chromatography on BioGel P-300 medium was used for all systems. HAI was added to a part of the sample to observe the transmission of CR and CR-Dox complex from the initial complexes with BSA.

3. Results

3.1. Characterization of CR-Dox Co-Micelles

3.1.1. DLS Analysis of CR-Dox Complexes—The Effect of Concentration on the Complex Size

DLS size analysis revealed well-defined peaks for CR-Dox complexes. The results indicate that the CR-Dox complex is bigger than free CR and free Dox. The size of the CR-Dox complex depends upon the initial concentrations of the components. The samples containing CR-Dox complexes at 2:1 CR:Dox molar ratio were compared with complexes formed at different component concentrations (lower: CR: 0.22 mg/mL, Dox: 0.09 mg/mL and higher: CR: 1 mg/mL, Dox: 0.83 mg/mL). In the case of free Dox and free CR, no effect of concentration on the hydrodynamic diameter was observed—for Dox it was 0.6 nm and for CR 2.3 nm for both concentrations. In the case of the CR-Dox complex, the size increased with increasing concentration, amounting to 3.6 nm for lower to 4.85 nm for the higher concentration (Figure 2).

The above results show that Dox alone does not form supramolecular structures and that the dimensions of the supramolecular CR ribbon are not influenced by the concentration while in the case of CR-Dox complex, the size increases with concentration (for samples with the same proportion of the components).

To explain the above results, it is necessary to clarify how the hydrodynamic diameter is read in the DLS analysis. In the case of spheroidal particles, it is just the diameter, but in the case of supramolecular CR, it is rather the width of the ribbon-like structure it creates in the solution. Both CR and Dox molecules contain charged groups, negative in the case of CR and positive in the case of Dox. As a result, supramolecular ribbons created by CR repel each other and the measured value of the hydrodynamic diameter is the same, even if the length of the ribbon differs, being higher at higher concentrations. A similar effect concerns Dox. In the case of CR-Dox complexes, the electrostatic interactions may stabilize the complex and the effect of electrostatic repulsion between separate ribbon-like structures disappears. This allows the individual ribbons to create more complex, tangled structures, registered in DLS analysis as the ones with higher hydrodynamic diameters.

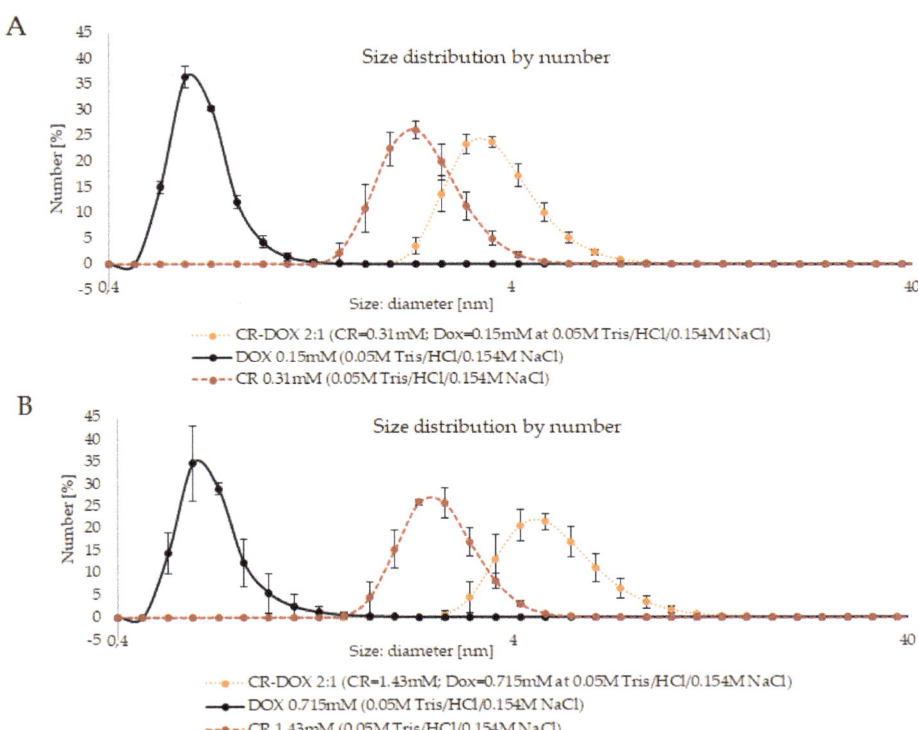

Figure 2. DLS analysis of CR-Dox complexes. The effect of concentration on the hydrodynamic diameter of CR, Dox, and CR-Dox complex. (**A**). lower concentration probes (final concentration of CR: 0.31 mM, Dox: 0.15 mM); (**B**). higher concentration probes (final concentration of CR: 1.43 mM, Dox: 0.715 mM).

3.1.2. DLS Analysis of CR-Dox Complexes—The Effect of Ionic Strength on Complex Size

The effect of ionic strength of the solution on hydrodynamic diameter of CR, Dox, and CR-Dox complex (2:1) was analyzed. Complexes were prepared in 0.05 M tris/HCl buffer with the addition of 0.154 M or 0.3 M NaCl. Higher ionic strength led to significantly increased size of CR-Dox complexes (3.6 vs. 78.82 nm) while it only slightly influenced the hydrodynamic diameter of CR ribbons (2.3 vs. 2.7 nm) and did not affect the results for Dox (0.6 nm for both NaCl concentrations, Figure 3).

To interpret the result, as in the case of the previous experiment, we need to note that in DLS analysis the hydrodynamic diameter value read by the DLS instrument reflects the diameter of the supramolecular ribbon formed by the analyzed molecules in the solution. As the ionic strength increases more charged groups become shielded due to the interaction with salt ions. For free CR the observed effect is rather small, but in the case of the CR-Dox complex, the neutralization of charges promotes the interaction between individual ribbon-like structures which produce bundles characterized by significantly higher hydrodynamic diameter.

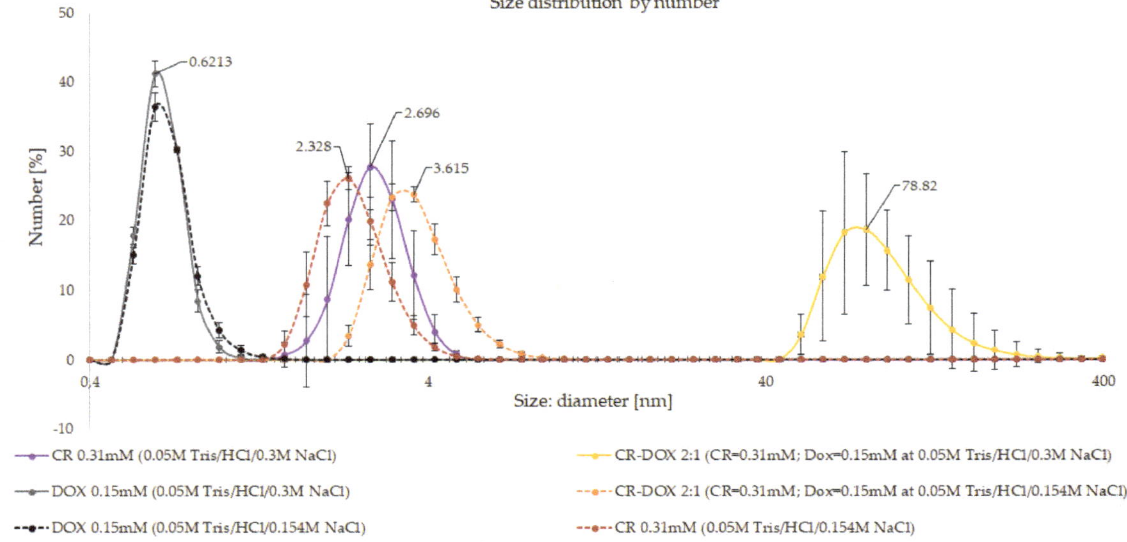

Figure 3. DLS analysis of CR-Dox complexes. The effect of ionic strength on the hydrodynamic diameter of CR, Dox, and CR-Dox complex. Lower concentration probes (final concentration of CR: 0.31 mM, Dox: 0.15 mM) at two different 0.05 M Tris/HCl buffers supplemented with: 0.154 M or 0.3 M NaCl.

3.1.3. Gel-Filtration Chromatography (BioGel P-10): Elution Volume (Ve) of CR, Dox, and CR-Dox

Gel filtration chromatography on BioGel P-10 columns was used to estimate the sizes of the complexes. Elution volumes (V_e) for CR, Dox, and CR-Dox (2:1 molar ratio) complex were compared (Table 1).

Table 1. Elution volume for CR, Dox, and CR-Dox complex.

	CR	Dox	CR-Dox (2:1 Molar Ratio)
Elution volume (V_e) (mL)	2.3	0.9	0.3

CR flows through the column very slowly and its elution volume is practically 100%, which may suggest adsorption to the BioGel bed. CR-Dox complex migrates faster than Dox alone indicating the formation of stable mixed SRLS created by CR and Dox. CR-Dox elution volume is 13% and Dox elution volume is 40%.

3.2. Interaction of Free CR and CR-Dox Co-Micelles with Proteins

3.2.1. Agarose Gel Electrophoresis and Chromatographic Analysis

1. CR-Dox co-micelles form complexes with BSA (BSA-CR-Dox)

Albumin, a universal transporter protein, has the binding site for CR [10]. To check its ability to bind CR-Dox supramolecular ligand, the agarose gel electrophoresis employed as binding for CR-Dox (2:1 molar ratio) changes the net charge of the protein. As seen in Figure 4, the BSA-CR complex migrates faster than BSA (but slower than free CR). The fastest migration was observed for the CR-Dox complex. This can be explained by a strong interaction and high cohesion between molecules, leading to the increased dissociation of CR sulfonic groups and thus more acidic properties and higher electrophoretic mobility. Migration of BSA-CR-Dox complex is similar to that of BSA-CR, but two-dimensional separation (electrophoresis in the direction pointed by the arrow 1, followed by chromatography of the filter paper replica (arrow 2) reveals the presence of Dox in the complex. This confirms the presence of CR-mediated binding of Dox to BSA. Under the above experimental conditions, formation of stable complexes between BSA and Dox was not observed.

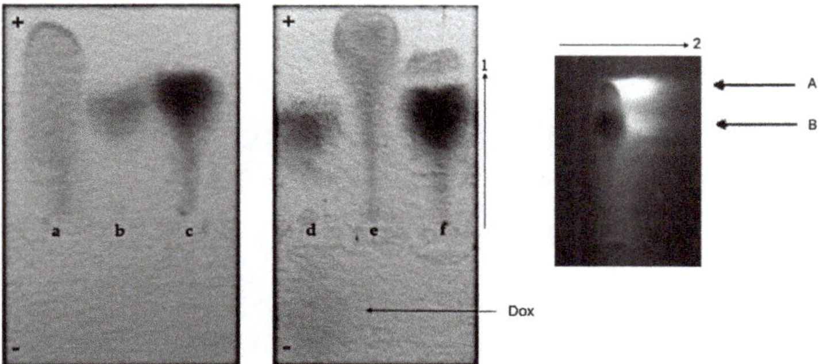

Figure 4. Complexes between BSA and CR-Dox (2:1)—replica on filter paper applied to bromophenol blue-stained gel (after agarose gel electrophoresis at pH 8.6); migration towards the anode (+): (**a**) CR, (**b**) BSA, (**c**) BSA-CR complex, (**d**) BSA and Dox (Dox can be seen as migrating towards the cathode (−), pointed by arrow), (**e**) CR-Dox complex, (**f**) BSA-CR-Dox (2:1) complex. The presence of Dox in the complex (BSA-CR-Dox) was confirmed chromatographically. The separation of CR-DOX mixtures was performed by Whatman 3 paper chromatography in butanol:acetic acid:water (5:1:4) solvent. Dox is seen as bright-orange fluorescence. For semiquantitative evaluation, DOX was eluted and the fluorescence was measured (emission signal at 550 nm upon excitation with a 470 nm laser beam) (the picture on the right; arrow no. 1 shows the direction of electrophoresis, and arrow 2 the direction of chromatography)—arrow (**A**) points to Dox released during chromatography from its complex with CR, while arrow (**B**) points to Dox released from the BSA-CR-Dox complex.

This experiment shows that this type of drug can be bound to albumin (BSA) via CR–a model SRLS.

2. CR-Dox co-micelles do not form complexes with Lλ

Conformational changes in the N-terminal loop of the immunoglobulin light chain λ (Lλ) heated up to 45 °C create the binding site for free CR [24]. Complexes with a defined stoichiometry (4 CR molecules per L chain monomer) can be visualized by agarose gel electrophoresis.

The possibility of creating complexes between Lλ and CR-Dox co-micelles was analyzed. In agarose gel electrophoresis (pH 8.6), Dox migrates towards the cathode and free CR (a), Lλ chain (b), Lλ-CR complex (c), CR-Dox complexes (e) migrate towards the anode. CR forms a strong complex with Lλ, which migrates towards the anode twice as fast as free protein. For CR-Dox molar ratio 2:1, the formation of CR-Lλ complex is not observed, and all the CR is incorporated in CR-Dox faster-migrating complex, while Lλ migrates at the same speed as free unbound protein as in lane "b". It suggests a strong competition between Lλ and doxorubicin for CR binding. All in all, preferentially the CR-Dox complex (migrating at the front) is formed first. This complex binds all the CR and thus Lλ-CR complex do not assembled, which is confirmed by the same migration distance of the protein as that of CR-free protein (f) (Figure 5).

We conclude that Lλ binding site for SRLS created in the Lλ-chain at 45 °C can accommodate CR but not CR-Dox, suggesting that the binding site is specific and the incorporation of Dox changes the supramolecular properties of the CR.

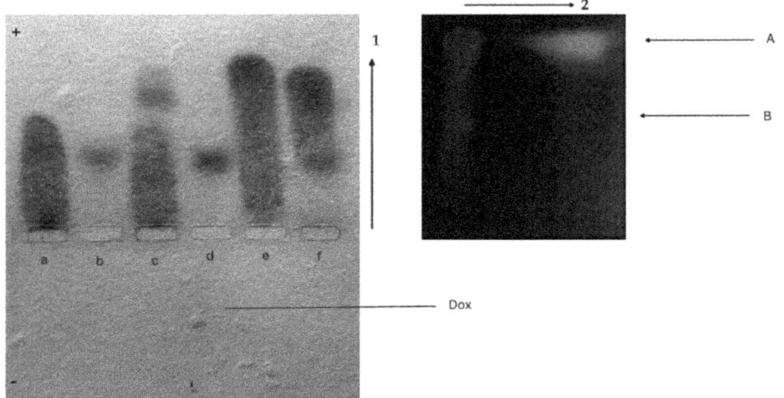

Figure 5. The possibility of creating complexes between Lλ and CR-Dox (2:1 molar ratio) was analyzed—replica on filter paper applied to bromophenol blue stained gel (after agarose gel electrophoresis at pH 8.6); migration towards the anode (+): (**a**) CR, (**b**) Lλ, (**c**) Lλ-CR complex, (**d**) Lλ-Dox mixture (Dox can be seen as migrating towards the cathode (−), pointed by arrow), (**e**) CR-Dox (molar ratio 2:1), (**f**) Lλ-CR-Dox (CR-Dox molar ratio 2:1); The absence of Dox in the Lλ-CR-Dox mixture was confirmed chromatographically. Arrow (**A**) points to Dox eluted from CR-Dox complex (fastest migration), arrow (**B**) points to the position of Lλ−CR complex (where Dox is absent).

3.2.2. Gel-Filtration Chromatography (BioGel P-300): Elution Volumes of CR and CR-Dox Complexes with Proteins

Elution volumes (V_e) of protein (BSA, Lλ) complexes with CR and CR-Dox were analyzed using gel filtration chromatography on BioGel P-300 columns. V_e values of free ligands and proteins (CR, Dox, BSA, Lλ), two-component complexes (BSA-CR, Lλ-CR), and three-component (triple) complex of BSA with CR-Dox co-micelle (BSA-CR-Dox) or a mixture of Lλ with CR-Dox co-micelle were compared (Table 2).

A low value of the elution volume for some complexes comparable to that of protein indicates the formation of large and stable co-micelles by e.g., CR and DOX.

Table 2. Gel-filtration chromatography: elution volume (V_e) of BSA, Lλ, CR, and Dox in mixtures of BSA-CR-Dox, Lλ-CR-Dox, CR-Dox, BSA-CR, Lλ-CR, and in free BSA and Lλ (Biogel P-300). The molar ratio of CR:Dox in all mixtures was 2:1. $V_{e\ DOX\ free}$ = 2.1 mL (not shown).

	Mixtures						
Elution Volume (V_e)	BSA-CR-Dox	Lλ-CR-Dox	CR-Dox	BSA-CR	Lλ-CR	BSA	Lλ
BSA (mL)	0.3	-	-	0.6	-	0.7	-
Lλ (mL)	-	0.8	-	-	0.9	-	1
CR (mL)	0.3	0.4	0.3	0.6	0.8	-	-
Dox (mL)	0.3	0.4	0.3	-	-	-	-
Complex formation:	**YES**	**NO**	**YES**	**YES**	±		

Free CR flows very slowly through the column, with tailing, which results from its adsorption to the BioGel. The elution volume of free BSA ($V_{e\ BSA\ free}$) was 0.7 mL, while for BSA-CR it was $V_{e\ BSA-CR}$ = 0.6 mL (both for separately determined CR and BSA). The BSA-CR complex shows a higher flow rate than ligand-free BSA. Simultaneous elution points to the formation of the BSA-CR complex. Elution volume for CR-Dox was $V_{e\ CR-Dox}$ = 0.3 mL (both for separately determined CR and Dox). Simultaneous elution points to the formation of the CR-Dox co-micelle. So, we can conclude that CR-Dox at 2:1 molar ratio forms a stable, fast-migrating, and large complex which shows the migration rate twice as fast as that of BSA-CR and free BSA. The same elution volume for separately determined BSA, CR, and Dox was $V_{e\ BSA\text{-}CR\text{-}Dox}$ = 0.3 mL. Simultaneous elution points to the binding of the CR-Dox co-micelle by BSA and formation of a stable, ternary complex with BSA.

The elution volume of free Lλ ($V_{e\ L\lambda\ free}$) was 1 mL. In the case of Lλ-CR, the V_e value differs for Lλ and CR ($V_{e\ L\lambda}$ = 0.9; $V_{e\ CR}$ = 0.8). The separate elution from the column, but with increased flow rate for CR (as compared to the control, which is free CR), suggests that the complex is formed but dissociates during the flow through the column (which adsorbs CR).

In the sample containing the mixture of Lλ and CR-Dox, different V_e values for CR-Dox co-micelle and Lλ were determined ($V_{e\ L\lambda}$ = 0.8; $V_{e\ CR\text{-}Dox}$ = 0.4). Separate elution points to the inability of Lλ to bind CR-Dox co-micelle.

Lλ migrates more slowly through the column than BSA due to its lower molecular weight. The differences in flow rates of Lλ-CR complex vs. free protein and free CR were observed. The mixture of Lλ and CR-Dox (2:1) did not produce a ternary complex (as in the case of BSA)–CR-Dox and Lλ were eluted separately from the column with V_e for CR-Dox equal to 0.4 and V_e for Lλ equal to 0.8.

3.2.3. CR Binds to Heat Aggregated Immunoglobulins G

HAI was dissolved with CR contrasted by $AgNO_3$. The formed complexes were isolated using gel filtration chromatography on Sephadex G-200 thin layer. HAI surrounded by clusters of supramolecular Congo red were visible using transmission electron microscopy. HAI are more diffused in the presence of CR and form smaller clusters than HAI aggregated without CR. (Figure 6).

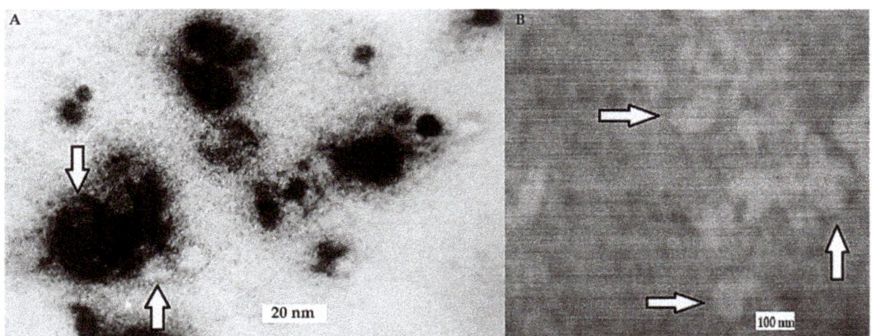

Figure 6. Complexes observed using TEM. (**A**). Heat aggregated immunoglobulins (HAI) with CR; (**B**). HAI without CR. Clusters of supramolecular Congo red can be distinguished around antibodies (darker spots). The images of immunoglobulins (brighter spots) are indicated by arrows.

3.2.4. Competition of BSA and HAI for Binding of the CR-RhoB Complexes (Congo Red-Rhodamine B) or CR-Dox (Congo Red-Doxorubicin) Complexes

Albumin has been shown to compete with HAI for binding of the CR-RhoB (or CR-Dox) and that it may eventually donate it to the immune complex model.

BSA-CR-RhoB complexes are formed when CR-RhoB is added to albumin. In electrophoresis, there is a clear difference in the location of BSA band (at the front, running faster) and HAI band (runs much slower, close to the loading site). After the addition of HAI to BSA-CR-RhoB complexes, CR-RhoB dissociates from BSA-CR-RhoB complexes and simultaneous binding of CR-RhoB to HAI is visible (Figure 7). Similar results were obtained with the drug-doxorubicin. These results indicate that albumin can be used as a carrier that binds other compounds, including CR-associated drugs. Such a drug can target the immune complex.

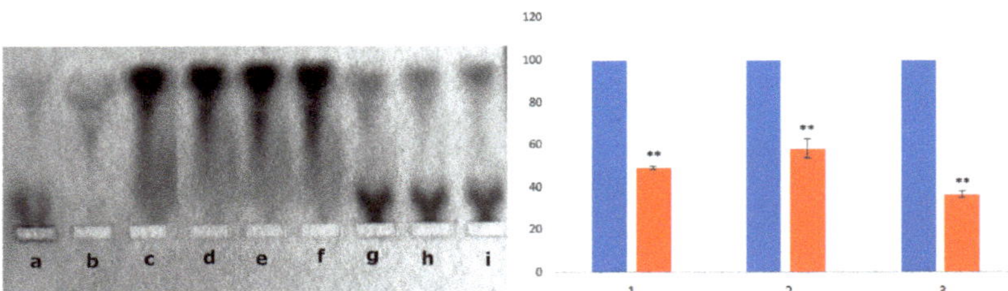

Figure 7. (**a**) HAI, (**b**) BSA, (**c**) BSA-CR, (**d–f**) BSA-CR-RhoB (1:1), (2:1), and (5:1) respectively, (**g–i**) BSA-CR-RhoB (1:1), (2:1), and (5:1) respectively after adding the same amount of HAI. The data in the graphs compare controls (blues bars) with results (red bars) and represent the mean ± SEM; test Student T. ** $p < 0.01$.

A loss of the CR-RhoB complex initially bound to albumin was observed after the addition of HAI in the amount of respectively: 50.6% in the case of the initially added CR:RhoB at molar ratio of 1:1 (compare "d" with "g" on the electrophoresis slab and with (1) on the histogram); 41.7% (for 2:1; compare "e" with „h" on the electrophoresis slab and (2) on the histogram) and 63.3% (for 5:1; compare "f" with "i" on the electrophoresis slab and (3) on the histogram). Complex binding to HAI is visible. The data in the graphs compare controls (blues bars) with results (red bars) and represent the mean ± SEM; test Student T. * $p < 0.05$, ** $p < 0.01$, *** $p < 0.001$.

4. Discussion

The design of safe and reliable carriers for drugs is an important goal in the development of new therapies. The safe delivery of drugs using albumin, heat aggregated immunoglobulins, or antibodies as carriers in targeted immunotherapy, is already used as a solution which either supplements or substitutes the hitherto used therapies [36,37].

Earlier studies showed that stable mixed supramolecular assemblies can be formed by CR and other molecules characterized by planar, polyaromatic ring structure [30–33]. Supramolecular structure of Congo red dissociated by heating (80 °C) was found to form chaotic, frozen organization after rapid cooling, but it reorganizes upon gradual heating at about 25 °C deg forming standard ribbon-like structure and dissociates again above 60 °C [38]. In addition, the interaction of CR, as a model supramolecular ligand, with plasma albumin [10], partly unfolded immunoglobulin light chain [9], with HAI [6], and with antigen-complexed antibodies [25] was described.

In this study, the analysis of the CR-drug complex (using DOX as an example) was extended, based on the previously conducted research [8,34,35,39,40]. Attempts were also made to introduce co-micelles formed from Congo red bound to Dox into all analyzed proteins. Moreover, the latest research using oil drop modeling has shown that not only free CR but also its complex with the drug can be bound to albumin. However, only free CR binds to the light chain while the CR-Dox complex is not bound [35].

The presented results provide new information about the properties of the SRLS-drug complexes and the capability of their interaction with proteins, and thus about their possible role in the delivery of drugs. It was shown that the L chain λ forms complexes with supramolecular CR but not with CR-Dox complexes (created at the 2:1 molar ratio). In the case of plasma, albumin complexes can be formed with both CR and CR-Dox. CR-Dox co-micelles were also bound by HAI, and additionally, the transfer of a part of the co-micelle-bound drug from the BSA-CR-Dox complex to HAI was observed.

The interaction of positively charged Dox with negatively charged CR molecules probably changes the regular, ribbon-like architecture of the supramolecular CR assembly.

The strong interaction between CR and Dox results in the increased electrophoretic mobility of CR-Dox as compared to CR. The explanation for this phenomenon is the face-to-face alignment of Congo red molecules in supramolecular ribbon-like structures. In the electric field, these systems become uniquely oriented dipoles (due to the delocalization of pi electrons from stacked aromatic rings). Electron relocation affects the polar groups of Congo red, changing their dissociation constants and consequently their charge. Hence, accelerated electrophoretic migration towards the anode upon increasing CR concentration is observed, which is indicative of the concentration-dependent rise in acidity of Congo red. CR-Dox complexes are formed by the intercalation of doxorubicin between the Congo red molecules forming the ribbon. The mechanism of the dipole formation is the same and hence the acceleration of the CR-Dox complex migration towards the anode [34,41].

The CR-Dox complex is probably too large, and perhaps too compact, to enter the binding site in a partly unfolded L chain. This binding site is capable of accommodating the supramolecular CR micelle composed of just 4 CR molecules [6] and requires certain plasticity of SRLS which is a characteristic of CR alone and is limited (or absent) in CR-Dox complex.

Therefore, the full analysis of protein complexes with a large CR-Dox complex could not be achieved using the partly unfolded L chain as a model. Supramolecular ligands, such as CR and CR-Dox can interact with proteins producing complexes in which some ligand molecules interact with the protein directly, while others are bound indirectly, as components of the SRLS. Such type of interaction would be possible in the case of HAI, where CR or CR-Dox could be localized also in-between immunoglobulins. It is thus possible that CR creates a kind of a scaffold structure capable of accommodating CR-Dox, and could work in the same way as in antigen-bound antibodies.

The potential of albumin to complex large-sized ligands (such as the CR-Dox complex) is much greater than that of the light chain model. Albumin, which has a ligand binding pocket, has greater ability to adapt this pocket to large ligands [35]. The structure of the ordered Congo red micelle changes significantly upon intercalation of other molecules (doxorubicin). The linear arrangement of the Congo red tape structure is likely to change, as indicated by the obtained results of the analysis using DLS and molecular filtration.

Albumin is one of the most often used drug carriers approved by the FDA, due to its high biocompatibility, availability, and accumulation in tissues that show high metabolic rate. It can also protect tissues against the harmful effects of the drugs it transports. Nab albumin nanoparticles (American Bioscience) have a diameter of 130–150 nm [42]. Albumin-based drug carriers are neither cytotoxic nor immunogenic, present optimum pharmacokinetics, are biodegradable, and easy to prepare [43–45]. Cancerous tissue presents increased capillary permeability and retention [EPR] and thus is more readily infiltrated by large particles, including albumin-based drug carriers. Here we show that native albumin can bind large ligands and can serve as a transporter for transport of SRLS-drug complexes.

Small molecules, like free Dox, can easily penetrate both healthy and diseased tissues while macromolecules (including albumin-drug conjugates or complexes) can easily infiltrate cancerous tissue but do not pass the endothelium of healthy ones. It increases the selective action of albumin-drug conjugates [46,47].

Albumin-CR-Dox complexes presented in this paper could serve as an alternative drug carrier. Due to the positive charge of Dox, its interaction with albumin is weak, and not observed in experimental systems presented here. Spectroscopy and docking results obtained by Agudelo et al. demonstrated that doxorubicin was able to bind to BSA and HSA via hydrophilic and hydrophobic interactions with more stable complexes created with human serum albumin than with bovine serum albumin. They also showed that drug-protein binding engaged several amino acid groups which were stabilized by a network of hydrogen bonds. The drug-protein interaction changed secondary structure of both bovine and human albumin causing partial destabilization of the protein. It can explain weak binding or no binding between BSA and Dox observed by us earlier [48].

Supramolecular systems, due to a wide range of their reactions with proteins and a variety of functional effects, become a new research tool in biology and pharmacology. The presented systems can be used as an alternative to the Nab technology (American Bioscience), consisting in mixing a hydrophobic drug suspended in an oil phase with an aqueous albumin solution and homogenization of the resulting mixture. The resulting albumin nanoparticles have a drug portion locked inside [42]. However, this technology does not work with hydrophilic, positively charged doxorubicin. The simplicity and speed of complex formation in the proposed system is also an interesting alternative to the previously tested combinations of albumin and doxorubicin [49]. On the other hand, drug delivery via CR that interacts with immune complexes is important due to the enhancing effect described in the literature [6].

Function-dependent Congo red binding to a protein like in the case when complexed with antigen-bound antibodies, consequently has a strong influence on this function. For instance, the affinity of antibodies forming immune complexes markedly increases in the presence of Congo red. Increasing the affinity of antibodies allows for the use of low-affinity antibodies in immune complexes, and an additional advantage is that an increased amount of drugs can be delivered in this way. The presented system offers a wide range of biomedical applications including drugs delivery to cancer cells.

5. Conclusions

To conclude, our results confirms that Congo red type supramolecular, self-assembled ribbon-like structures form complexes with the chemotherapeutic agent doxorubicin. CR-Dox are large-sized structures with properties different from the free CR. A model system composed of heated immunoglobulin light chain Lλ capable of CR binding, did not bind CR-Dox complexes. Heat aggregated immunoglobulins (HAI) and albumin were able to bind both free CR and CR-Dox complexes. Additionally albumin-bound CR-Dox complexes were transferred from albumin to HAI upon addition of HAI. This kind of interaction between CR-Dox and the described proteins, may in future become an important therapeutic system with the possibility of targeted drug transport and delivery. Supramolecular ribbon-like CR complexed with doxorubicin is a promising system in the treatment of cancers and may open new avenues for novel treatment strategies.

Author Contributions: Conceptualization, A.J. and I.R.; methodology, K.C., G.Z., I.K. and A.J.; software, G.Z. and K.C.; validation, A.J. and I.R.; formal analysis, K.C., G.Z., I.K., A.J. and I.R.; investigation, A.J.; resources, A.J.; data curation, K.C., G.Z., I.K., A.J. and I.R.; writing—original draft preparation, A.J.; writing—review and editing, I.K. and I.R.; visualization, A.J. and R.I.; supervision, A.J. and I.R.; project administration, A.J.; funding acquisition, A.J. All authors have read and agreed to the published version of the manuscript.

Funding: This research was funded by the National Science Centre, Poland, grant number 2016/21/D/NZ1/02763 and by the Polish Ministry of Science and Higher Education, grant number N41/DBS/000715.

Institutional Review Board Statement: Not applicable.

Informed Consent Statement: Not applicable.

Data Availability Statement: The data presented in this study are available on request from the corresponding author.

Acknowledgments: Many thanks to Leszek Konieczny for his valuable comments. Many thanks to Barbara Piekarska, Barbara Stopa and to Radoslawa Wrobel for helped edit the manuscript. Many thanks to Dorota Duraczynska from Jerzy Haber Institute of Catalysis and Surface Chemistry, Polish Academy of Sciences for transmission imaging of the samples.

Conflicts of Interest: The authors declare no conflict of interest. The funders had no role in the design of the study; in the collection, analyses, or interpretation of data; in the writing of the manuscript, or in the decision to publish the results.

References

1. Riley, R.S.; June, C.H.; Langer, R.; Mitchell, M.J. Delivery Technologies for Cancer Immunotherapy. *Nat. Rev. Drug Discov.* **2019**, *18*, 175–196. [CrossRef]
2. Milling, L.; Zhang, Y.; Irvine, D.J. Delivering safer immunotherapies for cancer. *Adv. Drug Deliv. Rev.* **2017**, *15*, 79–101. [CrossRef] [PubMed]
3. Vanneman, M.; Dranoff, G. Combining immunotherapy and targeted therapies in cancer treatment. *Nat. Rev. Cancer.* **2012**, *12*, 237–251. [CrossRef] [PubMed]
4. Conniot, J.; Silva, J.M.; Fernandes, J.G.; Silva, L.; Gaspar, R.; Ebrocchini, S.; Florindo, H.F.; Barata, T.S. Cancer immunotherapy: Nanodelivery approaches for immune cell targeting and tracking. *Front. Chem.* **2014**, *2*, 105. [CrossRef]
5. Cai, S.S.; Li, T.; Akinade, T.; Zhu, Y.; Leong, K.W. Drug delivery carriers with therapeutic functions. *Adv. Drug Deliv. Rev.* **2021**, *176*, 113884. [CrossRef] [PubMed]
6. Roterman, I.; Konieczny, L. Self-assembled molecules–New kind of protein ligands. In *Supramolecular Ligands*, 1st ed.; Springer: Cham, Switzerland, 2017. [CrossRef]
7. Piekarska, B.; Skowronek, M.; Rybarska, J.; Stopa, B.; Roterman, I.; Konieczny, L. Congo red-stabilized intermediates in the lambda light chain transition from native to molten state. *Biochimie* **1996**, *78*, 183–189. [CrossRef]
8. Jagusiak, A.; Panczyk, T. Interaction of Congo Red, Evans Blue and Titan Yellow with doxorubicin in aqueous solutions. A molecular dynamics study. *J. Mol. Liquids* **2019**, *279*, 640–648. [CrossRef]
9. Piekarska, B.; Konieczny, L.; Rybarska, J.; Stopa, B.; Zemanek, G.; Szneler, E.; Król, M.; Nowak, M.; Roterman, I. Heat-induced formation of a specific binding site for self-assembled Congo red in the V domain of immunoglobulin L chain lambda. *Biopolymers* **2001**, *59*, 446–456. [CrossRef]
10. Stopa, B.; Rybarska, J.; Drozd, A.; Konieczny, L.; Król, M.; Lisowski, M.; Piekarska, B.; Roterman, I.; Spólnik, P.; Zemanek, G. Albumin binds self-assembling dyes as specific polymolecular ligands. *Int. J. Biol. Macromol.* **2006**, *40*, 1–8. [CrossRef] [PubMed]
11. Peters, T., Jr. *All about Albumin: Biochemistry, Genetics, and Medical Applications*; Academic Press: San Diego, CA, USA, 1996. [CrossRef]
12. Bertucci, C.; Domenici, E. Reversible and covalent binding of drugs to human serum albumin: Methodological approaches and physiological relevance. *Curr. Med. Chem.* **2002**, *9*, 1463–1481. [CrossRef]
13. Yamasaki, K.; Chuang, V.T.; Maruyama, T.; Otagiri, M. Albumin-drug interaction and its clinical implication. *Biochim. Biophys. Acta.* **2013**, *1830*, 5435–5443. [CrossRef]
14. Otagiri, M. A molecular functional study on the interactions of drugs with plasma proteins. *Drug Metab. Pharmacokinet.* **2005**, *20*, 309–323. [CrossRef] [PubMed]
15. Ascenzi, P.; Bocedi, A.; Notari, S.; Fanali, G.; Fesce, R.; Fasano, M. Allosteric modulation of drug binding to human serum albumin. *Mini Rev. Med. Chem.* **2006**, *6*, 483–489. [CrossRef]
16. Yang, F.; Zhang, Y.; Liang, H. Interactive Association of Drugs Binding to Human Serum Albumin. *Int. J. Mol. Sci.* **2014**, *15*, 3580–3595. [CrossRef]
17. Atwood, J.L.; Steed, J.W. *Encyclopedia of Supramolecular Chemistry*; CRC Press Tylor&Francis Group: Boca Raton, FL, USA, 2004. [CrossRef]
18. Chattopadhyay, D.P. Chapter-Chemistry of dyeing. In *Handbook of Textile and Industrial Dyeing*; Woodhead Publishing Series in Textiles; Woodhead Publishing: Sawston, Cambridge, UK, 2011; Volume 1, pp. 150–183. [CrossRef]
19. Brown, M.A.; Mitar, D.A.; Whitworth, J.A. Measurement of Plasma Volume in Pregnancy. *Clin. Sci.* **1992**, *83*, 29–34. [CrossRef] [PubMed]
20. Natascha, G.A.; Zeinab, Y.M.; Sarah, Y.Y.; Jerome, W.B. *Chapter Four-Endothelial Protrusions in Junctional Integrity and Barrier Function*; Belvitch, P., Dudek, S., Eds.; Current Topics in Membranes; Academic Press: Cambridge, MA, USA, 2018; Volume 82, pp. 93–140. [CrossRef]
21. Grohe, M.E. *Chapter 9—Clinical Evaluation of Adults. Dysphagia Clinical management in Adults and Children*, 3rd ed.; Grohe, M.G., Crary, M.A., Eds.; Mosby; Elsevier: St. Louis, MI, USA, 2021; pp. 149–178. [CrossRef]
22. Spolnik, P.; Stopa, B.; Piekarska, B.; Jagusiak, A.; Konieczny, L.; Rybarska, J.; Król, M.; Roterman, I.; Urbanowicz, B.; Zieba-Palus, J. The use of rigid, fibrillar Congo red nanostructures for scaffolding protein assemblies and inducing the formation of amyloid-like arrangement of molecules. *Chem. Biol. Drug Des.* **2007**, *70*, 491–501. [CrossRef]
23. Ji, W.; Yuan, C.; Chakraborty, P.; Gilead, S.; Yan, X.; Gazit, E. Stoichiometry-controlled secondary structure transition of amyloid-derived supramolecular dipeptide co-assemblies. *Commun. Chem.* **2019**, *2*, 65. [CrossRef]
24. Krol, M.; Roterman, I.; Piekarska, B.; Konieczny, L.; Rybarska, J.; Stopa, B.; Spolnik, P. Analysis of Correlated Domain motions in IgG light chain reveals possible mechanisms of immunological signal transduction. *Proteins Struct. Funct. Bioinform.* **2005**, *59*, 545–554. [CrossRef]
25. Piekarska, B.; Drozd, A.; Konieczny, L.; Krol, M.; Jurkowski, W.; Roterman, I.; Spolnik, P.; Stopa, B.; Rybarska, J. The indirect generation of long-distance structural changes in antibodies upon their binding to antigen. *Chem. Biol. Drug Des.* **2006**, *68*, 276–283. [CrossRef]
26. Krol, M.; Roterman, I.; Drozd, A.; Konieczny, L.; Piekarska, B.; Rybarska, J.; Spólnik, P.; Stopa, B. The increased flexibility of CDR loops generated in antibodies by Congo red complexation favors antigen binding. *J. Biomol. Struct. Dyn.* **2006**, *23*, 407–416. [CrossRef] [PubMed]

27. Weiss, R.B. The anthracyclines: Will we ever find a better doxorubicin? *Semin. Oncol.* **1992**, *19*, e670–e686.
28. Zhu, J.; Liao, L.; Bian, X.; Kong, J.; Yang, P.; Liu, B. pH-controlled delivery of doxorubicin to cancer cells, based on small mesoporous carbon nanospheres. *Small* **2012**, *8*, 2715–2720. [CrossRef] [PubMed]
29. Yang, F.Y.; Teng, M.C.; Lu, M.; Liang, H.F.; Lee, Y.R.; Yen, C.C.; Liang, M.L.; Wong, T.T. Treating glioblastoma multiforme with selective high-dose liposomal doxorubicin chemotherapy, induced by repeated focused ultrasound. *Int. J. Nanomed.* **2012**, *7*, 965–974. [CrossRef]
30. Konieczny, L.; Piekarska, B.; Rybarska, J.; Skowronek, M.; Stopa, B.; Tabor, B.; Dabros, W.; Pawlicki, R.; Roterman, I. The use of Congo Red as a lyotropic liquid crystal to carry stain in a model immunotargeting system–microscopic studies. *Folia Histochem. Cytobiol.* **1997**, *35*, 203–210.
31. Schellman, J.A.; Reese, H.R. Extensions to the theory of intercalation. *Biopolymers* **1995**, *39*, 161–171. [CrossRef]
32. Tuite, E.; Kelly, J.M. The interaction of methylene blue, azure B, and thionine with DNA: Formation of complexes with polynucleotides as model systems. *Biopolymers* **1994**, *35*, 419–433. [CrossRef]
33. Konieczny, L.; Piekarska, B.; Rybarska, J.; Stopa, B.; Krzykwa, B.; Noworolski, J.; Pawlicki, R.; Roterman, I. Bis azo dye liquid crystalline micelles as possible drug carriers in immunotergeting technique. *J. Physiol. Pharmacol.* **1994**, *45*, 441–454.
34. Jagusiak, A.; Chlopas, K.; Zemanek, G.; Jemiola-Rzeminska, M.; Piekarska, B.; Stopa, B.; Panczyk, T. Self-assembled supramolecular ribbon-like structures complexed to single-walled carbon nanotubes as possible anticancer drug delivery systems. *Int. J. Mol. Sci.* **2019**, *20*, 2064. [CrossRef]
35. Ptak-Kaczor, M.; Kwiecinska, K.; Korchowiec, J.; Chlopas, K.; Banas, M.; Roterman, I.; Jagusiak, A. Structure and Location of Protein Sites Binding Self-Associated Congo Red Molecules with Intercalated Drugs as Compact Ligands-Theoretical Studies. *Biomolecules* **2021**, *11*, 501. [CrossRef]
36. Laginha, K.M.; Verwoert, S.; Charrois, G.J.R.; Allen, T.M. Determination of doxorubicin levels in whole tumor and tumor nuclei in murine breast cancer tumors. *Clin. Cancer Res.* **2005**, *11*, 6944–6949. [CrossRef]
37. Sparreboom, A.; Scripture, C.D.; Trieu, V.; Williams, P.J.; De, T.Y.A.; Beals, B.; Figg, W.D.; Hawkins, M.; Desai, N. Comparative preclinical and clinical pharmacokinetics of a cremophor-free, nanoparticle albumin-bound paclitaxel (abi-007) and paclitaxel formulated in cremophor (taxol). *Clin. Cancer Res.* **2005**, *11*, 4136–4143. [CrossRef] [PubMed]
38. Skowronek, M.; Stopa, B.; Konieczny, L.; Rybarska, J.; Piekarska, B.; Szneler, E.; Bakalarski, G.; Roterman, I. Self-assembly of Congo Red—A theoretical and experimental approach to identify its supramolecular organization in water and salt solutions. *Biopolymers* **1998**, *46*, 267–281. [CrossRef]
39. Kwiecińska, K.; Stachowicz-Kuśnierz, A.; Jagusiak, A.; Roterman, I.; Korchowiec, J. Impact of Doxorubicin on Self-Organization of Congo Red: Quantum Chemical Calculations and Molecular Dynamics Simulations. *ACS Omega* **2020**, *5*, 19377–19384. [CrossRef]
40. Jagusiak, A.; Chlopas, K.; Zemanek, G.; Wolski, P.; Panczyk, T. Controlled Release of Doxorubicin from the Drug Delivery Formulation Composed of Single-Walled Carbon Nanotubes and Congo Red: A Molecular Dynamics Study and Dynamic Light Scattering Analysis. *Pharmaceutics* **2020**, *12*, 622. [CrossRef] [PubMed]
41. Spólnik, P.; Król, M.; Stopa, B.; Konieczny, L.; Piekarska, B.; Rybarska, J.; Zemanek, G.; Jagusiak, A.; Piwowar, P.; Szoniec, G. Influence of the electric field on supramolecular structure and properties of amyloid-specific reagent Congo Red. *Eur. Biophys. J.* **2011**, *40*, 1187–1196. [CrossRef] [PubMed]
42. Elzoghby, A.O.; Samy, W.M.; Elgindy, N.A. Albumin-based nanoparticles as potential controlled release drug delivery systems. *J. Control. Release* **2012**, *157*, 168–182. [CrossRef]
43. Karimi, M.; Bahrami, S.; Ravari, S.B.; Zangabad, P.S.; Mirshekari, H.; Bozorgomid, M.; Shahreza, S.; Sori, M.; Hamblin, M.R. Albumin nano-structures as advanced drug delivery systems. *Expert Opin. Drug Deliv.* **2016**, *13*, 1609–1623. [CrossRef] [PubMed]
44. Elsadek, B.; Kratz, F. Impact of albumin on drug delivery–new applications on the horizon. *J. Control. Release* **2012**, *157*, 4–28. [CrossRef] [PubMed]
45. Kouchakzadeh, H.; Safavi, M.S.; Shojaosadati, S.A. Efficient delivery of therapeutic agents by using targeted albumin nanoparticles. *Adv. Protein Chem. Struct. Biol.* **2015**, *98*, 121–143. [CrossRef] [PubMed]
46. Duan, X.; Li, Y. Physicochemical characteristics of nanoparticles affect circulation, biodistribution, cellular internalization, and trafficking. *Small* **2013**, *9*, 1521–1532. [CrossRef]
47. Greish, K. Enhanced permeability and retention (EPR) effect for anticancer nanomedicine drug targeting. In *Cancer Nanotechnology*; Grobmyer, S., Moudgil, B., Eds.; Humana Press: New York, NY, USA, 2010; pp. 624–625. [CrossRef]
48. Agudelo, D.; Bourassa, P.; Bruneau, J.; Bérubé, G.; Asselin, E.; Tajmir-Riahi, H.A. Probing the binding sites of antibiotic drugs doxorubicin and *N*-(trifluoroacetyl) doxorubicin with human and bovine serum albumins. *PLoS ONE* **2012**, *7*, e43814. [CrossRef] [PubMed]
49. Lee, H.; Park, S.; Kim, J.B.; Kim, J.; Kim, H. Entrapped doxorubicin nanoparticles for the treatment of metastatic anoikis-resistant cancer cells. *Cancer Lett.* **2013**, *332*, 110–119. [CrossRef] [PubMed]

MDPI
St. Alban-Anlage 66
4052 Basel
Switzerland
Tel. +41 61 683 77 34
Fax +41 61 302 89 18
www.mdpi.com

Pharmaceutics Editorial Office
E-mail: pharmaceutics@mdpi.com
www.mdpi.com/journal/pharmaceutics